Advanced MEMS/NEMS Fabrication and Sensors

Zhuoqing Yang
Editor

Advanced MEMS/NEMS Fabrication and Sensors

Springer

Editor
Zhuoqing Yang
National Key Laboratory of Science and Technology on Micro/Nano Fabrication
School of Electronic Information and Electrical Engineering
Shanghai Jiao Tong University
Shanghai, China

ISBN 978-3-030-79748-5 ISBN 978-3-030-79749-2 (eBook)
https://doi.org/10.1007/978-3-030-79749-2

This Springer imprint is published by the registered company Springer Nature Switzerland AG
The registered company address is: Gewerbestrasse 11, 6330 Cham, Switzerland

Contents

Chapter 1
Tip-Based Nanofabrication for NEMS Devices

Dong Pu and Huan Hu

1.1 Introduction to NEMS

Nano-electro-mechanical systems (NEMS) are a class of devices evolved from microelectromechanical systems (MEMS). MEMS have been developed for more than half a century with the potential to radically transform commercial markets and people's lives. MEMS applications are multiple and have ranged from uses as accelerometers for vehicles to general-purpose uses in smartphones as MEMS accelerometers, gyroscopes, microspeakers, radiofrequency (RF) filters, and oscillators. MEMS technology provides functionality that has contributed extensively to society. The MEMS industry is still expected to grow continually in view of the establishment of the Internet of Things, and requires miniaturization and high precision to meet the demands.

NEMS, the next logical miniaturization step beyond MEMS, integrates electrical and mechanical functionalities at the nanoscale with at least one dimension below 1 μm. NEMS have been developed for more than two decades. At this scale, the device is often considered to enter the mesoscale from the macroscale (Schmid et al., 2016). As an emerging technology, NEMS have been used as tools devoted to fundamental studies to probe mesoscopic physical mechanisms (Duraffourg & Arcamone, 2015). Furthermore, as the dimension shrinks, their minuscule size makes them extremely sensitive to any external stimuli that allow them to serve as extremely good sensors. The fundamental sciences and metrology are the two driving forces for NEMS research (Schmid et al., 2016).

At this mesoscale, the nanodevices have been commonly utilized as experimental platforms to study many intriguing natural phenomena. For instance, the exotic

D. Pu · H. Hu (✉)
ZJU-UIUC Institute (ZJUI), Zhejiang University, Haining, China

School of Micro-nano Electronics, Zhejiang University, Hangzhou, China
e-mail: huanhu@intl.zju.edu.cn

Z. Yang (ed.), *Advanced MEMS/NEMS Fabrication and Sensors*,
https://doi.org/10.1007/978-3-030-79749-2_1

states of synchronization with highly complex dynamics have been documented—including weak chimeras, decoupled states, traveling waves, and inhomogeneous synchronized states based on eight coupled nanomechanical oscillators (Matheny et al., 2019). The parametric system breaking, as a control methodology, has also been investigated to assess potential applications that range from ultrasensitive force detection to low-energy computing memory units (Eichler et al., 2018; Papariello et al., 2016). Energy transfer between different modes in a single nanomechanical resonator (Güttinger et al., 2017), noise-induced signal amplification (Guerra et al., 2009), the strong coupling of mechanical motion and electron tunneling (Steele et al., 2009), parametric amplification (De Alba et al., 2016; Mahboob & Yamaguchi, 2008; Mathew et al., 2016; Prasad et al., 2017) and noise squeezing (Almog et al., 2007; Mahboob et al., 2014; Singh et al., 2018) have been investigated to enhance the resolutions for sensing applications.

In particular, NEMS constitute high-frequency mechanical systems that are rapid and extremely sensitive. A few examples of applications of NEMS will be described in detail below. Correspondingly, the mass and the position of a single protein can be measured by the multiple eigenmode frequencies of nanomechanical resonators (Dohn et al., 2007; Hanay et al., 2012; Naik et al., 2009; Olcum et al., 2015; Sader et al., 2018; Sage et al., 2018), as shown in Fig. 1.1a. The resolution of the mass spectrometer can be extremely high toward a single nanoparticle level (Chaste et al.,

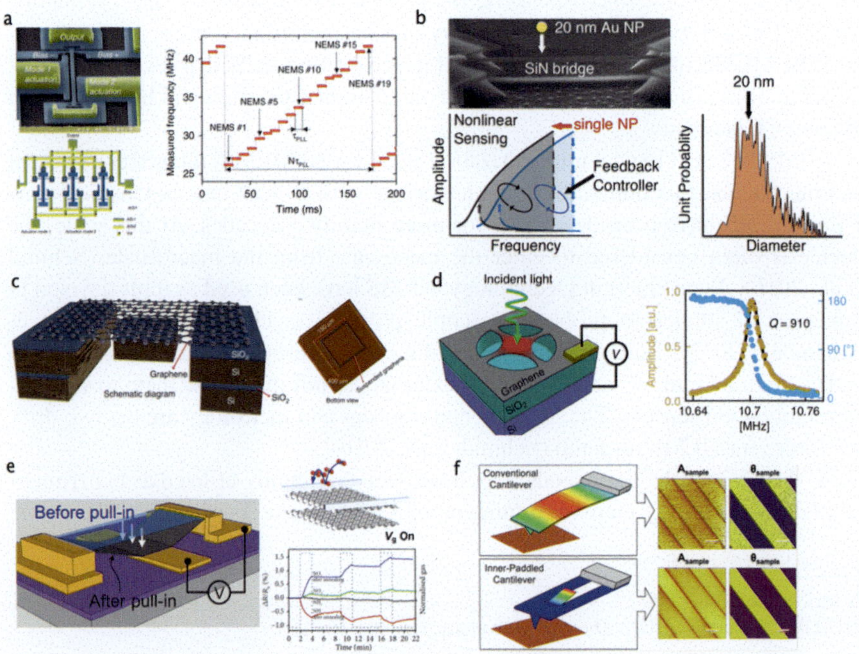

Fig. 1.1 Examples of nano-electro-mechanical systems (NEMS) and their electromechanical behaviors in sensing applications

2012; Dominguez-Medina et al., 2018; Yuksel et al., 2019), shown in Fig. 1.1b. The nanometer-scale resonant charge detector could ultimately reach sensitivities of the order of 10^{-6} e $Hz^{-0.5}$ (Cleland & Roukes, 1998). Another physical sensing application is the accelerometer. The silicon proof mass is suspended on a double-layer graphene ribbon. Based on the piezoresistive effect of graphene, this nanoscale spring–mass system is capable of detecting acceleration (Fan et al., 2019, 2020), as shown in Fig. 1.1c. In addition, the graphene resonator can be applicable for bolometers on account of the light absorption and induced frequency shift owing to the thermal tension (Blaikie et al., 2019), as shown in Fig. 1.1d. Additionally, with nanoscale graphene devices, gas and water can be electromechanically sensed (Meireles et al., 2020; Sun et al., 2016), as shown in Fig. 1.1e (the gas sensor). A key application of NEMS is in atomic force microscopy tips (Alsteens et al., 2017) shown in Fig. 1.1f. The increased sensitivity achieved by NEMS leads to smaller and more efficient sensors used to detect stresses, vibrations, forces at the atomic level, and chemical signals (Ventra et al., 2006). Atomic force microscopy (AFM) tips at the nanoscale rely heavily on NEMS.

Attributed to the microfabrication that is fully compatible with the microelectronics process, MEMS advances have been numerous; accordingly, MEMS have been extensively applied to numerous fields and have impacted the consumer market. Lessons can be learned such that a standardized fabrication process is imperative to foster the applications of NEMS. The remaining part of this chapter will provide a brief review of the existing fabrication methods of NEMS and will then highlight promising nanofabrication techniques based on tips.

1.2 NEMS Fabrication Methods

Nanofabrication is referred to fabricated structures with dimensions <100 nm, which is developed from microfabrication. Generally, the nanofabrication methods can be classified as either top-down or bottom-up. In the top-down methods, the nanostructures are made after patterning and the materials are removed to achieve the desired structure. The top-down approaches can fabricate nanostructures precisely in terms of position, orientation, and size. However, the top-down approach requires complex equipment and expensive maintenance. In addition, many challenges exist when the top-down method approaches its resolution limits. Bottom-up approaches rely on the material grown or the self-assembly, and employs manipulation or synthesis of biochemistry in assembling nanometer-scale building blocks, such as atoms, molecules, deoxyribonucleic acids (DNAs), and nanoparticles into desired nanostructures The bottom-up approach is low-cost and scalable, but it suffers from lack of precision and the low degree of control over the fabricated nanostructures (Hu, 2014). They play two complementary roles in practical nanofabrication.

Technologies in the top-down approach for NEMS device fabrication can be classified into various subtypes, including photon, electron beam (Cleland &

Roukes, 1998), ion beam (Chang et al., 2013), nano-imprint (Chou et al., 1996), and scanning probes (Tseng, 2011). In the optical lithography process, the photon energy is focused on the photoresist and the mask pattern is delineated. In this process, the dimension limit depends on the wavelength of the illumination source. Recent advances have promoted the technique into the extreme ultraviolet regime and the dimensional scale has been reduced to 7 nm.

Electron-beam lithography (EBL) evolved from scanning electron microscopy (SEM) and appeared half a century ago owing to the discovery of polymethyl methacrylate (PMMA), an electron-sensitive material (Haller et al., 1968). The Craighead group fabricated NEMS resonators using EBL combined with surface micromachining (Craighead, 2000). The process started with the nanopatterning of the PMMA resist layer on top of a silicon-on-insulator (SOI) wafer. Metal was then deposited, and lift-off was used to form metal nanopatterns. The metal nanopatterns were then transferred to silicon through etching to define resonator beams. After silicon etching, the buried oxide layer was etched to release the silicon resonator beams to form suspended beams. Finally, metal electrodes were deposited for the application of oscillating electric bias to generate electrostatic forces for the actuation of the NEMS resonator (Hu, 2014). As is known, the higher the electron energy, the shorter the wavelength (Grivet et al., 2013). The wavelength can be thousands of times shorter than the optical wavelength and enables ultrahigh resolution.

Focused ion beam (FIB) lithography is fundamentally the same as the electron beam lithography systems. The emission source (electron or ion source), lens column, work stage, and control systems are all included. Ion beam lithography is much more sensitive than the electron beam (by a factor of multiple hundreds of times) because of the efficient energy transfer between ions and atoms compared with the electron. This makes FIB perfect as a high-precision nanofabrication tool. For example, a three-terminal NEMS resonator was fabricated using FIB (Sulkko et al., 2010). First, aluminum patterns were prepared on the silicon substrate using conventional lithography and wet etching. Then, SF_6 isotropic dry etching of silicon undercuts the silicon underneath the aluminum film. Following the SF_6 etching, FIB cuts a 9 nm wide gap to form a suspended aluminum beam. FIB was able to cut a gap as small as 9 nm as shown in the SEM image. Other than FIB milling, FIB can also implant impurities to silicon that renders the implanted silicon region much less susceptible to chemical etching. NEMS resonators have also been demonstrated using this FIB method (Hu, 2014; Sievilä et al., 2010).

Nanoimprint lithography (NIL) was first proposed by Chou in 1995 who achieved the manufacturing of holes with diameters less than 25 nm in PMMA (Chou et al., 1995, 1996). It was originally proposed to replace the expensive optical lithography or EBL. Albeit its simple and high-resolution capability, NIL struggled to implement reliable technology routes as the technology nodes for integrated circuit (IC) manufacturing evolved to 22 nm (Cui, 2016). An example of NEMS resonators fabricated using NIL (Janzen et al., 2012) can be demonstrated below. First, bilayers of PMMA and lift-off resist (LOR) 3A polymers were spin-coated on a silicon carbonitride (SiCN) layer atop a silicon wafer. A NIL mold pressed on the polymer and replicated the pattern to the two polymer films. After the NIL process, oxygen plasma etched the remaining films. Chrome was then deposited and lifted to form an etch mask for silicon. Dry etching of silicon removed silicon unprotected by the chrome mask, and thus fabricated the SiCN beam. Finally, the SiCN beam was released from the substrate by the anisotropic etching of silicon (Hu, 2014).

Scanning probe-based lithography could be achieved with the tip of an AFM. Tip-based nanofabrication (TBN) has emerged for decades and has demonstrated many nanoscale devices. TBN uses a nanometer-scale tip to fabricate nanometer-scale features through a variety of mechanisms, including mechanical, electrical, electrochemical, thermal–mechanical, optical, plasma, etc. Among the nanomanufacturing methods, TBN is scalable, low-cost, and fast. Furthermore, TBN can fabricate arbitrary patterns. Therefore, TBN holds a great potential for mass-producing NEMS resonators for commercialization. In the following section, we will give a detailed demonstration of TBN.

Bottom-up approaches have also demonstrated fabrication of NEMS resonators but with the help of top-down approach equipment used for the identification and alignment of the nanowires or carbon nanotubes (Chaste et al., 2012) to the anchors. An extensively used method for application as a bottom-up approach involves the preparation of the nanowire or nanotube using synthesis or growth, the use of EBL to locate them, and the fabrication of electrodes to make connections. This bottom-up method has to rely on top-down equipment, such as EBL for alignment which is difficult to scale up and not scalable. There also exist scalable methods used to fabricate NEMS resonators with a bottom-up approach (Li et al., 2008). However, this method can only fabricate NEMS resonators with very limited geometry. The resonator shape is limited to the straight shape predetermined by the nanowire or nanotube shape.

Other methods have also emerged for fabricating NEMS resonators. One example is nanostencil lithography (Arcamone et al., 2009) that uses EBL-fabricated nanostencils to define nanopatterns functioning as a shadow mask. However, it can only replicate predefined patterns in a serial process and requires expensive EBL to fabricate the stencils.

Overall, there is no method that can satisfy all the requirements for the commercialization of NEMS resonators. For the commercialization of NEMS resonators, the fabrication method must be rapid, simple, scalable, low-cost, and flexible in design. Therefore, it is important to explore new fabrication methods for NEMS resonators. Table 1.1 shows the comparison of nanomanufacturing technologies (Hu, 2014; Malshe et al., 2010; Papadopoulos, 2016).

1.3 Tip-Based Nanofabrication for NEMS Device

As one of the emerging nanofabrication methods, tip-based nanofabrication (TBN) has been successfully applied in NEMS device fabrication. According to the nanofabrication mechanism, TBN can be divided into mechanical, electrochemical, thermal, and diffusion-based types. According to the formation type, it can be divided

Table 1.1 Comparison of nanomanufacturing technologies

Nanomanufacturing technologies	Resolution and materials	Defects and contamination	Type and processing speed	Scalability	Cost	Design flexibility
Photolithography	<10 nm with extensions	Photoresist/development residual	Machining, fast	Yes	Expensive	Fixed
EBL	5 nm; all	Ebeam resist/development residual	Machining, slow	No	High	Arbitrary
FIB	30 nm, all	Ion implantation and material redeposition	Machining, slow	No	High	Arbitrary
NIL	10 nm; polymers and silicon	Residual/contamination	Molding, fast	Yes	Low in volume	Fixed
Direct growth/self- assembly	Atomic level	Difficult to control	Assembly, fast	Yes	Inexpensive	Arbitrary
TBN	<10–80 nm, all materials	Occasional pile-up	Deposition, machining, assembly, fast	Yes	Low	Arbitrary

into material transfer, material change, and material removal subtypes. Figure 1.2 summarizes the existing TBN approaches.

In the material transfer category, there are dip-pen nanolithography (DPN) approaches that use the water meniscus in between the tip and substrate to transfer the molecule based on diffusion (Piner et al., 1999). Thermal dip-pen nanolithography (tDPN) uses thermal gradients and shear forces to achieve material transfer from the tip to the substrate (Sheehan et al., 2004).

In the material change category, there are two approaches. The first one is based on electrochemical reactions, wherein an applied bias between a conductive tip and the substrate could induce an electrochemical reaction within the water meniscus formed between the tip and the substrate (Liu et al., 2016). The electrochemical reaction converts the metal or semiconductor material to its oxide form or vice versa. The second approach uses a laser beam to point at the tip to induce a local reaction excited by photons (Kirsanov et al., 2003).

In material removal types, there are three approaches. The first one is to directly use an AFM tip to mechanically scratch the substrate to form nanopatterns (Hu et al., 1998). The second one is to use a heated AFM tip to induce local sublimation of substrate materials (Szoszkiewicz et al., 2007). The third one is to use a heated AFM tip to locally deform the substrate material through both heating and indentation (Vettiger et al., 2000).

Although there are many types of TBN approaches for fabricating nanopatterns as aforementioned, two additional requirements have to be fulfilled to fabricate NEMS devices. First, the TBN approach should be capable of fabricating nanostructures of semiconductor materials, such as silicon, or two-dimensional (2-D) materials, such as graphene. This is because NEMS devices are mostly made of

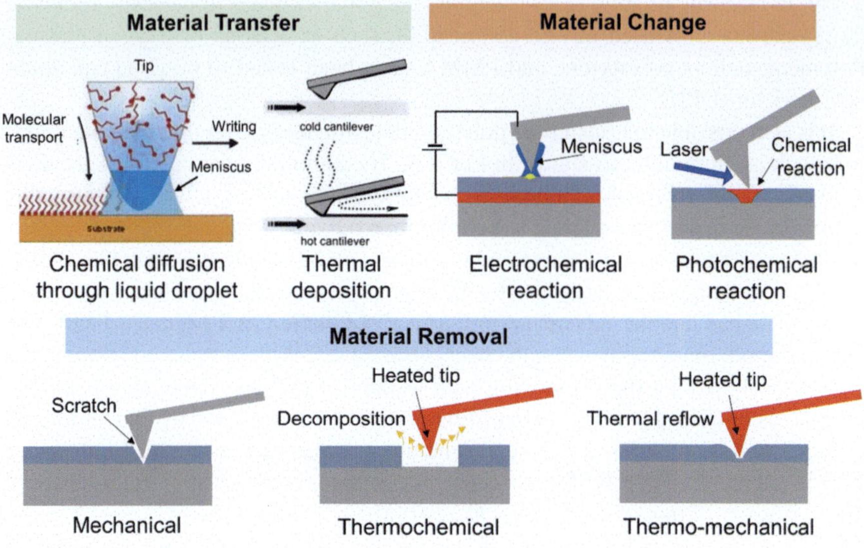

Fig. 1.2 Different types of tip-based nanofabrication (TBN) approaches

semiconductor materials and 2-D materials. Second, the TBN approach needs to be compatible with existing microfabrication technology. This is because NEMS devices consist of not only nanoscale patterns but also micropatterns, such as anchors for supporting suspended nanobeams. It is very difficult to use a TBN approach to fabricate both nanopatterns and micropatterns because of the long time it requires for TBN to finish both the micropatterns and nanopatterns.

Mainly because of the aforementioned two requirements, there are only three TBN approaches that have demonstrated successful NEMS device applications.

1.3.1 Thermal Dip-Pen Nanolithography

tDPN, first demonstrated in 2014 (Sheehan et al., 2004), was capable of depositing nanometer-scale polymers and low-melting temperature metals. This technology originated from the famous Millipede project at IBM that started in the 1990s, in which an array of thermal AFM makes indentations on the polymer substrate with tens of nanometer diameters. By producing nanoscale pits using thermal AFM tips, reading and erasing them, high-density memory technology was achieved (Vettiger et al., 2000). Although no commercial memory products were released, this thermal AFM technology has made a significant impact on many research fields.

One of the advantages of tDPN is that it can deposit relatively thick nanopatterns of polymers. To fabricate nanopatterns of semiconductor materials such as silicon, these polymer nanopatterns can serve as etching masks, and the nanopatterns can be transferred to the underlying silicon substrates through one or two etching steps (Hu et al., 2013), as shown in Fig. 1.3. Although many polymers can be deposited by tDPN, to fabricate pristine nanostructures, the polymer needs to be easily removed by regular solvents used in microfabrication, such as acetone or oxygen plasma. Polymers such as polystyrene and PMMA have been tested to be good candidate polymers for tDPN.

It is also possible to transfer the polymer nanostructures to an intermediate material, such as aluminum, as shown in Fig. 1.4. The aluminum nanopatterns are used as masks for the fabrication of high-aspect-ratio silicon nanostructures because aluminum has much higher etching selectivity than polymer materials.

Once nanostructures of silicon and graphene can be readily fabricated by tDPN, the next challenge is to integrate the TBN process with the microfabrication process. tDPN has another advantage in that it can function as a normal AFM tip to obtain sample topography when the heat is turned off. With sample topography, we can precisely design the tip moving path to deposit polymer nanopatterns on the sample to combine the nanopatterns with micropatterns fabricated by regular microfabrication. Figure 1.5 shows a representative process that successfully produces various types of silicon NEMS mechanical resonators, showcasing the robustness and versatility of this process (Hu et al., 2014a).

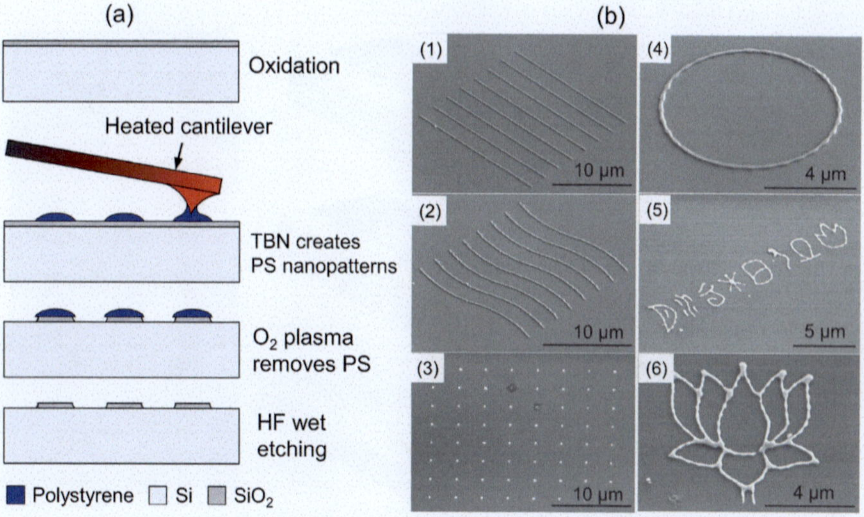

Fig. 1.3 (**a**) Process flow of fabricating silicon oxide nanostructures using deposited polymer nanopatterns as etching masks. (**b**) Scanning electron microscopic images of six different types of nanopatterns including straight nanolines, curved nanolines, nanodots, circular nanolines, ancient Chinese oracle characters, and a lotus flower

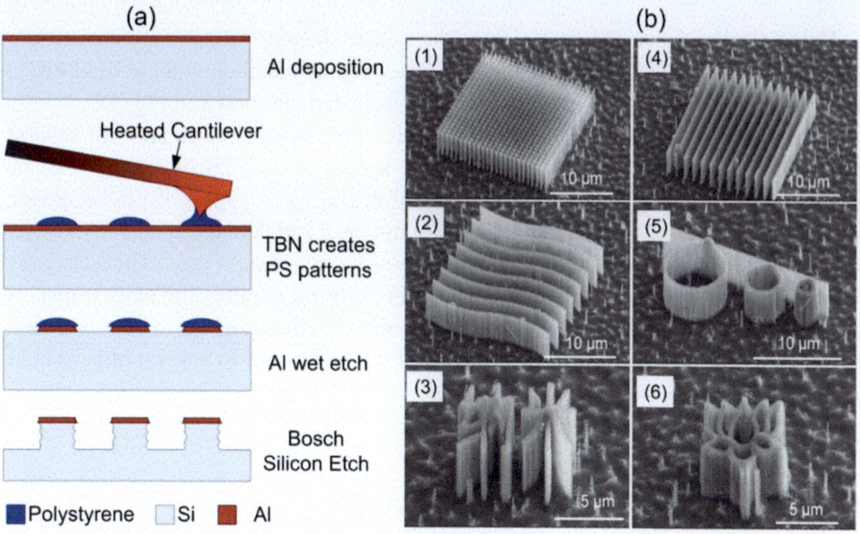

Fig. 1.4 (**a**) Process flow of fabricating high-aspect-ratio silicon nanostructures using thermal dip-pen nanolithography (tDPN). (**b**) Scanning electron microscopy (SEM) images of six different high-aspect-ratio silicon nanostructures fabricated by tDPN

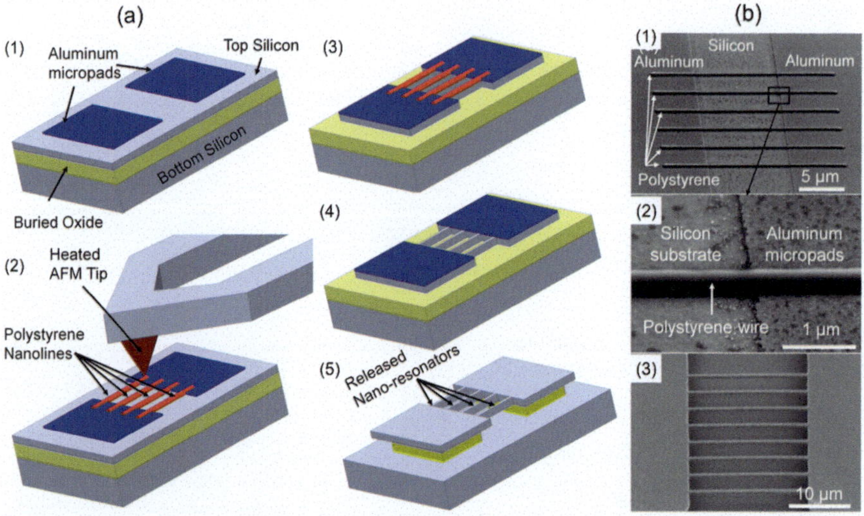

Fig. 1.5 (**a**) Fabrication process of a silicon NEMS resonator using tDPN combined with regular microfabrication process. (**b1**) SEM images of deposited polystyrene nanolines across aluminum micropads defined by regular microfabrication. (**b2**) Magnified view of polystyrene nanoline across the silicon surface and the aluminum micropad. (**b3**) Suspended silicon nanoresonators produced

Using similar fabrication processes, the spider-shaped mechanical nanoresonator, and PDMS nanofluidic channels, a silicon metal-semiconductor field-effect transistor (MESFET) was also successfully fabricated, as shown in Fig. 1.6.

Although tDPN is very good in nanodevice prototyping, there are still several challenges that need to be overcome. First, a special thermal probe that can be heated up controllably to several hundreds of degrees is required. Both the design and the fabrication of the thermal probes are not easy and demand special expertise. Second, the placement of the ink on the tip is still a manual process. This seriously limits the throughput and reliability. Third, the printing requires the mass transport of the molten ink from the tip to the substrate, which requires a considerable amount of time and limits the printing speed in the range of several to several hundreds of micrometers per second.

1.3.2 Thermal–Chemical Decompositions

Another type of TBN approach employs also a thermal probe or a heated AFM tip to decompose materials to form nanopatterns, also termed thermochemical nanolithography (TCNL). Under heating, the materials can change into either a liquid phase or a gas phase. For nanofabrication applications, the gas phase is typically

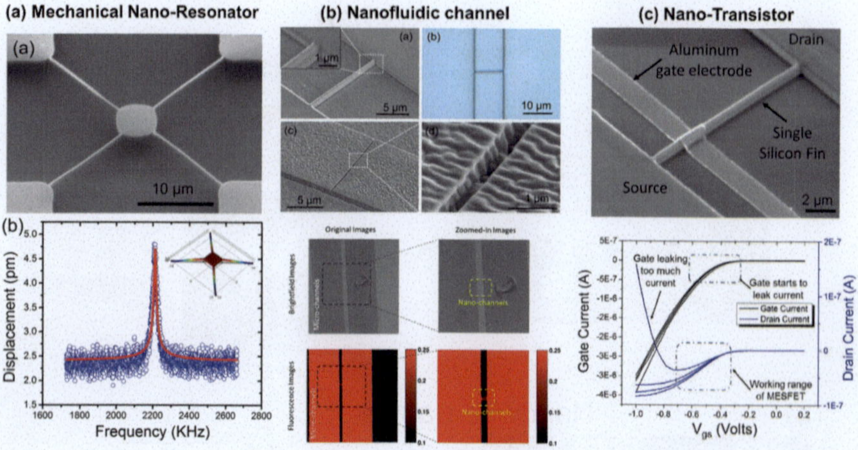

Fig. 1.6 Three types of NEMS devices fabricated by tDPN: (**a**) A spider-shaped mechanical nanoresonator and its resonance characterizations (Hu et al., 2014a). (**b**) A PDMS nanofluidic channel and its fluorescence imaging showing that the nanofluidic channel is connecting two other microfluidic channels (Hu et al., 2014b). (**c**) Silicon metal-semiconductor field-effect transistor (MESFET) with characterizations (Hu, 2014)

preferred because the liquid phase has surface tension that dominates in micro–/nanoscale, and is likely to damage the fabricated nanostructures.

The first advantage is the capability of patterning three-dimensional structures. By controlling the dwelling time, temperature, and force, a different thickness of materials can be removed to form a three-dimensional (3-D) topography or gray lithograph. As a result, rachet structures with different depths can be directly fabricated by removing different material thicknesses (Skaug et al., 2018). These nanodevices can transport nanoparticles with different diameters at different velocities.

The second advantage relates to the compatibility with other microfabrication processes. Transfer structures to different materials provide more material choices. Aligning with nanoscale accuracy is particularly preferred for nanodevice prototyping. For example, they can locate the device region and fabricate an indentation for fabricating a memristor device, as shown in Fig. 1.7a (Cheng et al., 2019). A nanotransistor can also be fabricated by taking advantage of the built-in capability of alignment as shown in Fig. 1.7b (Wolf et al., 2015). Wolf et al. applied TCNL, metal deposition, and lift-off to fabricate nanoscale metal connections precisely on top of the InAs nanowire. Zheng et al. fabricated the MoS_2 transistor using TCNL (Zheng et al., 2019). Unlike the Ebeam lithography that uses high-energy electron beams to expose photoresist that can deteriorate 2-D materials, TCNL only uses heat to decompose materials and will less likely cause defects on 2-D materials. Therefore, a very high on/off current ratio of 10^9 was achieved setting a record.

The third advantage is that no mass transport is required to form the nanopatterns. Therefore, the speed of TCNL can be quite high and reaches 5×10^4 µm²/h for a resolution that is approximately equal to 10 nm (Paul et al., 2011). The thermal tip

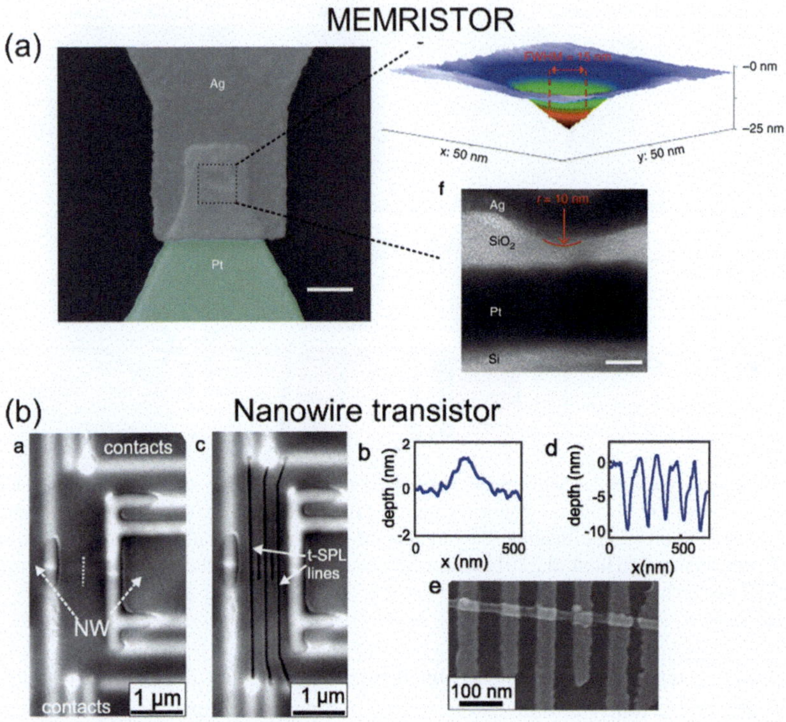

Fig. 1.7 (**a**) Memristor device consists of four layers of materials from the bottom to the top: silicon, Pt, SiO_2, and Ag (Cheng et al., 2019). (**b**) An InAs nanowire transistor with nanoscale metal connections fabricated by using TCNL, metal deposition, and lift-off (Wolf et al., 2015)

just transforms the material locally from the solid phase to the gas phase. Transforming it into the liquid phase will not work because the surface tension of the liquid material can deteriorate the nanopatterns (Hua et al., 2007).

Readers are referred to a recent comprehensive review paper on TCNL covering the heat transfer, resist material development, and comprehensive applications of TCNL (Howell et al., 2020).

1.3.3 Electrochemical SPL

TBN based on pure electrochemical reaction not enabled by heating was first demonstrated using a scanning tunneling microscope (STM) platform in the 1990s on silicon (Dagata et al., 1990) and tantalum surfaces (Thundat et al., 1990). Later, with the invention of AFM, high-vacuum environment was no longer required. Thus, it became much easier and faster to implement electrochemical SPL on AFM platforms.

When a tip is close to a substrate in certain humidity, a water meniscus forms between the tip and the substrate, which acted as an electrochemical cell at the nanometer scale. When a sufficient electrical bias is applied between the tip and the substrate, a local electrochemical reaction occurs within the water meniscus, thus resulting in a nanoscale material change. Therefore, the tip and the substrate should be conductive materials. The majority of the electrochemical reactions are substrate material oxidations. Therefore, we will focus on the oxidation SPL (o-SPL) approach.

o-SPL does not require a vacuum environment and is robust. Moreover, the o-SPL approach has demonstrated a successful nanopatterning of various types of materials. These include traditional materials, such as silicon and silicon oxide (Kyoung Ryu et al., 2014); metal materials, such as TiOx (Gotoh et al., 2000), Ni oxide nanopatterns (Shirakashi & Takemura, 2004), and Ta_2O_5 (Coronado et al., 2013); 2-D materials, such as graphene oxide (Neubeck et al., 2010), reduced graphene oxide (GO) (Benumof, 1982), and MoS_2 films (Zhang et al., 2012).

In addition, o-SPL is compatible with regular microfabrication processes and has a built-in nanoscale precision alignment function that makes it suitable for NEMS devices or nanodevice prototyping. Ryu et al. applied o-SPL to produce silicon oxide nanowire as a mask connecting two metal pads, and then etched to silicon using RIE (Ryu et al., 2013) (Fig. 1.8).

Some challenges of electrochemical SPL are the following (Ryu & Garcia, 2017). First, a conductive probe is required. Second, the environmental temperature and humidity can both have an impact over the size of the meniscus. Thus, for the size of the nanostructures, normally humidity control is required, and a sealed environment is needed.

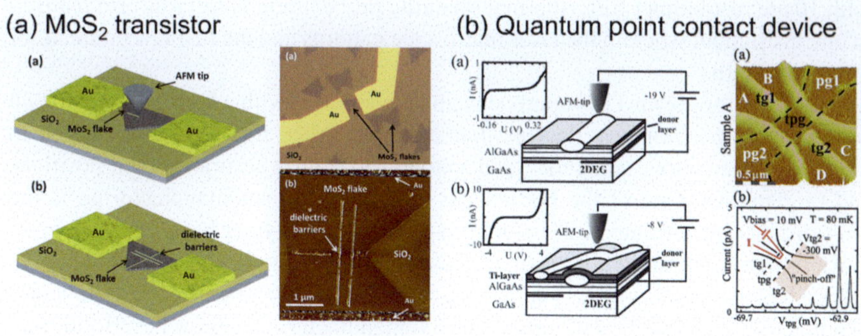

Fig. 1.8 (**a**) MoS_2 transistor (Espinosa et al., 2015). (**b**) Quantum point contact device (Sigrist et al., 2004)

1.4 Future Trends and Conclusions

We have reviewed the application and fabrication methods of NEMS. We then introduced three major types of TBN methods: material transfer, removal, and transformation. TBN technologies of each category have demonstrated successful fabrication of NEMS devices or nanodevices.

We foresee three major trends of current TBN technology development, especially for NEMS device fabrication.

First, with the shrinking of transistors and nanodevices with quantum electronics, smaller features are required to allow the validity of quantum physics. In addition, there is a trend of integrating TBN with more external sources. For example, the thermal probe has been combined with reacting gases to induce defects on 2-D materials that can potentially achieve atomic precision manufacturing (Zheng et al., 2020). Introducing external sources or excitations can provide additional control or functions.

Second, the throughput is increased that achieves fast turnaround fabrication. One disadvantage of TBN compared with optical beam lithography, such as DUV or EUV, is its low throughput. As a result, currently, it can only limit itself in nanolithography mask generation or fixing, as well as some small number of nanodevice prototyping that does not have a high demand for throughput. One promising approach may involve the use of lasers together with a micromirror array to direct laser beams to individual tips for heating for the purpose of nanofabrication. The confined geometry of the tip can further enhance heating (Milner et al., 2008). A recent development related to the combination of laser lithography with TCNL was achieved that could allow the fabrication of a quantum dot transistor in a fast turnaround (Rawlings et al., 2018). The integration of high-throughput laser writing with 10 nm resolution t-SPL (thermal scanning probe lithography) was also realized in one instrument wherein the laser microscale patterns and the t-SPL accounted for the nanopatterns.

Third, the top-down TBN approach can be combined with a bottom-up approach to build high-performance devices. For example, AFM tips can be employed to manipulate nanowires, nanotubes, and nanodots to control accurately their positions and orientations. One of the bottlenecks in bottom-up nanomanufacturing is the long-distance patterning. Self-assembly and block copolymer have good nanopattern properties but are within a small distance. Recent work of using AFM tips to induce better capability is very intriguing. It combines the best of bottom-up nanomanufacturing and the best of top-down nanomanufacturing approaches. Bottom-up manufacturing can provide atom precision crystals or nanomaterials, while TBN provides accurate control of position and orientation.

With the capability of controlling temperature, chemical reaction, force, and magnetic field in a highly confined region down to a few nanometers or a few atoms together with the built-in nanoimaging and super accurate alignment function, TBN technology will play more important roles in NEMS devices, 2-D material nanoelectronics, and quantum devices.

References

Almog, R., Zaitsev, S., Shtempluck, O., & Buks, E. (2007). Noise squeezing in a nanomechanical duffing resonator. *Physical Review Letters, 98*, 1–4. https://doi.org/10.1103/PhysRevLett.98.078103

Alsteens, D., Gaub, H. E., Newton, R., Pfreundschuh, M., Gerber, C., & Müller, D. J. (2017). Atomic force microscopy-based characterization and design of biointerfaces. *Nature Reviews Materials, 2*, 17008. https://doi.org/10.1038/natrevmats.2017.8

Arcamone, J., Sansa, M., Verd, J., Uranga, A., Abadal, G., Barniol, N., et al. (2009). Nanomechanical mass sensor for spatially resolved ultrasensitive monitoring of deposition rates in stencil lithography. *Small, 5*, 176–180.

Benumof, R. (1982). Momentum propagation by traveling waves on a string. *American Journal of Physics, 50*, 20–25. https://doi.org/10.1119/1.12980

Blaikie, A., Miller, D., & Alemán, B. J. (2019). A fast and sensitive room-temperature graphene nanomechanical bolometer. *Nature Communications, 10*, 1–8. https://doi.org/10.1038/s41467-019-12562-2

Chang, J., Koh, K., Min, B.-K., Lee, S. J., Kim, J., & Lin, L. (2013). Synthesis and bidirectional frequency tuning of cantilever-shape nano resonators using a focused ion beam. *ACS Applied Materials & Interfaces, 5*, 9684–9690.

Chaste, J., Eichler, A., Moser, J., Ceballos, G., Rurali, R., & Bachtold, A. (2012). A nanomechanical mass sensor with yoctogram resolution. *Nature Nanotechnology, 7*, 301–304. https://doi.org/10.1038/nnano.2012.42

Cheng, B., Emboras, A., Salamin, Y., Ducry, F., Ma, P., Fedoryshyn, Y., et al. (2019). Ultra compact electrochemical metallization cells offering reproducible atomic scale memristive switching. *Communications on Physics, 2*, 1–9.

Chou, S. Y., Krauss, P. R., & Renstrom, P. J. (1995). Imprint of sub-25 nm vias and trenches in polymers. *Applied Physics Letters, 67*, 3114–3116.

Chou, S. Y., Krauss, P. R., & Renstrom, P. J. (1996). Imprint lithography with 25-nanometer resolution. *Science (80-), 272*, 85–87.

Cleland, A. N., & Roukes, M. L. (1998). A nanometre-scale mechanical electrometer. *Nature, 392*, 160–162. https://doi.org/10.1038/32373

Coronado, E., Forment-Aliaga, A., Navarro-Moratalla, E., Pinilla-Cienfuegos, E., & Castellanos-Gomez, A. (2013). Nanofabrication of TaS 2 conducting layers nanopatterned with ta 2 O 5 insulating regions via AFM. *Journal of Materials Chemistry C, 1*, 7692–7694.

Craighead, H. G. (2000). Nanoelectromechanical systems. *Science (80-), 290*, 1532–1535.

Cui, Z. (2016). *Nanofabrication: Principles, capabilities and limits*. Springer.

Dagata, J. A., Schneir, J., Harary, H. H., Evans, C. J., Postek, M. T., & Bennett, J. (1990). Modification of hydrogen-passivated silicon by a scanning tunneling microscope operating in air. *Applied Physics Letters, 56*, 2001–2003.

De Alba, R., Massel, F., Storch, I. R., Abhilash, T. S., Hui, A., McEuen, P. L., et al. (2016). Tunable phonon-cavity coupling in graphene membranes. *Nature Nanotechnology, 11*, 741–746. https://doi.org/10.1038/nnano.2016.86

Dohn, S., Svendsen, W., Boisen, A., & Hansen, O. (2007). Mass and position determination of attached particles on cantilever based mass sensors. *The Review of Scientific Instruments, 78*, 1–4. https://doi.org/10.1063/1.2804074

Dominguez-Medina, S., Fostner, S., Defoort, M., Sansa, M., Stark, A., Halim, M. A., et al. (2018). Neutral mass spectrometry of virus capsids above 100 megadaltons with nanomechanical resonators. *Science (80-), 362*, 918–922. https://doi.org/10.1126/science.aat6457

Duraffourg, L., & Arcamone, J. (2015). *Nanoelectromechanical systems*. Wiley.

Eichler, A., Heugel, T. L., Leuch, A., Degen, C. L., Chitra, R., & Zilberberg, O. (2018). A parametric symmetry breaking transducer. *Applied Physics Letters, 112*, 233105. https://doi.org/10.1063/1.5031058

Espinosa, F. M., Ryu, Y. K., Marinov, K., Dumcenco, D., Kis, A., & Garcia, R. (2015). Direct fabrication of thin layer MoS2 field-effect nanoscale transistors by oxidation scanning probe lithography. *Applied Physics Letters, 106*, 103503.

Fan, X., Forsberg, F., Smith, A. D., Schröder, S., Wagner, S., Rödjegård, H., et al. (2019). Graphene ribbons with suspended masses as transducers in ultra-small nanoelectromechanical accelerometers. *Nature Electronics, 2*, 394–404. https://doi.org/10.1038/s41928-019-0287-1

Fan, X., Smith, A. D., Forsberg, F., Wagner, S., Schröder, S., Akbari, S. S. A., et al. (2020). Manufacture and characterization of graphene membranes with suspended silicon proof masses for MEMS and NEMS applications. *Microsystems & Nanoengineering, 6*. https://doi.org/10.1038/s41378-019-0128-4

Gotoh, Y., Matsumoto, K., Maeda, T., Cooper, E. B., Manalis, S. R., Fang, H., et al. (2000). Experimental and theoretical results of room-temperature single-electron transistor formed by the atomic force microscope nano-oxidation process. *Journal of Vacuum Science & Technology A: Vacuum, Surfaces, and Films, 18*, 1321–1325.

Grivet, P., Hawkes, P. W., & Septier, A. (2013). *Electron optics*. Elsevier.

Guerra, D. N., Dunn, T., & Mohanty, P. (2009). Signal amplification by 1/f noise in silicon-based nanomechanical resonators. *Nano Letters, 9*, 3096–3099.

Güttinger, J., Noury, A., Weber, P., Eriksson, A. M., Lagoin, C., Moser, J., et al. (2017). Energy-dependent path of dissipation in nanomechanical resonators. *Nature Nanotechnology, 12*, 631–636. https://doi.org/10.1038/nnano.2017.86

Haller, I., Hatzakis, M., & Srinivasan, R. (1968). High-resolution positive resists for electron-beam exposure. *IBM Journal of Research and Development, 12*, 251–256.

Hanay, M. S., Kelber, S., Naik, A. K., Chi, D., Hentz, S., Bullard, E. C., et al. (2012). Single-protein nanomechanical mass spectrometry in real time. *Nature Nanotechnology, 7*, 602–608. https://doi.org/10.1038/nnano.2012.119

Howell, S. T., Grushina, A., Holzner, F., & Brugger, J. (2020). Thermal scanning probe lithography—A review. *Microsystems & Nanoengineering, 6*, 1–24.

Hu, H. (2014). *Nano-electro-mechanical systems fabricated by tip-based nanofabrication*. University of Illinois at Urbana-Champaign.

Hu, S., Hamidi, A., Altmeyer, S., Köster, T., Spangenberg, B., & Kurz, H. (1998). Fabrication of silicon and metal nanowires and dots using mechanical atomic force lithography. *Journal of Vacuum Science & Technology, B: Microelectronics and Nanometer Structures—Processing, Measurement, and Phenomena, 16*, 2822–2824.

Hu, H., Mohseni, P. K., Pan, L., Li, X., Somnath, S., Felts, J. R., et al. (2013). Fabrication of arbitrarily shaped silicon and silicon oxide nanostructures using tip-based nanofabrication. *Journal of Vacuum Science & Technology, B: Nanotechnology & Microelectronics: Materials, Processing, Measurement, & Phenomena, 31*, 06FJ01.

Hu, H., Cho, H., Somnath, S., Vakakis, A. F., & King, W. P. (2014a). Silicon nano-mechanical resonators fabricated by using tip-based nanofabrication. *Nanotechnology, 25*, 275301.

Hu, H., Zhuo, Y., Oruc, M. E., Cunningham, B. T., & King, W. P. (2014b). Nanofluidic channels of arbitrary shapes fabricated by tip-based nanofabrication. *Nanotechnology, 25*, 455301.

Hua, Y., Saxena, S. R., Henderson, C. L., & King, W. P. (2007). Nanoscale thermal lithography by local polymer decomposition using a heated atomic force microscope cantilever tip. *Journal of Micro/Nanolithography, MEMS and MOEMS, 6*, 23012.

Janzen, A., Poshtiban, S., Singh, A., & Evoy, S. (2012). Fabrication of nanoresonator biosensing arrays using nanoimprint lithography. *Journal of Micro/Nanolithography, MEMS and MOEMS, 11*, 23013.

Kirsanov, A., Kiselev, A., Stepanov, A., & Polushkin, N. (2003). Femtosecond laser-induced nanofabrication in the near-field of atomic force microscope tip. *Journal of Applied Physics, 94*, 6822–6826.

Kyoung Ryu, Y., Aitor Postigo, P., Garcia, F., & Garcia, R. (2014). Fabrication of sub-12 nm thick silicon nanowires by processing scanning probe lithography masks. *Applied Physics Letters, 104*, 223112.

Li, M., Bhiladvala, R. B., Morrow, T. J., Sioss, J. A., Lew, K.-K., Redwing, J. M., et al. (2008). Bottom-up assembly of large-area nanowire resonator arrays. *Nature Nanotechnology, 3*, 88–92.

Liu, H., Hoeppener, S., & Schubert, U. S. (2016). Nanoscale materials patterning by local electrochemical lithography. *Advanced Engineering Materials, 18*, 890–902.

Mahboob, I., & Yamaguchi, H. (2008). Piezoelectrically pumped parametric amplification and Q enhancement in an electromechanical oscillator. *Applied Physics Letters, 92*, 1–4. https://doi.org/10.1063/1.2903709

Mahboob, I., Okamoto, H., Yamaguchi, H., Onomitsu, K., & Yamaguchi, H. (2014). Two-mode thermal-noise squeezing in an electromechanical resonator. *Physical Review Letters, 113*, 167203. https://doi.org/10.1103/PhysRevLett.113.167203

Malshe, A. P., Rajurkar, K. P., Virwani, K. R., Taylor, C. R., Bourell, D. L., Levy, G., et al. (2010). Tip-based nanomanufacturing by electrical, chemical, mechanical and thermal processes. *CIRP Annals, 59*, 628–651.

Matheny, M. H., Emenheiser, J., Fon, W., Chapman, A., Salova, A., Rohden, M., et al. (2019). Exotic states in a simple network of nanoelectromechanical oscillators. *Science (80-), 363*, eaav7932. https://doi.org/10.1126/science.aav7932

Mathew, J. P., Patel, R. N., Borah, A., Vijay, R., & Deshmukh, M. M. (2016). Dynamical strong coupling and parametric amplification in mechanical modes of graphene drums. *Nature Nanotechnology, 11*, 747–751. https://doi.org/10.1038/nnano.2016.94

Meireles, L. M., Neto, E. G. S., Ferrari, G. A., Neves, P. A. A., Gadelha, A. C., Silvestre, I., et al. (2020). Graphene electromechanical water sensor: The Wetristor. *Advanced Electronic Materials, 6*, 1–6. https://doi.org/10.1002/aelm.201901167

Milner, A. A., Zhang, K., & Prior, Y. (2008). Floating tip nanolithography. *Nano Letters, 8*, 2017–2022.

Naik, A. K., Hanay, M. S., Hiebert, W. K., Feng, X. L., & Roukes, M. L. (2009). Towards single-molecule nanomechanical mass spectrometry. *Nature Nanotechnology, 4*, 445–450. https://doi.org/10.1038/nnano.2009.152

Neubeck, S., Ponomarenko, L. A., Freitag, F., Giesbers, A. J. M., Zeitler, U., Morozov, S. V., et al. (2010). From one electron to one hole: Quasiparticle counting in graphene quantum dots determined by electrochemical and plasma etching. *Small, 6*, 1469–1473.

Olcum, S., Cermak, N., Wasserman, S. C., & Manalis, S. R. (2015). High-speed multiple-mode mass-sensing resolves dynamic nanoscale mass distributions. *Nature Communications, 6*, 7070. https://doi.org/10.1038/ncomms8070

Papadopoulos, C. (2016). *Nanofabrication: Principles and applications*. Springer.

Papariello, L., Eichler, A., Zilberberg, O., Leuch, A., Degen, C. L., & Chitra, R. (2016). Parametric symmetry breaking in a nonlinear resonator. *Physical Review Letters, 117*, 1–5. https://doi.org/10.1103/physrevlett.117.214101

Paul, P. C., Knoll, A. W., Holzner, F., Despont, M., & Duerig, U. (2011). Rapid turnaround scanning probe nanolithography. *Nanotechnology, 22*, 275306.

Piner, R. D., Zhu, J., Xu, F., Hong, S., & Mirkin, C. A. (1999). "Dip-pen" nanolithography. *Science (80-), 283*, 661–663.

Prasad, P., Arora, N., & Naik, A. (2017). Parametric amplification in MoS_2 drum resonator. *Nanoscale, 9*, 18299–18304. https://doi.org/10.1039/C7NR05721K

Rawlings, C., Ryu, Y. K., Rüegg, M., Lassaline, N., Schwemmer, C., Duerig, U., et al. (2018). Fast turnaround fabrication of silicon point-contact quantum-dot transistors using combined thermal scanning probe lithography and laser writing. *Nanotechnology, 29*, 505302.

Ryu, Y. K., & Garcia, R. (2017). Advanced oxidation scanning probe lithography. *Nanotechnology, 28*, 142003.

Ryu, Y. K., Chiesa, M., & Garcia, R. (2013). Electrical characteristics of silicon nanowire transistors fabricated by scanning probe and electron beam lithographies. *Nanotechnology, 24*, 315205.

Sader, J. E., Hanay, M. S., Neumann, A. P., & Roukes, M. L. (2018). Mass spectrometry using nanomechanical systems: Beyond the point-mass approximation. *Nano Letters, 18*, 1608–1614. https://doi.org/10.1021/acs.nanolett.7b04301

Sage, E., Sansa, M., Fostner, S., Defoort, M., Gély, M., Naik, A. K., et al. (2018). Single-particle mass spectrometry with arrays of frequency-addressed nanomechanical resonators. *Nature Communications, 9*, 3283. https://doi.org/10.1038/s41467-018-05783-4

Schmid, S., Villanueva, L. G., & Roukes, M. L. (2016). *Fundamentals of nanomechanical resonators*. Springer International. https://doi.org/10.1007/978-3-319-28691-4

Sheehan, P. E., Whitman, L. J., King, W. P., & Nelson, B. A. (2004). Nanoscale deposition of solid inks via thermal dip pen nanolithography. *Applied Physics Letters, 85*, 1589–1591.

Shirakashi, J.-I., & Takemura, Y. (2004). Ferromagnetic ultra-small tunnel junction devices fabricated by scanning probe microscope (SPM) local oxidation. *IEEE Transactions on Magnetics, 40*, 2640–2642.

Sievilä, P., Chekurov, N., & Tittonen, I. (2010). The fabrication of silicon nanostructures by focused-ion-beam implantation and TMAH wet etching. *Nanotechnology, 21*, 145301.

Sigrist, M., Fuhrer, A., Ihn, T., Ensslin, K., Driscoll, D. C., & Gossard, A. C. (2004). Multiple layer local oxidation for fabricating semiconductor nanostructures. *Applied Physics Letters, 85*, 3558–3560.

Singh, R., Nicholl, R. J. T., Bolotin, K. I., & Ghosh, S. (2018). Motion transduction with thermomechanically squeezed graphene resonator modes. *Nano Letters, 18*, 6719–6724. https://doi.org/10.1021/acs.nanolett.8b02293

Skaug, M. J., Schwemmer, C., Fringes, S., Rawlings, C. D., & Knoll, A. W. (2018). Nanofluidic rocking Brownian motors. *Science (80-), 359*, 1505–1508.

Steele, G. A., Huttel, A. K., Witkamp, B., Poot, M., Meerwaldt, H. B., Kouwenhoven, L. P., et al. (2009). Strong coupling between single-electron tunneling and nanomechanical motion. *Science (80-), 325*, 1103–1107. https://doi.org/10.1126/science.1176076

Sulkko, J., Sillanpaa, M. A., Hakkinen, P., Lechner, L., Helle, M., Fefferman, A., et al. (2010). Strong gate coupling of high-Q nanomechanical resonators. *Nano Letters, 10*, 4884–4889.

Sun, J., Muruganathan, M., & Mizuta, H. (2016). Room temperature detection of individual molecular physisorption using suspended bilayer graphene. *Science Advances, 2*, 1–8. https://doi.org/10.1126/sciadv.1501518

Szoszkiewicz, R., Okada, T., Jones, S. C., Li, T.-D., King, W. P., Marder, S. R., et al. (2007). High-speed, sub-15 nm feature size thermochemical nanolithography. *Nano Letters, 7*, 1064–1069.

Thundat, T., Nagahara, L. A., Oden, P. I., Lindsay, S. M., George, M. A., & Glaunsinger, W. S. (1990). Modification of tantalum surfaces by scanning tunneling microscopy in an electrochemical cell. *Journal of Vacuum Science & Technology A: Vacuum, Surfaces, and Films, 8*, 3537–3541.

Tseng, A. A. (2011). *Tip-based nanofabrication: Fundamentals and applications*. Springer.

Ventra, M., Evoy, S., & Heflin, J. R. (2006). *Introduction to nanoscale science and technology*. Springer Science & Business Media.

Vettiger, P., Despont, M., Drechsler, U., Durig, U., Haberle, W., Lutwyche, M. I., et al. (2000). The "millipede"—More than thousand tips for future AFM storage. *IBM Journal of Research and Development, 44*, 323–340.

Wolf, H., Rawlings, C., Mensch, P., Hedrick, J. L., Coady, D. J., Duerig, U., et al. (2015). Sub-20 nm silicon patterning and metal lift-off using thermal scanning probe lithography. *Journal of Vacuum Science & Technology, B: Nanotechnology & Microelectronics: Materials, Processing, Measurement, & Phenomena, 33*, 02B102. https://doi.org/10.1116/1.4901413

Yuksel, M., Orhan, E., Yanik, C., Ari, A. B., Demir, A., & Hanay, M. S. (2019). Nonlinear nanomechanical mass spectrometry at the single-nanoparticle level. *Nano Letters, 19*, 3583–3589. https://doi.org/10.1021/acs.nanolett.9b00546

Zhang, K., Fu, Q., Pan, N., Yu, X., Liu, J., Luo, Y., et al. (2012). Direct writing of electronic devices on graphene oxide by catalytic scanning probe lithography. *Nature Communications, 3*, 1–6.

Zheng, X., Calò, A., Albisetti, E., Xiangyu, L., Alharbi, A. S. M., Arefe, G., et al. (2019). Patterning metal contacts on monolayer MoS 2 with vanishing Schottky barriers using thermal nanolithography. *Nature Electronics, 2*, 17–25.

Zheng, X., Calò, A., Cao, T., Liu, X., Huang, Z., Das, P. M., et al. (2020). Spatial defects nanoengineering for bipolar conductivity in MoS 2. *Nature Communications, 11*, 1–12.

Chapter 2
Dimensional-Nanopatterned Piezoresistive Silicon Microcantilever for Environmental Sensing

Jiushuai Xu and Erwin Peiner

2.1 Introduction

A microcantilever can be modeled as a cantilever beam, i.e., with one end fixed and the other free, which is typically composed of silicon, silicon derivatives, inorganic semiconductors, or polymers, in various dimensions and shapes (Fig. 2.1). Inspired by the micromachined cantilevers that have been developed and used in atomic force microscopes (AFM) for surface analysis, microcantilever-based chemical and biological sensing are relatively young research fields, which were initiated in the mid-1990s. Most of the early cantilever-based chemical detection works were performed in standard AFM setups using metal-coated normal AFM probes, where only one microcantilever can be measured at a time (Boisen et al., 2011). In 1998, an array of eight polymer-coated Si cantilevers was designed and micromachined by HP Lang and coworkers at IBM (1998). Besides, a corresponding system was developed to measure simultaneously the response of the cantilevers to different alcohols, through that the analytes were distinguished due to their different adsorption rates in the polymer coatings, referring to an uncoated cantilever. Almost in the same period, cantilever array with integrated readout for environmental sensing appeared, for example, piezoresistive cantilevers for surface stress sensing were reported (Boisen et al., 2000), utilizing its benefit for compact designs of cantilever-based sensing and probing systems initially developed for AFM (Tortonese et al., 1991).

Microcantilever-based sensors with different readout schemes have since then been extensively investigated and developed for chemical sensing applications

J. Xu · E. Peiner (✉)
Institute of Semiconductor Technology (IHT), Technische Universität Braunschweig, Braunschweig, Germany

Laboratory for Emerging Nanometrology (LENA), Technische Universität Braunschweig, Braunschweig, Germany
e-mail: e.peiner@tu-bs.de

Z. Yang (ed.), *Advanced MEMS/NEMS Fabrication and Sensors*,
https://doi.org/10.1007/978-3-030-79749-2_2

Fig. 2.1 Microcantilevers of different shapes (oblique view): (**a**) rectangular, (**b**) double-legged with end mass, (**c**) triangular. (Reprinted from (Xu et al., 2019), Copyright 2019, with permission from IOP)

requiring low-level detection capability, in which they benefit from their high surface/volume ratio despite miniature dimensions enabling high mass sensitivity. Compared to conventional sensing technologies, microcantilevers provide several prospective advantages, as they are highly sensitive and have rapid response time while following nonhazardous label-free procedures. In the end, they are inexpensive due to high-throughput manufacturing at minimal material consumption. As mechanical sensors, they are robust and consume little power (Goeders et al., 2008; Vashist et al., 2018). Moreover, unlike other resonant elements such as quartz crystals, microcantilevers as MEMS resonators have the potential for higher levels of integration with microelectronics at both die and package levels (Abdolvand et al., 2016).

Microcantilevers have been mostly fabricated from silicon, demonstrating in extensive investigations their diverse potentials in environmental monitoring and healthcare, whereby this rapid growth has been following the advancement in micromachining technologies (Bertke et al. 2021). Si-based materials such as Si itself, silicon nitride (Si_3N_4), and silicon dioxide (SiO_2) have well-known properties, can be operated at different temperatures and environmental conditions, and are stable over time. Besides, they can be fabricated at low cost by using microelectromechanical systems (MEMS) or printing processes derived from planar microelectronic technologies (Bertke et al. 2021), thus providing advantages such as miniaturization, batch fabrication, and integration of microelectronics circuitry for control and processing (Debeda & Dufour, 2020).

This chapter provides developments in microcantilever-based sensors, their working principle, readout systems, the fabrication of microcantilevers with dimensional nanostructure frameworks, and their environmental sensing applications, together with the future prospects.

2.2 Mechanism of Operation

Surface functionalization is generally required when microcantilevers are proposed for physical and chemical detection, which is achieved by coating a functional layer supposed to be highly sensitive and selective to the target analytes. Then the microcantilevers are used for sensing applications in two operating modes, which are known as static and dynamic modes. The static mode of cantilever deflection occurs when adsorption of target species on the cantilever causes differential surface stresses on its upper and lower surfaces, leading to bending either upward or downward depending on the adsorption effect (Goeders et al., 2008). Principally, a long and soft cantilever is required to achieve large deflections for the static detection mode. The static mode is straightforward because no actuation element is needed. Another mode is known as dynamic mode, where an actuation system to excite the microcantilever at resonance is a prerequisite. By measuring the eigen resonance frequency of the microcantilever, mass (or surface stress) changes due to adsorption or desorption of analytes can be monitored and calculated (Vashist et al., 2018; Oden et al., 1996). Similar to QCM sensors, the resonance frequency of a microcantilever decreases/increases correlating to the target analyte that is adsorbed/desorbed on/from the sensing material (Xu et al., 2018a), due to the interactions between the sensitive layer and the target species. Generally, a short and stiff cantilever is required for the dynamic method, where a high operation frequency is favorably needed. By measuring resonance-frequency shift, experimental detection of mass changes in yoctogram range (Chaste et al., 2012) has been demonstrated using nanomechanical resonators. Cantilever-based resonators are potential candidate for high-resolution detection. Si has high stability, high mechanical quality factor, and high degree of freedom for the geometrical cantilever design, and is therefore considered as a preferable base material for fabricating cantilever-based resonators. Large surface/volume ratio and conformal coverage with a thin functionalization layer are key issues to provide a maximum number of well accessible reaction/adsorption sites on a resonator for high sensitivity and selectivity as well as fast response and recovery.

2.3 Basics of Microcantilevers

2.3.1 *Resonance Frequency* ($\mathbf{f}_n$)

The dynamic behavior of a coated microcantilever is composed of an infinite number of harmonic flexural modes. It is commonly assumed that Young's modulus of the cantilever is independent of frequency, and the effect of the coating on elasticity in the vicinity of the considered resonance peak can be neglected. The *n*th-mode f (f_n) of a coated microcantilever can then be approximated by:

$$f_n = \frac{a_n^2}{2\pi}\sqrt{\frac{k}{m}} \tag{2.1}$$

where a_n is a number corresponding to the considered *n*th flexural mode, which is 1.875 for the fundamental mode (Brand et al., 2015).

The small variation of stiffness and mass change due to the adsorption of analytes adsorbed on the coated layers causes a resonance-frequency shift (Δf) of the

nth resonant-mode frequency f_n, and Δf is related to a cantilever-mass and a cantilever-stiffness change (assuming uniform changes Δm and Δk of mass m and stiffness k, respectively):

$$\Delta f_n = \frac{1}{2} f_n \left(\frac{\Delta k}{k} - \frac{\Delta m}{m} \right) \tag{2.2}$$

When the analytes are randomly adsorbed on the entire cantilever, the effect of mass change is dominant compared to the effect of stiffness change. Therefore, the stiffness term can be neglected and Eq. (2.2) can be simplified as:

$$\frac{\Delta f_n}{\Delta m} = -\frac{1}{2}\frac{f_n}{m} = -\frac{1}{2}\frac{f_n}{(\rho t + \rho_c t_c) wL} \tag{2.3}$$

where ρ, w, L and t are density, width, length, and thickness of the microcantilever, respectively (cf. Fig. 2.2), and assuming that a sensitive layer is uniformly coated on the cantilever with ρ_c and t_c as its density and thickness, respectively.

2.3.2 *Quality Factor*

In real environmental measurements, the operation of coated microcantilevers can suffer from drastic decrease in their Q factor. The Q factor is commonly used to quantify energy dissipation in mechanical systems, and is defined as 2π times the ratio of the maximum energy stored in a resonating system to the amount of energy dissipated in one cycle (Brand et al., 2015). However, when considering the

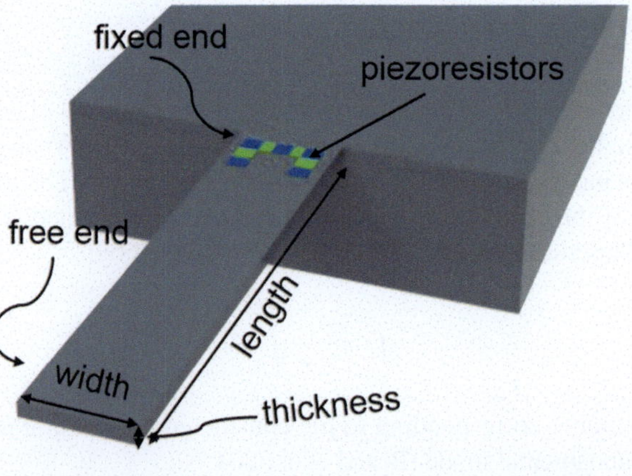

Fig. 2.2 3D schematic of a silicon microcantilever integrated with piezoresistors at its fixed end. (Reprinted from (Xu et al., 2019), Copyright 2019, with permission from IOP)

cantilever transfer function, the Q factor is commonly obtained using the 3-dB bandwidth Δf_{3dB} of the cantilever and is given by (Bargatin et al., 2007; Dufour et al., 2007):

$$Q = \frac{f_n}{\Delta f_{n,3dB}} \tag{2.4}$$

The Q factor quantifies the sharpness of the resonant peak and is influenced by damping. Too low a Q factor impacts on the noise of the resonance-frequency measurement, and consequently reduces the performance of microcantilever-based sensors. For instance, the minimum detectable mass of a gravimetric sensor is inversely proportional to Q (Johnson & Mutharasan, 2012; Fantner et al., 2017):

$$\Delta m_{\min} \propto \frac{m}{Q} \tag{2.5}$$

In resonant microcantilever-based sensors, the Q factor therefore determines the figure of merit of the sensor, i.e., for high resolution and efficiency, high Q factors are required (Lavrik et al., 2004; Bertke et al., 2017).

2.3.3 *Figure of Merit of Microcantilever-Based Gas Sensors*

2.3.3.1 Sensitivity

Sensitivity is a primary property to evaluate the performance of a gas sensor. In the scenario of a microcantilever-based sensor, the relative sensitivity (S_R), which means the relative frequency change $\Delta f/f_n$ related to the concentration change of the considered analyte (Δc_A) of a resonant sensor, is generally used instead of its (absolute) sensitivity (S):

$$S = \frac{\partial f_n}{\partial c_A}, \quad S_R = \frac{1}{f_n}\frac{\partial f_n}{\partial c_A}, \tag{2.6}$$

since using S_R we are able to better compare the performances of the sensors with different f_n. In the detection of gases, we assume that the mass change of the sensitive layer due to the adsorption of the target species is the only contribution to Δf_n. Alternatively, S_R can be expressed as the product of relative gravimetric sensitivity G_R (Brand et al., 2015) (i.e., Δf_n due to a change $\Delta \rho_c$ in the density of the sensing layer, cf. Eq. (2.3)):

$$G_R = \frac{1}{f_n} \times \frac{\Delta f}{\Delta \rho_c} = \frac{1}{f_n} \times \left(-\frac{1}{2} \frac{f_n}{(\rho t + \rho_c t_c) wL} \right) \times \frac{\Delta m}{\Delta \rho_c} = -\frac{1}{2} \frac{t_c}{(\rho t + \rho_c)} \tag{2.7}$$

and S_A (i.e., the sensitivity of this layer to adsorbing/desorbing the target gas given by the density change of the sensitive layer $\Delta\rho_c$ related to the analyte concentration change Δc_A):

$$S_A = \frac{\Delta\rho_c}{\Delta c_A}. \tag{2.8}$$

We then obtain:

$$S_R = G_R \times S_A = \frac{1}{f_n} \times \left(-\frac{f_n}{2} \frac{wLt_c}{wL(\rho t + \rho_c t_c)} \times S_A \right) = -\frac{t_c}{2(\rho t + \rho_c t_c)} \times S_A \tag{2.9}$$

Therefore, in the scenario of using mass-sensitive-based chemical gas sensors, for example, microcantilevers, with uniformly coated sensitive layers vibrating in lateral (in-plane) or transverse (out-plane) flexural mode, the resulting S_R is independent of the microcantilever's in-plane dimension and the used flexural eigenmode, but only depends on the layer thicknesses and densities of the microcantilever and the sensing film, respectively, as well as on the adsorption/desorption properties of the sensing film (Brand et al., 2015).

It should be noted that S_R in Eq. (2.9) is determined by the changes in the density of the sensing film. That is to say, S_R cannot be improved by reducing the microcantilever's in-plane dimensions, which are the width and length of the microcantilever (Fig. 2.2). This is a distinct difference compared to the general case of mass sensitivity of a resonant sensor, e. g., with respect to a material deposited on the cantilever instead of being taken up in a sensing layer (Eq. 2.3), which can be increased by simply decreasing the resonator mass (effective mass), in other words, by reducing its dimensions. Furthermore, the adsorption/desorption sensitivity S_A of a resonant sensor will not be improved by shrinking its size.

2.3.3.2 Limit of Detection (LOD)

One ultimate performance metric for any sensor is its LOD, in other words, the smallest detectable target analyte concentration of a gas sensor. In the case of a resonant mass-sensitive sensor, LOD is defined as three times the noise-equivalent analyte concentration (Brand et al., 2015), which can be approximated by the ratio of the minimal detectable frequency change $\Delta f_{n,\min}$ and the gas sensor sensitivity:

$$\mathrm{LOD} = 3\frac{\Delta f_{n,\min}}{S} \tag{2.10}$$

Therefore, a small value LOD can be achieved by combining high sensitivity and low measurement noise. While the sensitivity of the microcantilever sensor is independent on dimensions, as aforementioned, $\Delta f_{n,\min}$ is, however, dependent on resonator geometry via the Q factor. Thus, the LOD is enhanced by improving the

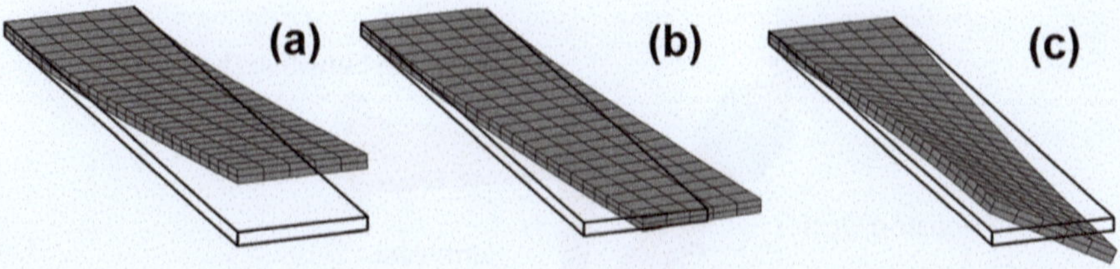

Fig. 2.3 Schematic of the first bending (**a**), lateral (**b**), and torsional (**c**) modes of a resonating cantilever. The heavy lines denote the undeformed cantilever; the shaded regions denote the deformed cantilever. (Reprinted from (Goeders et al., 2008), Copyright © 2008, with permission from ACS)

microcantilever's Q factor, such as choosing a higher-order resonance mode, which generally has a larger Q factor, using in-plane and torsional modes instead of out-of-plane flexural modes (Brand et al., 2015) (Fig. 2.3), miniaturized cantilever dimensions (Bertke et al., 2020), or special device geometries such as triangular (Bertke et al., 2017), disk (Seo & Brand, 2008), and hammerhead shapes (Beardslee et al., 2011).

2.3.3.3 Selectivity

Selectivity represents the ability of gas sensors to identify a specific target gas among many chemical molecules in the atmosphere. For a microcantilever-based gas sensor, selective sensing is achieved by a proper coating on the cantilever acting as sensing layer which can offer interaction specifically with the target gas. In case the coated layer is sensitive to specific groups of gases, usually these gases have to be separated before they arrive to the sensing layer by a gas chromatography column or other pre-conditioning approaches. This can be circumvented using a microcantilever array, where the individual cantilevers are coated with different materials sensitive to different gas species. Consequently, such cantilever arrays work as a system for detecting individual species from a mixture of gases, which is the normal case, e. g., for environmental air-quality monitoring applications.

2.3.4 Readout of the Microcantilever

The optical triangulation method (Meyer & Amer, 1988) is the most widely used technique to detect and characterize both static and dynamic deflections of microcantilevers. The method involves the reflection of a low-power optical beam off the microcantilever onto a segmented photodiode or a position-sensitive detector (PSD), where a light-emitting diode (LED) coupled with a lens or laser diode is typically used as a source to generate the low-power light beam (Fig. 2.4). The reflected laser beam falls at a certain position when the undeflected microcantilever is fixed, but

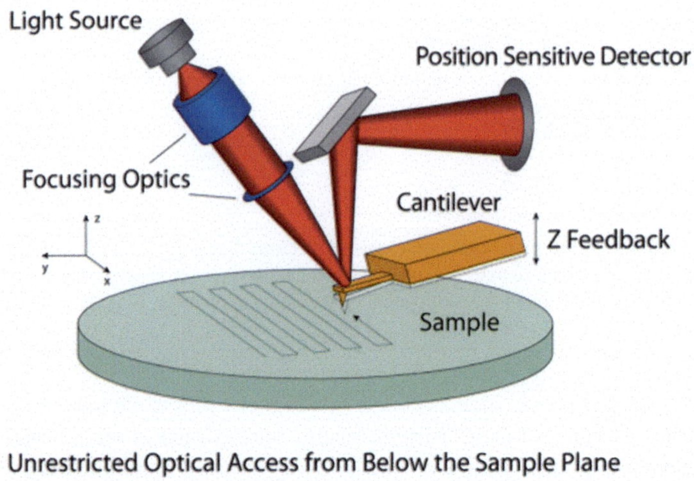

Fig. 2.4 Schematic depicting the optical-lever detection system used in AFMs. Light is reflected off the top of a cantilever and read on a detector as the cantilever/probe is raster scanned to measure the topography of the sample. Most of the AFM instrumentation can be located above the sample, allowing for unrestricted optical access from below. (Reprinted from (Geisse, 2009), Copyright 2009, with permission from Elsevier)

when the microcantilever deflects due to the adsorption of analytes, a proportional reflection-spot shift of the laser beam occurs. This shift will be recorded by the PSD, which possesses high spatial resolution and fast response speed. Even though the optical detection tends to be complicated in terms of its optical-tool alignment, it offers high-sensitive response and has been already commercially available, in which deflections of microcantilevers in the range of angstroms and nanometers can already be detected (e.g., AFM) (Vashist et al., 2018). However, this technique still has a distinct limitation as it is sensitive to the optical properties of the cantilever surface and the atmosphere around the microcantilever. Thus, it is not effective when it is employed with cantilevers having a light-absorbing surface as well as in smoky air or volatile organic compounds (VOCs), which can scatter light reflected from the cantilever on its path to the photodetector. Because of that, it is clear that optical readout systems cannot be used for measuring opaque liquids neither, such as blood. Besides, optical detection systems of high precision and resolution are expensive and not portable (Vashist et al., 2018).

Piezoresistive detection is another commonly used method to characterize cantilever deflection. A piezoresistive microcantilever (PMC) contains embedded piezoresistive materials designed as resistors (i.e., piezoresistors), which are normally arranged in either half or full Wheatstone bridge (WB) circuit configurations, to record the stress change along the cantilever axis induced by static or resonant cantilever deflections (Fig. 2.2). In case only a single piezoresistor is integrated directly on the microcantilever, the remaining three resistors will be normally placed on the unstressed frame close to the cantilever or directly on a printed circuit board using commercial electronic components. However, to obtain maximum sensitivity, the piezoresistor (or even four piezoresistors in case of a full WB integrated on the cantilever, Fig. 2.2) should be positioned as close to the cantilever fixed end as possible. Depending on operation mode, the resistance change can be measured as both static and dynamic signals. Doped single crystal silicon is one of the most common materials that shows a strong piezoresistive effect, in which silicon-based PMCs with a boron-doped full WB have a typical resistance of a few kilo-ohms (Wasisto et al., 2013, 2014).

The benefit of using piezoresistive detection is that the principle works well in both liquid and gas phases, where large arrays of such piezoresistive sensors can be operated and read out simultaneously. Besides, a PMC-based sensor can integrate the readout system on-chip, which consequently shrinks the size of a final system setup. Therefore, bulky and expensive instrumentation is not needed for electrical characterization of the devices (Boisen et al., 2000; Thaysen et al., 2000; Jensenius et al., 2000). However, current flow through the microcantilever is always required in piezoresistive detection, which may result in additional dissipation of heat inducing associated thermal drifts (Bertke et al., 2017, 2018). Thus, any variations in the thermal conductivity of the environment, which result in fluctuations of the cantilever temperature, in turn may lead to parasitic piezoresistance changes and cantilever deflections (Ma et al., 2008). In the reported works about PMCs for sensing applications, these thermal effects were eliminated, e.g., by using a bare unfunctionalized microcantilever as reference sensor (Vashist et al., 2018).

2.3.5 *Online Frequency Tracking*

The most straightforward method to detect the resonance frequency of a mechanical device is by sweeping the frequency of an AC excitation signal around the frequency of the corresponding amplitude maximum. This method is common especially in laboratory settings but is not suitable for online measurements due to its slow sweep rates needed for proper frequency resolution. Another technique is to excite the resonator with a fixed stable signal at a frequency that is slightly different (typically higher) than the resonance frequency of the device. In this case, the resonance-frequency shift will be converted to changes in amplitude of the output signal and can then be related back to the measurand (Abdolvand et al., 2016; Rodriguez et al., 2007). With this method, online monitoring can be achieved, but it leads to a short range of frequency tracking, since it just relies on the slope side of a resonance peak. Additional Q factor changes caused by the environment may lead to misinterpretations.

A more practical method is to place a two-port resonant device into the feedback loop of a properly designed electronic circuit to form an oscillating system, which follows the desired resonance frequency (Abdolvand et al., 2016). Corresponding feedback loops can be active or passive units. As feedback loop, a phase-locked loop (PLL) circuit can be used. This method is common for online resonance-frequency tracking using instruments like a lock-in amplifier (e.g., Zurich Instruments HF2LI and MFLI) (Pandit et al., 2017; Yang et al., 2015) or individual circuits (Wasisto et al., 2014) and can reach very high frequency stability. PLL circuits are active controlling circuits and consist of three core elements: a phase detector, a loop filter, and a voltage-controlled oscillator (VCO). The phase detector detects the phase difference θ_d between the input and output signals of the resonator and generates a θ_d-depended output signal (v_{out}). This is converted to an increasing or decreasing voltage by the loop filter and further results in an increasing or decreasing frequency of the output signal of the VCO. This can be realized by analogue and digital components or by software. Typically, software-based PLL circuits are slower than their hardware-based counterparts, but can be adapted to emerging, more complicated demands, e.g., nonlinear resonance behavior (Bertke et al., 2018).

Microcantilevers working in dynamic mode can also be placed in oscillators. A resonator-based oscillator consists of a positive-feedback loop circuit with an amplifier. Although the oscillator circuit does not offer superior accuracy or resolution with respect to the aforementioned PLL, the possibility of simple oscillator circuits can reduce the size, cost, and power consumption of the system (Manzaneque et al., 2014) and enables their potential application for integration into a single chip comprising the microcantilever sensors.

A very important criterion for comparing methods of the resonant PMC actuation is the impact of the chosen method on cross talk superimposed to the signal detection (Lee et al., 2017). In comparison to other resonance elements such as quartz crystals (e.g., quartz crystal microbalance, QCM), PMC-based resonators have the potential for higher degrees of integration with microelectronics at the die or package level. The actuator integration can lead to systems with enhanced performance and functionality. PMC-based devices show very suitable properties for mass-sensing applications, since the sensitivity increases with decreasing resonator mass and piezoresistivity enhances at shrinking scales, while other transduction methods lose efficiency (Abdolvand et al., 2016).

2.4 Fabrication of Si Microcantilever-Based Resonators

2.4.1 Fabrication of Microcantilever Structure

Microcantilever-based resonators have typically dimensions of few hundreds μm to millimeter in length, tens to hundreds μm in width, and few to tens μm in thickness (Fig. 2.2). For Si and its derivative-based dynamic microcantilevers, they are fabricated using well-mastered Si technologies, with Si wafers or Si on insulator (SOI) wafers as starting substrates. Thanks to the well-established MEMS Si micromachining techniques and microfabrication process developed from the IC industry in the 1960s, microcantilevers are already commercially available by many

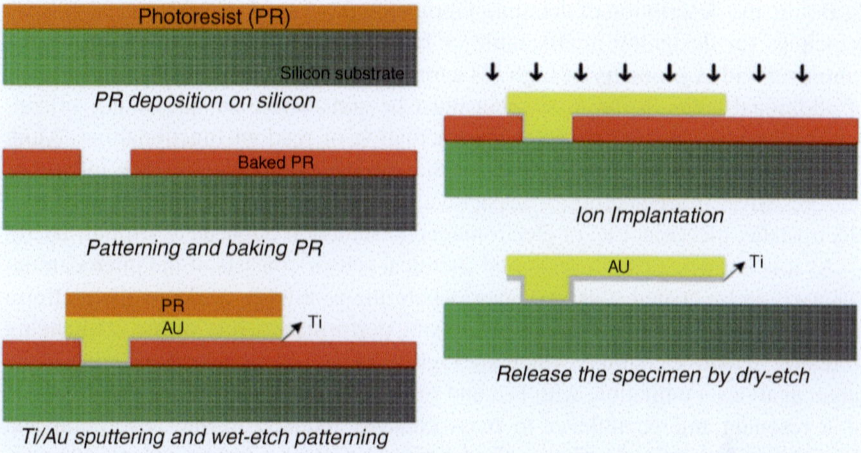

Fig. 2.5 Fabrication steps of a microcantilever by a surface-micromachining approach. (Reprinted from (Kang et al., 2005), Copyright 2005, with permission from IOP)

companies, such as SCL-Sensor.Tech. Fabrication GmbH, 1220 Vienna, Austria, Micromotive Mikrotechnik, Germany, and Concentris GmbH, Switzerland.

Si-microcantilever-based resonators are generally fabricated using cleanroom processes, either by surface-micromachining or bulk-micromachining techniques. Surface micromachining is considered as a bottom-up process, where a microcantilever functional substances of (poly)Si, Si_3N_4, SiO_2, or SiC are deposited on sacrificial layers of different natures (SiO_2, metal, etc.). Deposition can be realized by different techniques, for example, chemical vapor deposition, physical vapor deposition, spin coating, inkjet printing, spray coating, etc., and the microcantilevers are released by a final elimination of sacrificial layer. Using surface-micromachining technique, some metals, inorganic materials, and biocompatible materials can also be constructed as microcantilevers (Fig. 2.5).

Bulk micromachining is analogized as a top-down process. In a bulk-micromachining process, the microcantilever pattern is defined on the front side of the Si wafer by UV or electron-beam lithography (EBL), and then the cantilever thickness is successively etched by anisotropic cryogenic deep reactive ion etching (cryo-DRIE). The microcantilevers are released by doing another cryo-DRIE process from the backside. Different to bulk Si, SOI wafers are composed of a monocrystalline Si layer of fixed thickness and a buried SiO_2 layer on a bulk Si wafer. The buried SiO_2 layer can act as an etching stop layer for both front and backside etching; therefore, microcantilevers fabricated from SOI have reproducible thickness. To fabricate the microcantilevers with "big dimension," for example, tens μm in thickness, bulk Si is preferable compared to SOI, due to its lower price accompanied by higher stability, higher mechanical Q factor, and higher degree of freedom for cantilever design. For "small" microcantilevers, which have only few μm or even hundreds nm in thickness, SOI is a better candidate, since thicknesses in these low ranges are much more easily controllable.

Before the description of detailing fabrication process of microcantilevers, some principles for designing microcantilever resonators should be introduced: firstly, cantilever with a geometry of high *Q* factor is favorable. Besides, the compatibility of additional steps in the processing must be considered for fabricating silicon-based microcantilevers with integrated actuation or readout functionalities. More fabrication steps based on thin-film technologies are, hence, necessary in case of piezoresistive, piezoelectric, or capacitive readout, as well as for electrothermal, electrostatic, piezoelectric, or electromagnetic actuation (Debeda & Dufour, 2020).

As it is known, a cantilever-based chemical sensor consists of the microcantilever itself and a coated sensitive layer, where the cantilever works as the platform and the sensitive layer determines the sensing performance. It is always challenging to realize sensitive coatings which are highly stable, sensitive, and selective to the target analytes, compatible with fast and reversible response. For a strong response of a resonant microcantilever to mass changes induced by the adsorption and desorption of target molecules on the functional coating, a bigger-volume (thicker-layer) coating of a sensitive material is generally required corresponding to higher sensitivity. But the mechanical and physicochemical properties of the coated layer, and its thickness, will also affect the overall performance of the cantilever-based sensors, not only the sensitivity but also the response time and *Q* factor. The obtained *Q* factor determines further the limit of detection (LOD). Therefore, to fabricate high-performance environmental sensors, sensitive layers of high surface area are essential. This can be achieved either by depositing directly high surface area functional nanomaterials (bottom-up method) as sensing layers or by fabricating microcantilevers patterned with high surface-to-volume ratio nanostructure arrays (top-down method) as sensing materials or supporting frameworks for deposition of thin sensing layers, or the combination of both methods. Fabrication of cantilever-based environmental sensors using these aforementioned methods will be described in detail in this following section.

2.4.2 *Nanostructuration of Microcantilevers Using Bottom-Up Method*

Bottom-up synthesis of Si nanostructures relies on the sustained accumulations of Si atoms onto a metal catalyst (e.g., Au nanoparticle) located on the substrate surface (Morales, 2008). The epitaxial growth process ends till supersaturation is achieved, and phase separation subsequently results in nanostructure growth. Due to the concerns over high temperature, poor reproducibility, low crystallinity (Chang et al., 2009), and difficulty in orientation control and device integration (Schmidt et al., 2005; Huang et al., 2010), the bottom-up approach is not preferentially considered for synthesizing high-density and vertically aligned Si nanostructure arrays, although this method can produce high yields of nanostructures at low cost (Huang et al., 2010; de Boor et al., 2010).

Therefore, bottom-up approach-based nanostructures are generally referring to the deposition of sensitive layers. Various materials have been investigated as sensitive layer and deposited on resonant microcantilevers for gas detection, both organic and inorganic as well as composites of them. Nonconductive polymers are the most explored materials as sensitive coating for resonant microcantilevers. These polymer-based materials are the best candidates for gas detection, due to their abundant material availability, good time stability, and easy straightforward processing. In general, polymers have weak physisorption to interact with target analytes, which requires moderate reaction conditions, and offers reversible and reproducible fast response. Besides, the functional groups of the polymers can be modified or tailored to promote their selectivity, e. g., on arrays of resonant microcantilevers, which can also be built by depositing different polymeric films (Battiston et al., 2001; Possas-Abreu et al., 2017). Numerous films of synthetic polymers (polysiloxane [polyvinyl [PVA], PDMS], polyurethanes [PEUT], etc.) have been deposited on microcantilevers by drop-casting, spray- and spin-coating, or inkjet printing. In addition, the use of molecularly imprinted polymers (MIP) is also an interesting strategy, where these MIP can be either deposited on the cantilever as sensitive coating (Okan et al., 2017) or directly nanostructured into microcantilevers (Debeda & Dufour, 2020).

Inorganic porous materials with high specific area have also been studied to increase the adsorption of target gases. Dimensional nanomaterials of metal oxides (Xu et al., 2018a; Kilinc et al., 2014), carbon oxide, graphene oxide, and metal organic frameworks (MOF), which have large surface area and exhibit high sorption capabilities, have also attracted great attention. However, syntheses of these nanomaterials are complicated and require usually harsh conditions, making their preparation not compatible with the microcantilever fabrication. Conventionally, the combination of inorganic porous nanomaterials with microcantilevers requires a two-step method. Firstly, the nanomaterials are synthesized separately. Then the prepared nanomaterials are added into solvent (DI water, isopropanol, acetone, etc.) to form a suspension, which is used as ink in the following material deposition. Thereafter, the ink is printed on the top surface of the microcantilever near the free end using an inkjet printer and subsequently dried at a certain temperature.

A typical fabrication approach of a microcantilever with a selective area for sensing material deposition was reported by X. Li et al. (Fig. 2.6). A double-side polished (100) SOI silicon wafer with a 3 μm top silicon layer, 7000 Å buried oxygen (BOX) layer, and 380 μm silicon substrate is used as a raw material to fabricate the integrated microcantilevers using a MEMS machining technique. The fabrication process is detailed as follows: (1) 1000 Å low stress silicon nitride (LS SiNx) is deposited on the surface of the SOI wafer by low pressure chemical vapor deposition (LPCVD) and then patterned to form the region of resistors for heating (actuation) and strain sensing (piezoresistive Wheatstone bridge). (2) To fabricate these resistors, boron ions are implanted into the silicon, yielding a sheet resistance of 200 Ohm per square. (3) After opening the contact holes, 300 Å of titanium and 3000 Å of gold are sequentially sputter-deposited on the wafer surface and then patterned by lift-off for electrical contact. (4) 100 nm of parylene-C film is deposited on the resonant microcantilever using a Diener parylene coating system (model P6).

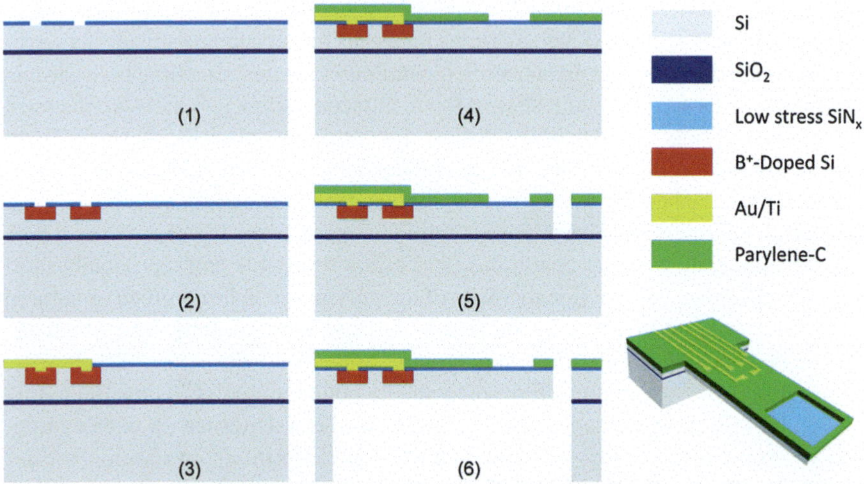

Fig. 2.6 Fabrication process of the parylene-C patterned resonant microcantilever. The bottom right image shows the 3D structure of the parylene-C patterned microcantilever. (Reprinted from (Cai et al., 2019), Copyright 2019, with permission from The Royal Society of Chemistry)

With the help of a photoresist mask, unwanted parylene-C at the free end of the microcantilever is removed by O_2 plasma etching. This way, a sample reservoir for corrosive substance loading is formed. (5) Using deep reactive-ion etching (DRIE), the silicon nitride and top silicon layer together with the parylene-C film on top are removed in a defined area, which defines the shape of the microcantilever. (6) Employing photoresist as a mask, the wafer is then deep-etched from the backside of the wafer to remove the silicon beneath the cantilever. Subsequently, the BOX layer is removed by buffered oxide etching (BOE) to free the cantilever. Finally, the wafer is immersed into acetone to remove the photoresist coating layer (Cai et al., 2019). The length, width, and thickness of the microcantilever are designed as 200 μm, 100 μm, and 3 μm (Fig. 2.7a), respectively (Lv et al., 2018).

Metal organic framework (MOF) crystals have been deposited on this resonant microcantilever platform as sensing material for CO detection, where the Ni-MOF-74 material with ultrahigh surface area and nanoporous structure was loaded on a microcantilever using a commercial GIX II Microplotter (SonoPlot, Inc.) and dried in an oven at 60 °C for 2 h. Accordingly, carboxyl group functionalized mesoporous silica nanoparticles (C-MSNs) were prepared and deposited on such microcantilevers (Fig. 2.7b). C-MSNs has ultrahigh-specific surface area as well, which promoted the resonant cantilever toward high performance for trace-level NH_3 detection. Similarly, a MOF material consisting of nitrogen-doped nanoporous carbon (NPC) was synthesized which was derived from a zeolitic imidazolate framework-8 (ZIF-8), which showed a nanoporous structure indicating a large surface area (Tang et al., 2020). The fabricated NPC nanomaterial and simultaneously prepared basic carbonate (BCC) are loaded onto two microcantilevers, respectively, through the described inkjet printing technology. Since these two materials show

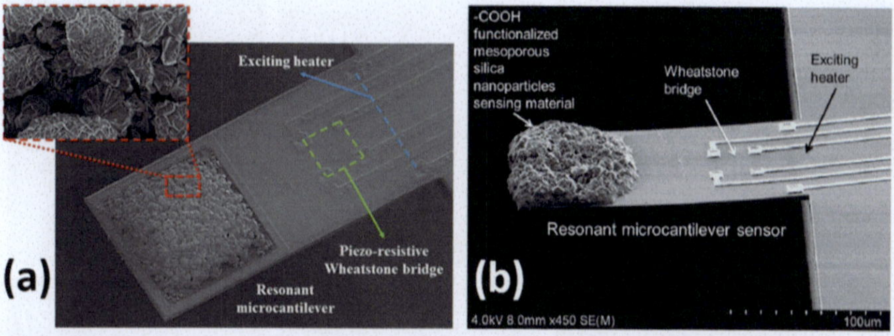

Fig. 2.7 (**a**) MOF-loaded resonant microcantilever sensor where Ni-MOF-74 is area-selectively deposited close to the free end of the microcantilever. (Reprinted from (Lv et al., 2018), Copyright 2018, with permission from Elsevier) (**b**) SEM image of a C-MSNs-loaded resonant microcantilever sensor, where the sensing material is inkjet-deposited close to the free end of the cantilever. (Reprinted from (Liu et al., 2018), Copyright 2018, with permission from Elsevier)

different adsorption and desorption behavior to H_2S, this dual-microcantilever array can selectively detect H_2S from a variety of common gases, with a LOD of below 1 ppb.

These aforementioned bottom-up approaches have successfully separated the synthesis of sensing nanomaterials and their addition to the microcantilever into individual steps, eliminating inevitable compatibility issues. The coating process was reported to be easy and reproducible, and the obtained microcantilever-based sensors showed good sensing performance assigned to the large surface area of the deposited sensing material. However, inkjet-printed materials are only physically deposited and thus weakly attached onto the cantilever surface. Their properties are prone to be influenced by environmental humidity and they may suffer from bad long-term stability. Besides, nanoporous structures lead to long response/recovery times.

Metal oxides are the most used materials for gas detection in resistive-type-based sensors, due to their high surface-to-volume ratio. However, the miniaturized dimension and slender geometry of microcantilevers hinder an area-selective deposition of metal oxide nanostructures on their surface. Recently, J. Xu et al. reported a two-step ZnO nanorod deposition method, they sputtered a zinc nanofilm using a direct-current sputtering technique, and the sputtered zinc was further oxidized at a certain temperature to transform into a ZnO nanofilm. The obtained ZnO nanofilm worked then as a seed layer for a subsequent growth of ZnO nanorods (NRs) using a chemical bath deposition (CBD) method. Area-selective deposition was realized on either the top or bottom surfaces of the microcantilever (Fig. 2.8), depending on the cantilever release order, with the help of photolithography (Xu et al., 2017a, 2018a, b). Such ZnO-NRs-patterned microcantilever gas sensing application was operated using both optical and piezoresistive detection methods. In addition, by changing the time of CBD, ZnO NRs can be obtained varying in length from few hundred nanometers to several micrometers, with corresponding diameters of tens of nanometers to around hundreds of nanometers.

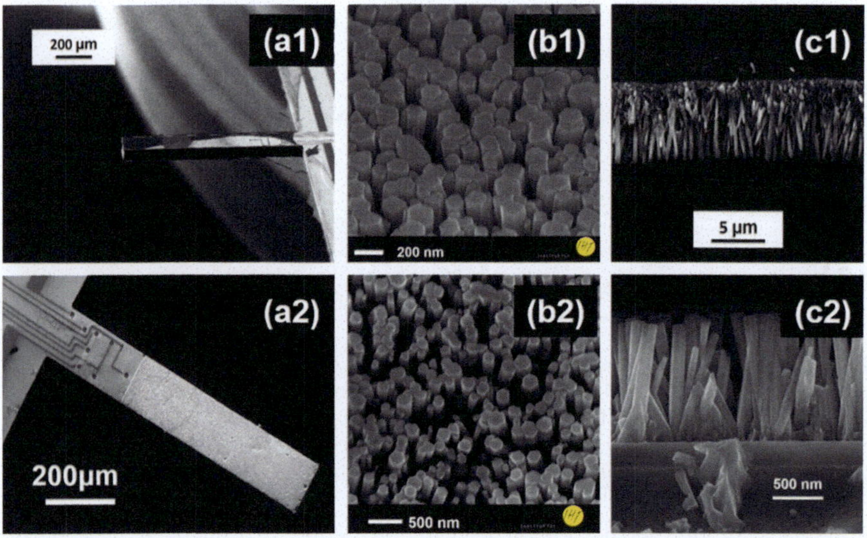

Fig. 2.8 SEM graphs of a microcantilever with area-selective ZnO deposition on its backside (1) and frontside (2): oblique view (**a1**, **a2**), magnified inclined overview (**b1**, **b2**), and magnified cross-sectional view (**c1**, **c2**). (Reprinted from (Xu et al., 2019), Copyright 2019, with permission from IOP)

L. Schlur et al. reported a similar method for the fabrication of ZnO nanostructures (NRs and nanotubes) on one side of commerical Tipless Force Modulation AFM cantilever (TL-FM cantilever). Using ZnO nanotubes instead of NRs, the surface area can be further increased. The TL-FM tipless cantilevers have a length of 225 ± 10 µm, a width of 28 ± 7.5 µm, a thickness of 3.0 ± 1 µm, and a resonant frequency between 45 and 115 kHz (Fig. 2.9). In their work, the ZnO seed layer was not sputtered area-selectively but on the whole surface of the AFM chip, the subsequent ZnO NRs deposition was realized also on the whole-chip surface. Besides, deposition of copper hydroxide ($Cu(OH)_2$) or copper oxide (CuO) nanostructures on TL-FM tipless microcantilevers was reported for NO_2 sensing application. The $Cu(OH)_2$ NRs were grown on a copper layer previously evaporated on the top surface of the cantilever. CuO NRs were obtained by the annealing of the copper hydroxide nanostructures (Schlur et al., 2018b). Similarly, vertically aligned rutile TiO_2 NRs, with an average length and width of 9.5 µm and 14.7 nm, respectively, were successfully grown on TL-FM tipless microcantilevers (Biapo et al., 2019).

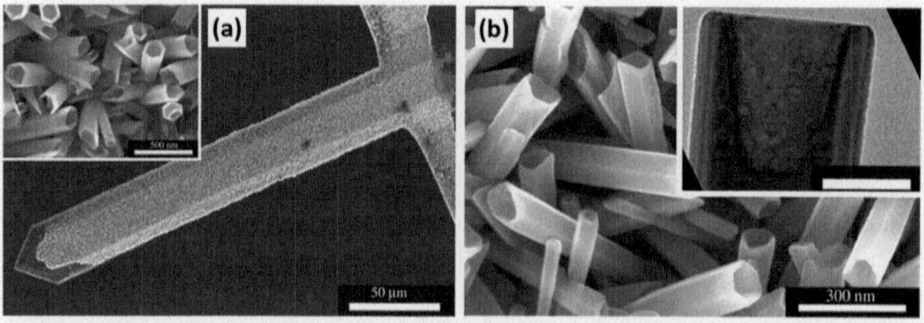

Fig. 2.9 (**a**) SEM top view of a cantilever covered by ZnO nanotubes. These nanotubes are obtained by the dissolution during 1.75 h of ZnO nanorods in a solution containing 0.5 wt% of cetyltrimethylammonium bromide (CTAB) and 0.5 wt% of $NH_{3(aq)}$. Inset: zoom of the obtained nanotubes. (**b**) SEM top view of nanotubes obtained after an etching time of 5.25 h. Inset: TEM images of the nanotubes (scale bar: 50 nm). (Reproduced from (Schlur et al., 2018a), CC BY 4.0)

2.4.3 *Nanostructuration of Microcantilevers Using Top-Down Method*

The top-down technique is a reducing approach, where an amount of Si defined by a processing mask is removed from the substrate during a final etching process, leaving back structures with certain lateral shapes and dimensions, and vertical heights. Considering the applied high temperature during Si epitaxial process (bottom-up approach) (Kanungo et al., 2008; Fuhrmann et al., 2005; Bauer et al., 2007), micromachining Si nanostructures using the top-down technique is preferable, due to its low price and good compatibility with the standard CMOS fabrication process. In the top-down approach, the mask with a precise nanopatterning step is crucial for realizing the desired nanostructure. The mask can be achieved by appropriate patterning technologies, for example, photolithography is the most utilized technique for processing Si-based materials. However, photolithography is elaborate for micromachining small device and structures with lateral sizes of below 100 nm, due to the limitation of its typical lowest-convenient UV wavelength (365 nm). The developments in deep ultraviolet illumination (DUV, UV wavelength of ~254 nm was applied) and the commercialization of extreme ultraviolet (EUV) lithography have overcome this limitation with resist exposure at a wavelength of 13.5 nm (EUV), which enables the realization of lateral patterns below 10 nm in mass-produced CMOS chips. However, it is hard to adopt these techniques for nanopatterning on microcantilevers, due to cost and additional issues such as suitable photoresist properties, mask defects, compatibility with fabrication process, etc. Therefore, other nonoptical lithography methods have been developed and adopted for nanopatterning and processing on Si materials.

2.4.3.1 Nanoimprint Lithography

Nanoimprint lithography (NIL) is a simple nanolithography technique with low cost, high resolution, and high throughput. The nanopattern is created by the mechanical deformation of imprint resist induced by pressing a predefined stamp (mold) during the replication process. The imprint resist is typically a low-viscosity monomer or polymer formulation that is cured either by a certain temperature or by UV irradiation (Hamdana et al., 2018). The mold is normally made of transparent material like fused silica or polydimethylsiloxane (PDMS). The cured imprint photoresist cannot be damaged during photolithography; therefore, after patterning, it can be employed either as direct etching mask or as a predefined image-reverse pattern for the subsequent deposition of hard mask materials. After an additional lift-off process, the imprint resist can then be removed, and the left deposited materials (e.g., chromium and SiO_2) will work as harder mask for etching Si micro/nanostructures with higher aspect ratios by a cryo-DRIE.

J. Xu et al. reported a full fabrication process, during which NIL was adopted for nanostructuring of microcantilevers. In their process, a heating resistor (for electrothermal actuation) and a full Wheatstone bridge (for strain sensing of deflected cantilever) were embedded by boron diffusion. An array of Si nanopillars of ~200 nm in diameter and ~2.5 μm in height was selectively engraved on the microcantilever by deep reactive-ion etching at cryogenic temperature (cryo-DRIE). In addition, ZnO NRs of ~25 nm in diameter and ~250 nm in length were grown on top and the sidewall surfaces of the Si nanopillars, using the aforementioned two-step ZnO NRs deposition method, building an array of "suction-cup cleaner-brushes." The details are illustrated in Fig. 2.10, and described in their publication (Xu et al., 2020).

2.4.3.2 Nanosphere Lithography

The nanosphere lithography (NSL) technique, which is also known as colloidal lithography or natural lithography (Park et al., 2018; Hamdana et al., 2017; Haynes & Van Duyne, 2001), has been widely used to prepare patterned masks for nanostructure etching, because of its simplicity, low cost, and high throughput compared to conventional nanolithography methods (Park et al., 2018; Xu et al., 2017b; Li et al., 2017). NSL utilizes periodic arrays of self-assembled closed-packed mono–/bilayer nanospheres (e.g., polystyrene, SiO_2, and others) as masks and could be an alternative choice for constructing uniform and well-ordered nanopatterns of minimum sub-10 nm underlying substrate materials (Park et al., 2018; Ji et al., 2017; Su et al., 2013). P. Weiss et al. reported a strategy called multiple-patterning nanosphere lithography (MP-NSL) for the fabrication of periodic 3D hierarchical nanostructures in a highly scalable and tunable manner. This nanofabrication technique exploits the selected and repeated etching of polymer nanospheres that serve as resists and that can be shaped in parallel for each processing step. The application of MP-NSL enables the fabrication of periodic, vertically aligned Si nanotubes at the wafer scale with nanometer-scale control in three dimensions including outer/

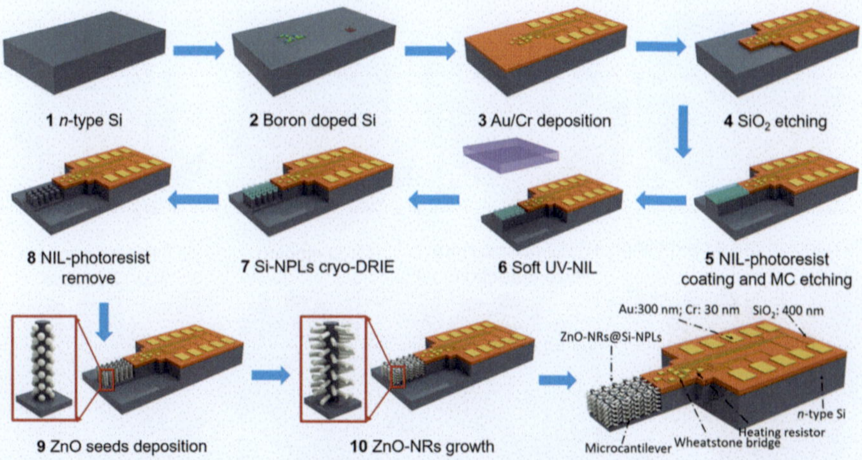

Fig. 2.10 Schematic diagram of the main steps to fabricate a piezoresistive microcantilever with a 3D hierarchical nanoframework. (1) An n-type bulk silicon wafer. (2) The heating and Wheatstone bridge resistors were embedded by boron diffusion (Xu et al., 2017a, 2018a). (3) A bilayer of 30-nm-thick Cr and 300-nm-thick Au was subsequently deposited as the contact pads on the up-surface by electron-beam physical vapor deposition. (4) The wafer upside was selectively etched using cryogenic deep reactive-ion etching (cryo-DRIE) to form a 10-μm-thick "pre-cantilever." (5) A bonding agent layer (mr-APS1) of and a layer of ~130-nm-imprint resist (mr-UVcur06 1:1) were spun-on the sample surface. (6) Then a fabricated polydimethylsiloxane (PDMS) stamp (pillar diameter, ~300 nm; pillar space, ~200 nm) was pressed on the coated substrate. (7) With the help of photolithography, an array of pillars (diameter, ~200 nm; length, ~1.5 μm) was selectively engraved on the "pre-cantilever" by cryo-DRIE. (8) Thereafter, the photoresists were removed, and (9) a layer of Zn film was selectively sputtered and oxidized to form a layer of ZnO nanoparticles on the "pre-cantilever." (10) Then the sample with ZnO seed layer was put in a chemical bath deposition (CBD) solution for 30 min, where the ZnO nanorods of ~25 nm in diameter and ~250 nm in length were self-organizedly grown. Finally, a microcantilever with triangular end shape (width, 170 μm; length, 300 μm; thickness, 12 μm (in the beam part); triangular side length, 700 μm (in the tip part)) was obtained after backside free releasing using cryo-DRIE. (Reprinted from (Xu et al., 2020), with permission from Copyright © 2020 ACS)

inner diameters, heights/hole depths, and pitches (Fig. 2.11) (Xu et al., 2017b). J. Xu et al. reported also the application of NSL technique for micromachining Si NRs on microcantilevers (Xu et al., 2019b).

2.4.3.3 Electron-Beam Lithography (EBL)

With electron-beam lithography (known as e-beam lithography, EBL) nanostructures down to ~2 nm can be realized by exposing a polymeric thin resist film under a highly focused electron beam (Chen, 2015). Similar to the photolithography process, the solubility of the resist was changed by irradiation (electrons here vs. photons in case of photolithography), enabling selective removal of either the exposed or the nonexposed area of the resist after developing. And the nanopatterned resist can be used either as a positive etching mask or negative etching mask for the

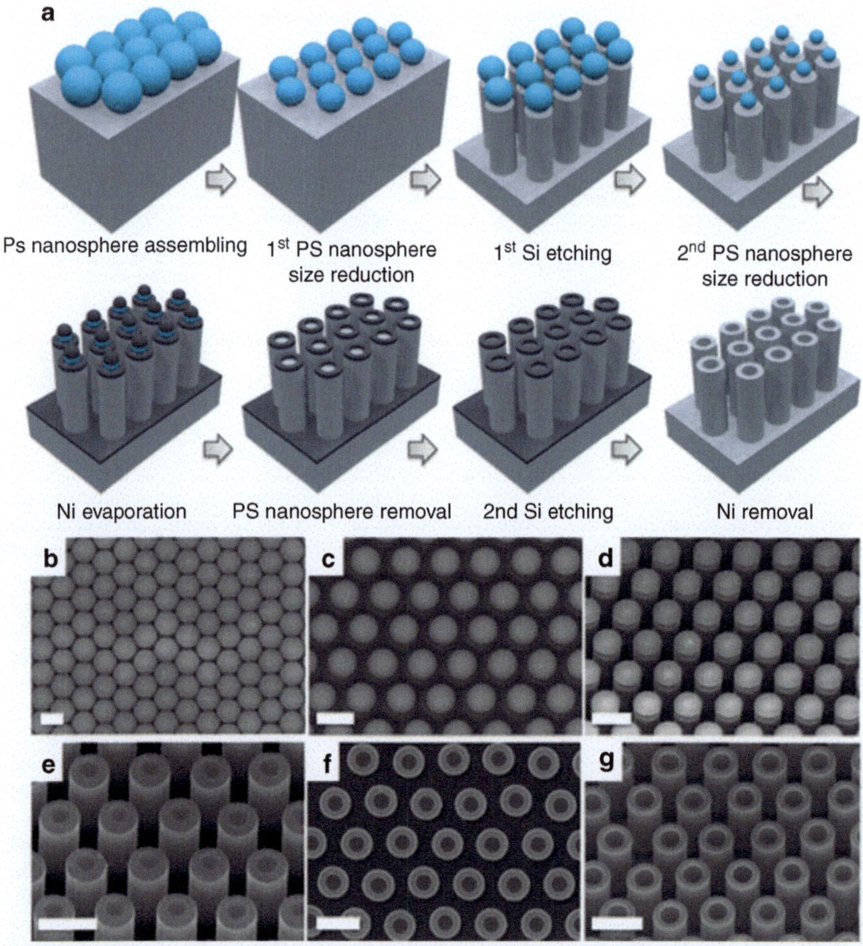

Fig. 2.11 Self-aligned multiple-patterning nanosphere lithography (MP-NSL) for Si nanotube arrays. (**a**) A schematic illustration of the process: Step 1: A monolayer of close-packed polystyrene (PS) nanospheres is formed at a water/air interface and then transferred onto a Si wafer. The size of the nanospheres defines the ultimate pitch (p) of the Si nanotubes. Step 2: Oxygen plasma RIE reduces the sphere size and defines the outer diameter (d_o) of Si nanotubes. Step 3: Deep reactive-ion etching (DRIE) etches Si into nanopillars by using the nanospheres as mask. The outer height (h_o) of the Si nanotubes is controlled by the etch time. Step 4: A second oxygen plasma RIE further reduces the size of the polymer nanoparticles and defines the inner diameter (d_i) of Si nanotubes. Step 5: Ni is deposited to form Ni nanorings on the top surface of the Si nanopillars, which functions as a DRIE mask. Step 6: The polystyrene nanoparticles are removed by 10 min oxygen plasma etching lifting the Ni on top and exposing the center part of the Si nanopillars to a second DRIE. Step 7: DRIE is performed to etch holes in the pillars to a defined depth (h_i). Step 8: The Ni is removed by HCl, and pristine Si nanotube arrays are obtained. Scanning electron microscope (SEM) images of key intermediates: (**b**) monolayer of closed-packed polystyrene nanospheres (diameter: 1 μm) formed on a Si wafer; (**c**) polystyrene nanoparticles on the Si wafer after first size reduction; (**d**) polystyrene nanoparticles on top of periodic Si nanopillar arrays; (**e**) second size reduction of polystyrene nanoparticles by oxygen plasma (step 4); and (**f**) Ni nanorings on top of the Si nanopillars. The region displayed as dark is Si and the region displayed as bright is Ni. (**g**) Etching of the inner regions by DRIE to form Si nanotubes (step 7). Images (**d, e, g**) were taken at a tilt of 30°. Scale bars: 1 μm. (Reprinted from (Xu et al., 2017b), Copyright © 2017, with permission from ACS)

subsequent metal deposition, where the metal works as hard mask for etching after a lift-off process.

The primary advantage of the EBL is that it is a maskless (direct write) process and it can draw nanopatterns with sub-10 nm resolution. Although it has high resolution, throughput is low limiting its application in high-volume production of semiconductor devices. Nevertheless, it is an eligible candidate for micromachining nanostructure on microcantilevers, and nanopatterns of hard masks prepared by EBL can combine well with the photolithography technique for the processing of Si.

These top-down approaches can realize microcantilevers with uniform nanostructure arrays, which enhance the cantilever's surface area and maintain its Q. Although this combination will increase the number of necessary fabrication steps, the nanostructure-patterned microcantilevers will benefit from better performance due to their large surface area, which enable their application for gas detection.

2.5 Environmental Sensing Applications

Microcantilever-based sensors have very high sensitivity with an LOD lower than most of the conventional advanced techniques (Vashist et al., 2018; Debeda & Dufour, 2020). Resonant microcantilevers of Si are the mostly used microcantilevers for gas detection requiring LOD of sub-ppm level, for example, volatile organic compounds (VOCs), chemical warfare agents, and explosives. There are many resonant cantilever-based sensors which have been reported and reviewed (Boisen et al., 2011; Goeders et al., 2008; Vashist et al., 2018; Debeda & Dufour, 2020; Lavrik et al., 2004). Many approaches have been explored to improve the performance of resonant microcantilever-based gas sensors, for example, increasing the resonant frequency of the device, by using shorter and stiffer microcantilevers or by selecting higher resonance modes. For increasing the microcantilever's Q factor, specific designs were considered, such as using in-plane and torsional modes instead of out-of-plane flexural modes (Fig. 2.3) (Brand et al., 2015), miniaturized cantilever dimensions Bertke et al. (2021), or special device geometries such as triangular (Bertke et al., 2017), disk (Seo & Brand, 2008), and hammerhead shapes (Beardslee et al., 2011). In view of gas-sensitive coatings, metallic, ceramics, polymers, etc. have been studied. However, the sensing applications of nanostructured microcantilevers are not reviewed before. In this chapter, we will focus only on resonant microcantilevers coated with 3D hierarchically nanostructured sensing material or surface, which are mostly fabricated to realize an LOD of sub-ppm or sub-ppb level.

Relative humidity sensing is one of the most investigated applications of microcantilever-based resonators. Resistive humidity sensors and capacitive humidity sensors are currently the most widely used commercial humidity sensors, because of their advantages of easy fabrication, simple readout circuits (resistive humidity sensor) and low power consumption, wide temperature range, and long-term stability (capacitive humidity sensor). However, there are still some corresponding drawbacks that limit their applications, such as long recovery times and low stability for the former, complex readout circuits required for highly precise detection for the latter (Fenner & Zdankiewicz, 2001; Lee & Lee, 2005; Farahani et al., 2014; Chen & Lu, 2005). In addition, the performance of both resistive and capacitive humidity sensors is directly affected by the electrical characteristics of the sensitive materials, which thus leads to nonlinear sensing responses at either low RH levels (typical for a resistive sensor) or high RH levels (capacitive sensor) (Farahani et al., 2014; Kim et al., 2010; Wang et al., 2010; Huang et al., 2015). As a type of gravimetric sensor, microcantilevers with micro/nanopatterns, such as SiO_2 microcantilever with controlled micro-patterns on their surface (Balasubramanian et al., 2020), and Si microcantilevers with chitosan SAMs-modified ZnO nanorods, etc. (Xu et al., 2019) have been demonstrated to be high-performance relative humidity sensors. These nanostructured-microcantilevers show high sensitivity and fast response/recovery (~1 s) to a wide range of relative humidity detection (0–97.3%RH).

One great advantage of resonant microcantilever-based gas sensors is their high sensitivity at moderate working conditions, i. e., they mostly work at room temperature; therefore, they are widely investigated as a promising portable device for chemical warfare agents and explosives, for example, the organophosphorus compounds (Ops). Ppb-level gas detection of H_2S, CO, and dimethyl methylphosphonate (DMMP, which is a well-known simulant of sarin (GB)) has also been demonstrated by the coating of a zeolitic imidazolate framework-8 (ZIF-8)-derived nitrogen-doped nanoporous carbon (NPC), a MOF nanomaterial based on UiO-66 (Cai et al., 2019) and Ni-MOF-74 (Lv et al., 2018) on a resonant cantilever, respectively. S. Cai reported a method for toxic organophosphorus molecule detection by using a typical MOF consisting of a UiO-66 film deposited on a resonant microcantilever (Cai et al., 2019). With the help of photolithography, parylene-C with hydrophobic character is deposited and patterned onto the microcantilever, to form area-selectively a sample reservoir at the free end of the microcantilever for deposition of UiO-66. UiO-66 is constructed in situ onto the sensing region (at the free end) of the microcantilever, by being dropped using an inkjet printer and annealed, realizing an experimentally observed LOD to DMMP of better than 5 ppb. Besides, the sensor exhibits a good selectivity and repeatability (Fig. 2.12).

Selective deposition of 1D CuO nanorods (Schlur et al., 2020) and TiO_2 nanorods (Biapo et al., 2019) on microcantilevers as well as 3D ZnO nanorods on a Si-nanopillar array (3D ZnO-NRs@Si-NPLs) (Xu et al., 2019c) on resonant microcantilevers has been reported, where sensitivity enhancement factors of tens to hundreds have been realized with respect to planar structures (Fig. 2.13a). Nevertheless, 1D metal oxide nanostructure-patterned cantilevers showed only ppm-level LODs, which are scant compared with cantilevers functionalized with ink printed MOF nanomaterials. The former cantilevers exhibit normally higher Q factor, thanks to the more uniformly distributed nanostructures. However, as reported by J. Xu et al. (2020), dimensional inorganic nanostructures can be further functionalized by a thin (organic) self-assembled monolayers (SAM) to realize a ppb-level gas-detection performances (Fig. 2.13b). This finding demonstrates that resonant microcantilevers with 3D hierarchical nanostructured framework are potential sensing platforms

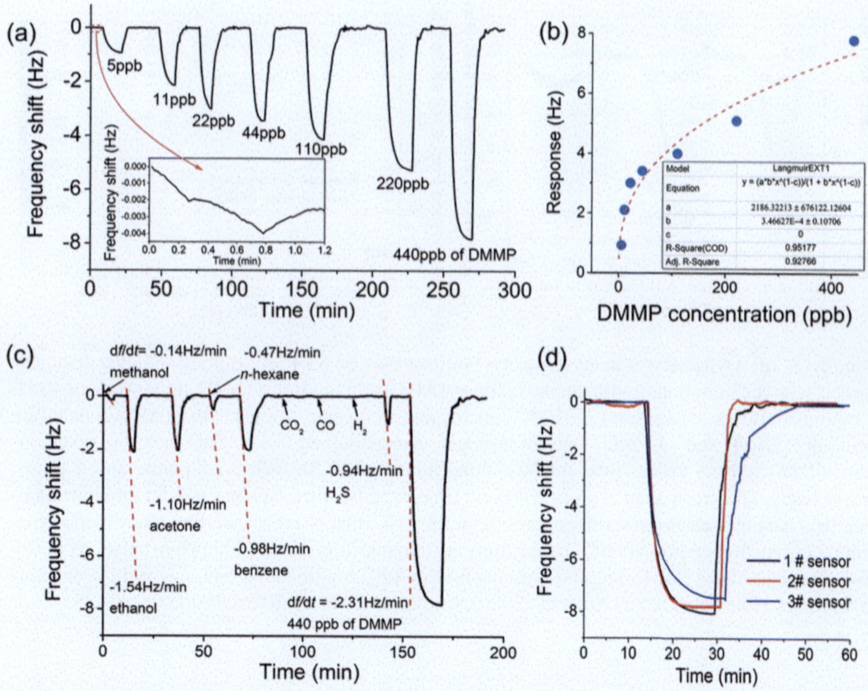

Fig. 2.12 (**a**) Gravimetric frequency shift of a microcantilever functionalized with a UiO-66 film as the sensitive material to DMMP vapor. The inset shows the short-term noise floor of ±0.002 Hz; (**b**) relationship between the sensing response (frequency shift) and the DMMP concentration, which can be well fitted as a function of the Langmuir equation; (**c**) selectivity of the sensor to nine kinds of interfering gases (each with a concentration of 50 ppm); (**d**) sensing response of three prototypes of the microcantilever sensor to DMMP vapor at an identical concentration of 440 ppb. (Reprinted from (Cai et al., 2019), Copyright 2019, with permission from The Royal Society of Chemistry)

of combining both top-down and bottom-up approaches, by modifying nanopatterned microcantilevers (with micromachined or grown nanostructures) with high surface area sensing materials, such as MOFs, SAMs, graphene, phosphorene, MXene, and other 2D sensing materials. Therefore, numerous possible combinations of these techniques with the resonant microcantilever platform will trigger future research for sub-ppb-level gas detections.

The gas sensing performance of most recently collected state-of-the-art nanostructure-patterned microcantilevers can be compared as shown in Table 2.1.

2.6 Conclusion and Future Trends

Microcantilever-based sensors have distinct advantages including high sensitivity, fast response, high throughput, miniaturized size, low cost, and low power consumption. The approaches of fabricating semiconductor and other nanostructures

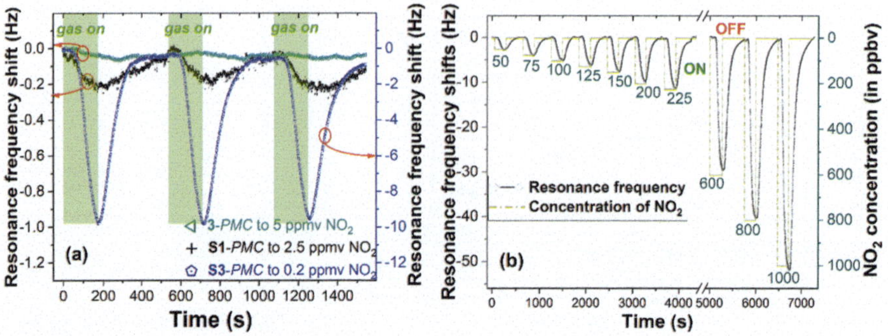

Fig. 2.13 (**a**) Dynamic resonant-frequency shift curves of S3-PMC (microcantilever nanopatterned with a (3-aminopropyl)trimethoxysilane (APTES) nanofilm on a 3D framework of ZnO-nanorods@silicon-nanopillars), S1-PMC (microcantilever nanopatterned with an APTES nanofilm on ZnO NRs) and 3-PMC (microcantilever nanopatterned with ZnO-nanorods@silicon-nanopillars) sensors versus time for NO_2 concentrations of 200 ppbv, 2.5 ppmv, and 5 ppmv, respectively. The green-shaded areas represent time periods, when NO_2 gas was let into the chamber. (**b**) Dynamic sensing-response curves of an S3-PMC, that is, resonance-frequency shift versus time for switched on and off NO_2 concentrations ranging from 50 to 1000 ppbv (balanced in dry air). The dashed-dot line (orange) represents the "on/off" conditions of NO_2 gas at different concentrations. (Reprinted from (Xu et al., 2020), Copyright 2020, with permission from ACS)

on the microcantilever's surface to increase their surface area have been systematically introduced and discussed in this chapter; this dimensional nanopatterning of sensing materials can notably improve the sensitivity and LOD of microcantilever-based sensors. Especially, we can build 3D nanostructures on microcantilevers by combining top-down method and bottom-up method, which can give them super-large surface area. Furthermore, the combined use of inorganic nanostructures with large surface area organic coating will substantially increase the surface area and reduce the thickness of sensitive layer, thereby resulting in increased number of bound receptor molecules that can detect an analyte with high sensitivity, good signal-to-noise ratio, and reduced response/recovery time, at room temperature. These properties enable the microcantilever-based sensors to be a perfect candidate as high-performance gas sensors for ppb- or even ppt-level gas detection. If fabricating an array of these dimensional microcantilevers and the successful integration of readout for such arrays can be realized, real-time selective analysis of gas mixes at room temperature will be feasible, which will be an attractive and the most dominant application of microcantilever-based sensors.

The rapid advances in complementary technologies would further facilitate the development of critically improved microcantilever-based sensors. Extensive research efforts in developing and optimizing integration of readout microcantilever arrays will improve further the real-time portable or wireless sensing of microcantilever-based sensors with high precision. Application of microcantilever-based sensors for fabricating portable detectors or integrating with smart sensing systems of Internet of Things can be expected.

Table 2.1 Sensing performances of microcantilevers with nanostructures

Cantilever-based sensors	Readout	*Q*	Surface area	Target gas	Reaction condition	LOD (ppb)	Selectivity
Dual Si PMC (chip size: 700 × 700 × 450 μm³) with BCC and NPC (Tang et al., 2020)	Piezoresistive	Not given	Not given	H_2S	RT, BCC irreversible, NPC reversible	1	Yes
Si PMC (200 × 100 × 3 μm³) with MOF of UiO-66 (Cai et al., 2019)	Piezoresistive	326.2	939 (m²/g)	DMMP	RT	<5	Yes
Si PMC (200 × 100 × 3 μm³) with MOF nanoporous material of Ni-MOF-74 (Lv et al., 2018)	Piezoresistive	Not given	1152 (m²/g)	CO	25 °C	<10	Yes
Si PMC with SAM-on-3D ZnO-NRs@ Si-NPL (Xu et al., 2020)	Piezoresistive	~1000	112 times	NO_2	RT, with light illumination	~2.1	Not given
Si PMC (225 × 28 × 3 μm³) with CuO NRs (Schlur et al., 2020)	Optical	Not given	27 times	DMMP	RT	16.4 ppm	Not given
Si PMC (225 × 38 × 7 μm³) with oxime-modified TiO_2 NRs (Biapo et al., 2019)	Not given	Not given	75 times	DMMP	RT	2.25 ppm	Not given
Si microcantilever (225 × 38 × 7 μm³) with double-sided coated TiO_2 nanotubes (Thomas et al., 2020)	Optical	Not given	343 times	DMMP	16 °C	2.8 ppm	Not given

BCC basic carbonate, *DMMP* dimethyl methylphosphonate, *MOF* metal organic frameworks, *NPC* nanoporous carbon, *NPL* nanopillar, *NRs* nanorods, *PMC* piezoresistive microcantilever, *SAM* self-assembled monolayer

Acknowledgements This project has received funding from the EMPIR program co-financed by the Participating States and from the European Union's Horizon 2020 research and innovation program under No. 19ENG05 Nanowires.

References

Abdolvand, R., Bahreyni, B., Lee, J. E. Y., & Nabki, F. (2016). Micromachined resonators: A review. *Micromachines, 7*, 160.

Balasubramanian, S., Polaki, S. R., & Prabakar, K. (2020). Ultrahigh sensitive and ultrafast relative humidity sensing using surface enhanced microcantilevers. *Smart Materials and Structures, 29*, 095006. https://doi.org/10.1088/1361-665x/ab9f1a

Bargatin, I., Kozinsky, I., & Roukes, M. L. (2007). Efficient electrothermal actuation of multiple modes of high-frequency nanoelectromechanical resonators. *Applied Physics Letters, 90*, 88–91.

Battiston, F. M., Ramseyer, J., Lang, H. P., Baller, M. K., & Gerber, C. (2001). A chemical sensor based on a microfabricated cantilever array with simultaneous resonance-frequency and bending readout. *Sensors and Actuators B: Chemical, 77*, 122–131.

Bauer, J., et al. (2007). Electrical properties of nominally undoped silicon nanowires grown by molecular-beam epitaxy. *Applied Physics Letters, 90*, 10–13.

Beardslee, L. A., et al. (2011). Selectivity enhancement strategy for cantilever-based gas-phase VOC sensors through use of peptide-functionalized carbon nanotubes. In *Proceedings of the IEEE International Conference on Micro Electro Mechanical Systems (MEMS)* (pp. 964–967). IEEE. https://doi.org/10.1109/MEMSYS.2011.5734587

Bertke, M., et al. (2017). Analysis of asymmetric resonance response of thermally excited silicon micro-cantilevers for mass-sensitive nanoparticle detection. *Journal of Micromechanics and Microengineering, 27*, 064001.

Bertke, M., et al. (2018). Contact resonance spectroscopy for on-the-machine manufactory monitoring. *Sensors and Actuators, A: Physical, 279*, 501–508.

Bertke, M., et al. (2020). Fabrication of a microcantilever-based aerosol detector with integrated electrostatic on-chip ultrafine particle separation and collection. *Journal of Micromechanics and Microengineering, 30*, 014001.

Bertke, M., et al. (2021). Ultrafine Aerosol Particle Sizer Based on Piezoresistive Microcantilever Resonators with Integrated Air-Flow Channel. *Sensors, 21*(3731), 19. https://doi.org/10.3390/s21113731.

Biapo, U., et al. (2019). Functionalized TiO2 nanorods on a microcantilever for the detection of organophosphorus chemical agents in air. *ACS Applied Materials & Interfaces, 11*, 35122–35131.

Boisen, A., Thaysen, J., Jensenius, H., & Hansen, O. (2000). Environmental sensors based on micromachined cantilevers with integrated read-out. *Ultramicroscopy, 82*, 11–16.

Boisen, A., Dohn, S., Keller, S. S., Schmid, S., & Tenje, M. (2011). Cantilever-like micromechanical sensors. *Reports on Progress in Physics, 74*, 036101.

Brand, O., Dufour, I., Heinrich, S., & Josse, F. (2015). *Resonant MEMS—Fundamentals, implementation and application*. Wiley.

Cai, S., et al. (2019). In situ construction of metal-organic framework (MOF) UiO-66 film on Parylene-patterned resonant microcantilever for trace organophosphorus molecules detection. *Analyst, 144*, 3729–3735.

Chang, S. W., Chuang, V. P., Boles, S. T., Ross, C. A., & Thompson, C. V. (2009). Densely packed arrays of ultra-high-aspect-ratio silicon nanowires fabricated using block-copolymer lithography and metal-assisted etching. *Advanced Functional Materials, 19*, 2495–2500.

Chaste, J., et al. (2012). A nanomechanical mass sensor with yoctogram resolution. *Nature Nanotechnology, 7*, 301–304.

Chen, Y. (2015). Nanofabrication by electron beam lithography and its applications: A review. *Microelectronic Engineering, 135*, 57–72.

Chen, Z., & Lu, C. (2005). Humidity sensors: A review of materials and mechanisms. *Sensor Letters, 3*, 274–295.

de Boor, J., Gösele, U., Huang, Z., Geyer, N., & Werner, P. (2010). Metal-assisted chemical etching of silicon: A review. *Advanced Materials, 23*, 285–308.

Debeda, H., & Dufour, I. (2020). Resonant microcantilever devices for gas sensing. In *Advanced nanomaterials for inexpensive gas microsensors* (pp. 161–187). https://doi.org/10.1016/B978-0-12-814827-3.00009-8

Dufour, I., Lochon, F., & Josse, F. (2007). Effect of coating viscoelasticity on quality factor and limit of detection of microcantilever chemical sensors. *IEEE Sensors Journal, 7*, 230–236.

Fantner, G. E., Odermatt, P. D., & Eskandarian, H. A. (2017). Applications of MEMS to cell biology. In B. Bhushan (Ed.), *Springer handbook of nanotechnology* (pp. 587–616). Springer. https://doi.org/10.1007/978-3-662-54357-3_19

Farahani, H., Wagiran, R., & Hamidon, M. N. (2014). Humidity sensors principle, mechanism, and fabrication technologies: A comprehensive review. *Sensors, 14*, 7881–7939.

Fenner, R., & Zdankiewicz, E. (2001). Micromachined water vapor sensors: A review of sensing technologies. *IEEE Sensors Journal, 1*, 309–317.

Fuhrmann, B., et al. (2005). Ordered arrays of silicon nanowires produced by nanosphere lithography and molecular beam epitaxy. *Nano Letters, 5*, 2524–2527.

Geisse, N. A. (2009). AFM and combined optical techniques. *Materials Today, 12*, 40–45.

Goeders, K. M., Colton, J. S., & Bottomley, L. A. (2008). Microcantilevers: Sensing chemical interactions via mechanical motion. *Chemical Reviews, 108*, 522–542.

Hamdana G., et al. (2017). Towards fabrication of 3D isotopically modulated vertical silicon nanowires in selective areas by nanosphere lithography. *Microelectronic Engineering, 179*, 74–82. http://dx.doi.org/10.1016/j.mee.2017.04.030.

Hamdana, G., et al. (2018). Nanoindentation of crystalline silicon pillars fabricated by soft UV nanoimprint lithography and cryogenic deep reactive ion etching. *Sensors and Actuators A: Physical, 283*, 65–78.

Haynes, C. L., & Van Duyne, R. P. (2001). Nanosphere lithography: A versatile nanofabrication tool for studies of size-dependent nanoparticle optics. *The Journal of Physical Chemistry. B, 105*, 5599–5611.

Huang, J., Chiam, S. Y., Tan, H. H., Wang, S., & Chim, W. K. (2010). Fabrication of silicon nanowires with precise diameter control using metal nanodot arrays as a hard mask blocking material in chemical etching. *Chemistry of Materials, 22*, 4111–4116.

Huang, J. Q., Li, F., Zhao, M., & Wang, K. (2015). A surface micromachined CMOS MEMS humidity sensor. *Micromachines, 6*, 1569–1576.

Jensenius, H., et al. (2000). A microcantilever-based alcohol vapor sensor-application and response model. *Applied Physics Letters, 76*, 2615–2617.

Ji, D., Li, T., & Fuchs, H. (2017). Nanosphere lithography for Sub-10-nm nanogap electrodes. *Advanced Electronic Materials, 3*, 1–5.

Johnson, B. N., & Mutharasan, R. (2012). Biosensing using dynamic-mode cantilever sensors: A review. *Biosensors & Bioelectronics, 32*, 1–18.

Kang, T. J., et al. (2005). Low-thermal-budget and selective relaxation of stress gradients in gold micro-cantilever beams using ion implantation. *Journal of Micromechanics and Microengineering, 15*, 2469–2478.

Kanungo, P. D., et al. (2008). Controlled in situ boron doping of short silicon nanowires grown by molecular beam epitaxy. *Applied Physics Letters, 92*, 263107.

Kilinc, N., et al. (2014). Fabrication of 1D ZnO nanostructures on MEMS cantilever for VOC sensor application. *Sensors and Actuators B: Chemical, 202*, 357–364.

Kim, J. H., Hong, S. M., Moon, B. M., & Kim, K. (2010). High-performance capacitive humidity sensor with novel electrode and polyimide layer based on MEMS technology. *Microsystem Technologies, 16*, 2017–2021.

Lang, H. P., Gerber, C., et al. (1998). A chemical sensor based on amicromechanical cantilever array for the identification of gases and vapors. *Applied Physics A, 66*, S61–S64.

Lavrik, N. V., Sepaniak, M. J., Datskos, P. G., & Lavrik, N. V. (2004). Cantilever transducers as a platform for chemical and biological sensors. *The Review of Scientific Instruments, 2229*, 2229–2253.

Lee, C.-Y., & Lee, G.-B. (2005). Humidity sensors: A review. *Sensor Letters, 3*, 1–15.

Lee, B., et al. (2017). Magnetoelectric versus thermal actuation characteristics of shear force AFM probes with piezoresistive detection. *Measurement Science and Technology, 28*, 034011.

Li, L., et al. (2017). Controlling the geometries of Si nanowires through tunable nanosphere lithography. *ACS Applied Materials & Interfaces, 9*, 7368–7375.

Liu, M., et al. (2018). Revealing humidity-enhanced NH3 sensing effect by using resonant microcantilever. *Sensors and Actuators B: Chemical, 257*, 488–495.

Lv, Y., Xu, P., Yu, H., Xu, J., & Li, X. (2018). Ni-MOF-74 as sensing material for resonant-gravimetric detection of ppb-level CO. *Sensors and Actuators B: Chemical, 262*, 562–569.

Ma, R.-H., Lee, C.-Y., Wang, Y.-H., & Chen, H.-J. (2008). Microcantilever-based weather station for temperature, humidity and flow rate measurement. *Microsystem Technologies Nanosystems and Storage Processing Systems, 14*, 971–977.

Manzaneque, T., et al. (2014). Piezoelectric MEMS resonator-based oscillator for density and viscosity sensing. *Sensors and Actuators, A: Physical, 220*, 305–315.

Meyer, G., & Amer, N. M. (1988). Novel optical approach to atomic force microscopy. *Applied Physics Letters, 53*, 2400–2402.

Morales, A. M. (2008). A laser ablation method for the synthesis of crystalline semiconductor nanowires. *Science (80-), 208.*

Oden, P. I., Chen, G. Y., Steele, R. A., Warmack, R. J., & Thundat, T. (1996). Viscous drag measurements utilizing microfabricated cantilevers. *Applied Physics Letters, 68*, 3814–3816.

Okan, M., Sari, E., & Duman, M. (2017). Molecularly imprinted polymer based micromechanical cantilever sensor system for the selective determination of ciprofloxacin. *Biosensors & Bioelectronics, 88*, 258–264.

Pandit, M., Zhao, C., Sobreviela, G., Mustafazade, A., & Seshia, A. A. (2017). Reduction of amplitude ratio dependence on drive level in mode localized resonant MEMS sensors. In *Sensors, 2017 IEEE* (pp. 1–3). IEEE.

Park, I., Han, H.-J., Jung, Y. S., Cho, M., & Gao, M. (2018). Hydrogen sensors: Palladium-decorated silicon nanomesh fabricated by nanosphere lithography for high performance, room temperature hydrogen sensing (small 10/2018). *Small, 14*, 1870041.

Possas-Abreu, M., et al. (2017). Development of diamond and silicon MEMS sensor arrays with integrated readout for vapor detection. *Sensors (Switzerland), 17*, 1–15.

Rodriguez, B. J., Callahan, C., Kalinin, S. V., & Proksch, R. (2007). Dual-frequency resonance-tracking atomic force microscopy. *Nanotechnology, 18*, 475504.

Schlur, L., Calado, J. R., & Spitzer, D. (2018a). Synthesis of zinc oxide nanorods or nanotubes on one side of a microcantilever. *Royal Society of Chemistry, 5*, 180510.

Schlur, L., et al. (2018b). Cu(OH)2 and CuO nanorod synthesis on piezoresistive cantilevers for the selective detection of nitrogen dioxide. *Sensors (Switzerland), 18*, 1108.

Schlur, L., Agostini, P., Thomas, G., Gerer, G., Grau, J., & Spitzer, D. (2020). Detection of organophosphorous chemical agents with CuO-nanorod-modified microcantilevers. *Sensors (Switzerland), 20*, 1061.

Schmidt, V., Senz, S., & Gösele, U. (2005). Diameter-dependent growth direction of epitaxial silicon nanowires. *Nano Letters, 5*, 931–935.

Seo, J. H., & Brand, O. (2008). High Q-factor in-plane-mode resonant microsensor platform for gaseous/ liquid environment. *Journal of Microelectromechanical Systems, 17*, 483–493.

Su, S., Lin, L., Li, Z., Feng, J., & Zhang, Z. (2013). The fabrication of large-scale sub-10-nm core-shell silicon nanowire arrays. *Nanoscale Research Letters, 8*, 1–7.

Tang, L., Xu, P., Li, M., Yu, H., & Li, X. (2020). H2S gas sensor based on integrated resonant dual-microcantilevers with high sensitivity and identification capability. *Chinese Chemical Letters, 31*, 2155–2158.

Thaysen, J., Boisen, A., Hansen, O., & Bouwstra, S. (2000). Atomic force microscopy probe with piezoresistive read-out and a highly symmetrical Wheatstone bridge arrangement. *Sensors and Actuators, A: Physical, 83*, 47–53.

Thomas, G., et al. (2020). Double side nanostructuring of microcantilever sensors with TiO2-NTs as a route to enhance their sensitivity. *Nanoscale, 12*, 13338–13345.

Tortonese, M., Yamada, H., Barrett, R. C., & Quate, C. F. (1991). Atomic force microscopy using a piezoresistive cantilever. In *TRANSDUCERS '91: 1991 International conference on solid-state sensors and actuators. Digest of Technical Papers* (pp. 448–451).

Vashist, S. K., Luong, J. H. T., Vashist, S. K., Chaudhary, M., & Gupta, A. (2018). Microcantilever-based sensors. In J. L. Sandeep Vashist (Ed.), *Handbook of immunoassay technologies* (Vol. 59, pp. 305–332). Elsevier.

Wang, Y., Park, S., Yeow, J. T. W. W., Langner, A., & Müller, F. (2010). A capacitive humidity sensor based on ordered macroporous silicon with thin film surface coating. *Sensors and Actuators B: Chemical, 149*, 136–142.

Wasisto, H. S., et al. (2013). Portable cantilever-based airborne nanoparticle detector. *Sensors and Actuators B: Chemical, 187*, 118–127.

Wasisto, H. S., et al. (2014). A phase—locked loop frequency tracking system for portable micro-electromechanical piezoresistive cantilever mass sensors. *Microsystem Technologies, 20*, 559–569.

Xu, J., et al. (2017a). Fabrication of ZnO nanorods on MEMS Piezoresistive silicon microcantilevers for environmental monitoring. *Proceedings, 1*, 290.

Xu, X., et al. (2017b). Multiple-patterning Nanosphere lithography for fabricating periodic three-dimensional hierarchical nanostructures. *ACS Nano, 11*, 10384–10391.

Xu, J., et al. (2018a). Fabrication of ZnO nanorods and chitosan@ZnO nanorods on MEMS piezoresistive self-actuating silicon microcantilever for humidity sensing. *Sensors and Actuators B: Chemical, 273*, 276–287.

Xu, J., et al. (2018b). Area-selective growth of aligned ZnO nanorod arrays for MEMS device applications. *Proceedings, 2*, 887.

Xu, J., Bertke, M., Wasisto, H. S., & Peiner, E. (2019). Piezoresistive microcantilevers for humidity sensing. *Journal of Micromechanics and Microengineering, 29*, 053003.

Xu, J., et al. (2019b). Silicon nanopillars with ZnO nanorods by nanosphere lithography on a piezoresistive microcantilever. In *2019 20th International Conference on Solid-State Sensors, Actuators Microsystems Eurosensors XXXIII (TRANSDUCERS EUROSENSORS XXXIII)* (pp. 2420–2423).

Xu, J., et al. (2019c). Piezoresistive microcantilevers 3D-patterned using ZnO-nanorods@silicon-nanopillars for room-temperature ethanol detection. In *2019 20th International Conference on Solid-State Sensors, Actuators Microsystems Eurosensors XXXIII (TRANSDUCERS EUROSENSORS XXXIII)* (pp. 1211–1214).

Xu, J., Setiono, A., & Peiner, E. (2020). Piezoresistive microcantilever with SAM-modified ZnO-nanorods@silicon-nanopillars for room-temperature parts-per-billion NO2 detection. *ACS Applied Nano Materials, 3*, 6609–6620.

Yang, Y., et al. (2015). Experimental investigation on mode coupling of bulk mode silicon MEMS resonators. In *Micro Electro Mechanical Systems (MEMS), 2015 28th IEEE International Conference on 1008–1011*. IEEE.

Chapter 3
Micromachining Based on Mask-Free Direct Writing: An Advanced Approach to Innovative MEMS Gas Sensors

Rui You, Wenshuai Lu, Dongdong Han, and Yonglai Zhang

3.1 Introduction

Over the past decade, microelectromechanical systems (MEMS), tiny systems that make a huge difference, have promoted the wave of electronics evolution. The potential utility of MEMS has been expanded from consumer electronics to various next-generation applications (Fischer et al., 2015; Lee et al., 2018; Ren et al., 2020; Voiculescu & Nordin, 2012). Environment monitoring by miniaturized devices, to date, is still an ongoing area of great academic and technological interest. Among all these devices, MEMS gas sensors have become increasingly evident to satisfy the requirement in the growing market of healthcare management, intelligent manufacturing, etc. Despite certain market share, conventional silicon-based MEMS gas sensors have limitations in material varieties and fabrication flexibility (Ando & Fu, 2019; DelRio et al., 2015; Qiu et al., 2019; Shin et al., 2009; Zheludev & Kivshar, 2012). Certain limitations are inevitably originated and inherited from semiconductor foundries. These limitations lead to great challenges for conventional silicon-based MEMS gas sensors to embrace emerging opportunities of application expansion. To guarantee competitiveness in the future market, disruptive innovation in materials is a long-cherished desire to improve and even innovate MEMS gas

R. You (✉)
National Key Laboratory of Science and Technology on Micro/Nano Fabrication, Institute of Microelectronics, Peking University, Beijing, China
e-mail: yourui19@pku.edu.cn

W. Lu
State Key Laboratory of Precision Measurement Technology and Instruments, Department of Precision Instrument, Tsinghua University, Beijing, China

D. Han · Y. Zhang
State Key Laboratory of Integrated Optoelectronics, College of Electronic Science and Engineering, Jilin University, Changchun, China

Z. Yang (ed.), *Advanced MEMS/NEMS Fabrication and Sensors*,
https://doi.org/10.1007/978-3-030-79749-2_3

sensors. For this purpose, massive efforts have been made to explore new functional materials for the development of advanced MEMS gas sensors. These functional materials, generally with tailoring nanoscale structures, are achieved by special treatment of binary composite oxides, polymers, carbon-based materials, etc. (Briand et al., 2006; Chang & Hsihe, 2016; Jian et al., 2020; Schroeder et al., 2019; Seshadri et al., 2019; Tan et al., 2019). It is noted that standard silicon micromachining, either bulk-silicon (top-down) or surface micromachining (bottom-up), is not fully competent to cope with these functional materials. To address this issue, the concept of mask-free direct writing has been introduced as a welcome addition to traditional MEMS fabrication (Farahani et al., 2016; Fischer & Wegener, 2013; Guo et al., 2019; Yang & Zhou, 2014; You et al., 2020).

Mask-free direct writing, widely utilized in printed electronics, enables additive and subtractive manufacturing of micro/nano−/mix-scale structures. Similar to traditional MEMS fabrication, mask-free direct writing includes both "bottom-up" and "top-down" approaches (Heule et al., 2003; Ugrinic et al., 2019; Wu et al., 2016). Bottom-up approaches are designed to build micro/nano-components onto complex assemblies, while top-down approaches aim at fabricating micro/nanodevices from bulky materials (Altmann et al., 2012; Brauns et al., 2013). This creates a common basis to match mask-free direct writing and traditional MEMS fabrication in a complementary manner. Although many original direct writing MEMS gas sensors have already been presented in published literature, there is still plenty of room for improvement, especially for mix-scale cases (Arlett et al., 2011; Farahani et al., 2016; Sima et al., 2018; Xiao et al., 2015; Ye et al., 2019). Due to the variety of involved functional materials, more appropriate fabrication strategies are required to satisfy the increasing demand of fabrication accuracy and compatibility.

To facilitate the innovation of MEMS gas sensors, this chapter provides a special subject of two typical mask-free direct writing methods, namely, inkjet printing and femtosecond laser direct writing (FsLDW). These two methods are integrated in MEMS fabrication focusing on the treatment of binary composite oxides, polymers, and carbon-based materials for advanced MEMS gas sensors. Equipment, principle, standard process, and application examples of both inkjet printing and FsLDW are described in sequence in the following sections. It is expected that this chapter would be an essential guidance for researchers on the way forward to discover the beauty of mask-free direct writing in MEMS gas sensors.

3.2 Inkjet Printing for the Fabrication of Advanced MEMS Gas Sensor

3.2.1 Functional Materials Constructing and Patterning by In Situ Inkjet Printing

Micro inkjet printing is a controlled process to fabricate electronic devices with flow phase materials (Singh et al., 2010). By inkjet printing, trace metal or oxide material solution is programmatically printed from the micro nozzle to the substrate without mask patterning. Inkjet printing has high efficiency in material utilization by quantitative and precise control. In addition, the noncontact deposition process of inkjet printing minimizes the pollution and detrimental effect on the substrate (Tekin et al., 2008).

The essences of in situ inkjet printing for MEMS gas sensor are in situ synthesis and fabrication. In situ synthesis can be demonstrated by the following material preparation process. The first phase material is directly synthesized on the surface of the original structure, and the bi-materials are closely combined to avoid damage to the microstructure and introduction of pollution. To see from the big picture of the complete catalyst system, the support material is transferred to the substrate, and then the catalytic material is precisely grown on the surface of the support. In situ fabrication is to create micro/nanostructures at certain marked locations. For typical in situ inkjet printing, material preparation and patterning are completed simultaneously without the traditional chemical synthesis reaction, material transfer, or graphic deposition. This leads to the following: (1) reducing the consumption of the original material, (2) simplifying the process of mask graphic, and (3) protecting the original appearance of the material during synthesis. In the in situ fabrication of MEMS gas sensors, the catalyst is accurately printed on the functional area the support material without additional synthesis and transfer. This effectively realizes the maintenance of graphical and morphological properties of the catalyst.

By in situ synthesis and fabrication, in situ inkjet printing has great capability to pattern and construct varieties of functional materials. Figure 3.1 illustrates a classic fabrication case of nanocomposites by in situ inkjet printing. The nanocomposites are achieved by automatic screen printing and then layer-by-layer in situ hybrid printing. A porous supporter (framework) is provided for the attachment of the precursor (liquid phase catalytic material). In this case, the porous, consistent, accurate, and controllable fabrication of nanocomposites is realized by the absorbability of the flow phase material and the accurate positioning of the inkjet printing.

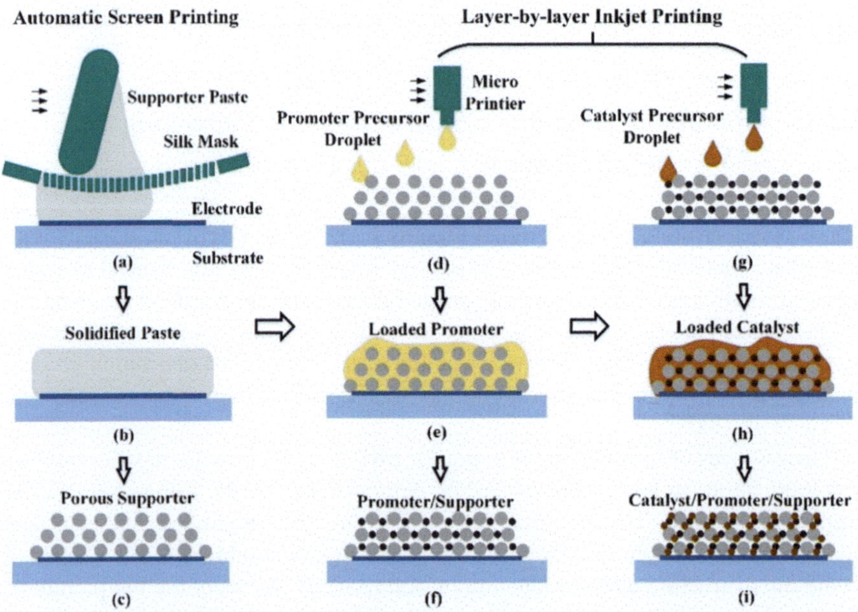

Fig. 3.1 Schematic diagrams of proposed screen-inkjet hybrid printing processes, including one-step screen printing and two-step inkjet printings

3.2.2 Application of Inkjet Printing in the Fabrication of Catalytic MEMS Methane Sensor

3.2.2.1 Principle and Method

The high-performance catalytic combustion sensor with a core-shell nanostructured catalyst is proposed here as an application example of inkjet printing. The catalyst was accurately in situ printed on the surface of the sensor by inkjet printing, with CeO_2 nanoparticles (NPs) and thermally decomposed palladium-platinum (Pd-Pt) bimetallic as core and shell material, respectively (CeO_2@Pd-Pt). In this nanostructure, CeO_2 NPs provide oxygen vacancies for PdO to stabilize its active phase, and sintering of Pd is suppressed by Pt. With the highly dispersed CeO_2 NPs promoter (framework cores and oxygen suppliers for Pd-Pt metal shell), the catalyst system shows good thermal and chemical stability at high working temperature up to 500 °C. Due to the unique core-shell structure catalyst of CeO_2@Pd-Pt, CeO_2 promoter shows chemical stability effect on Pd-Pt catalyst, significantly improving the signal-to-noise ratio of the sensor. The working temperature, response time, and robustness of the sensor are optimized by the increase of catalyst activity; the sensitivity and long-term stability of the sensor are enhanced by the special core-shell structure.

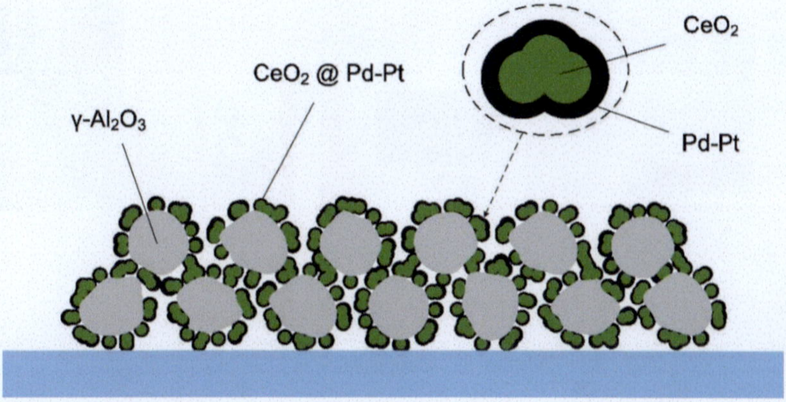

Fig. 3.2 Schematic diagram of proposed core-shell CeO_2@Pd-Pt/γ-Al_2O_3 catalyst system nanostructure, with γ-Al_2O_3 as a basic frame, CeO_2 nanoparticle cores as physical disperser and chemical stabilizer, Pd/PdO as the main catalyst, and Pt as co-catalyst for alloy effect against thermal deterioration

As shown in Fig. 3.2, the core-shell sphere cluster of CeO_2@Pd-Pt co-catalyst supported by γ-Al_2O_3 for catalytic combustion gas sensing was accurately deposited layer by layer. To start with, a porous γ-Al_2O_3 (colored in gray) film was selected as the thermal resistive frame of the catalyst with a micrometer particle and sub-micrometer internal pores. Then, CeO_2 nanoparticles (colored in green) were utilized as physical dispersants to form nanoscale surface, and monomer actives of noble metals were enhanced by CeO_2 NPs as chemical stabilizers and oxygen storage agents. Finally, Pd-Pt bimetallic alloy catalyst (colored in black) was discretely coated on deposited CeO_2 NPs. Such a mesoporous nanostructure and combustible gas can be fully contacted, while the interfacial sites between Pd and CeO_2 are fully maintained by individual CeO_2@Pd-Pt core-shell units. Thus, the redox potential of Pd at high temperature would be chemically stabilized and suppressed by CeO_2 NPs as oxygen suppliers.

The metal and oxide liquid materials were transferred by piezoelectric automatic inkjet instrument (Micro-Drop Company) with a nozzle diameter of 70 μm. The step resolution of horizontal positioning is 1 μm, and positioning accuracy is ±5 μm. The solution used for loading (including hex-hydrate chloroplatinic acid solution, palladium chloride hydrochloric acid solution, and CeO_2 NPs dispersion) and automatic inkjet equipment are shown in Fig. 3.3.

The repeatability of inkjet printing fabrication is determined by the stability of the printed micro droplets. A droplet is a microfluidic unit excited by shear or vibration piezoelectric mode, and the parameters of droplet generation need to be configured accurately and appropriately. Due to the difference of properties (such as viscosity, density, etc.) of solutions, the parameters of droplets (including diameter, drop speed, etc.) are controlled by driving voltage, pulse width, air pressure, and temperature of the equipment. The criteria of stable micro droplets are as follows: (1) only one droplet can be observed in the view of a synchronous CCD camera and

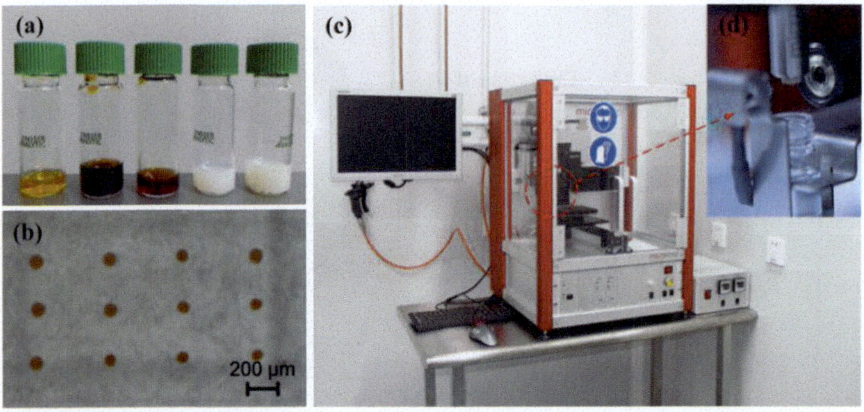

Fig. 3.3 Precursor solution and piezoelectric micro inkjet printer (**a**) solution or dispersion, left to right H_2PtCl_6 aqueous solution, $PdCl_2$ hydrochloric acid aqueous solution, Pd $(NO_3)_2$ nitric acid aqueous solution, CeO_2 aqueous dispersion, CeO_2 acetic acid dispersion, (**b**) graphic array of liquid printing, (**c**) piezoelectric automatic micro inkjet printer, (**d**) inkjet nozzle model of md-k-130

(2) the trajectory of the droplet is straight without any small droplets or drift. According to these two criteria, the driving parameters of the micro inkjet instrument are configured to obtain optimized structured print droplets, and the formation procedures of the droplets were recorded by a simultaneous CCD camera (Fig. 3.4).

The accurate quantitative control of loaded material is reflected in the unit of micro droplet. The volume, mass, and material quantity of the droplet unit are calculated as shown in Eqs. (3.1) to (3.3).

$$V = \frac{4}{3}\pi \cdot \left(\frac{d}{2}\right)^3 \tag{3.1}$$

$$m_{\text{cat,drop}} = m_{\text{sol}} \cdot \frac{M_{\text{cat}}}{M_{\text{sol}}} = \frac{4}{3}\pi \cdot \left(\frac{d}{2}\right)^3 \cdot \rho \cdot \omega \frac{M_{\text{cat}}}{M_{\text{sol}}} \tag{3.2}$$

$$n_{\text{cat,drop}} = \frac{m_{\text{sol}}}{M_{\text{sol}}} = \frac{4}{3}\pi \cdot \left(\frac{d}{2}\right)^3 \cdot \frac{\rho \cdot \omega}{M_{\text{sol}}} \tag{3.3}$$

where V is the volume of droplet unit, $m_{\text{cat,drop}}$ represents the mass of catalytic metal or promoter contained in droplet unit, $n_{\text{cat,drop}}$ indicates the amount of substance containing catalytic metal or promoter in the droplet, d is the diameter of the droplet, ρ is the density of the solution, and ω is the mass fraction of solute, while m_{sol}, M_{sol}, and M_{cat} are the mass of solute containing precursor in the droplet, the relative molecular mass of solute in the droplet, and relative atomic/molecular mass of catalytic metal or promoter, respectively.

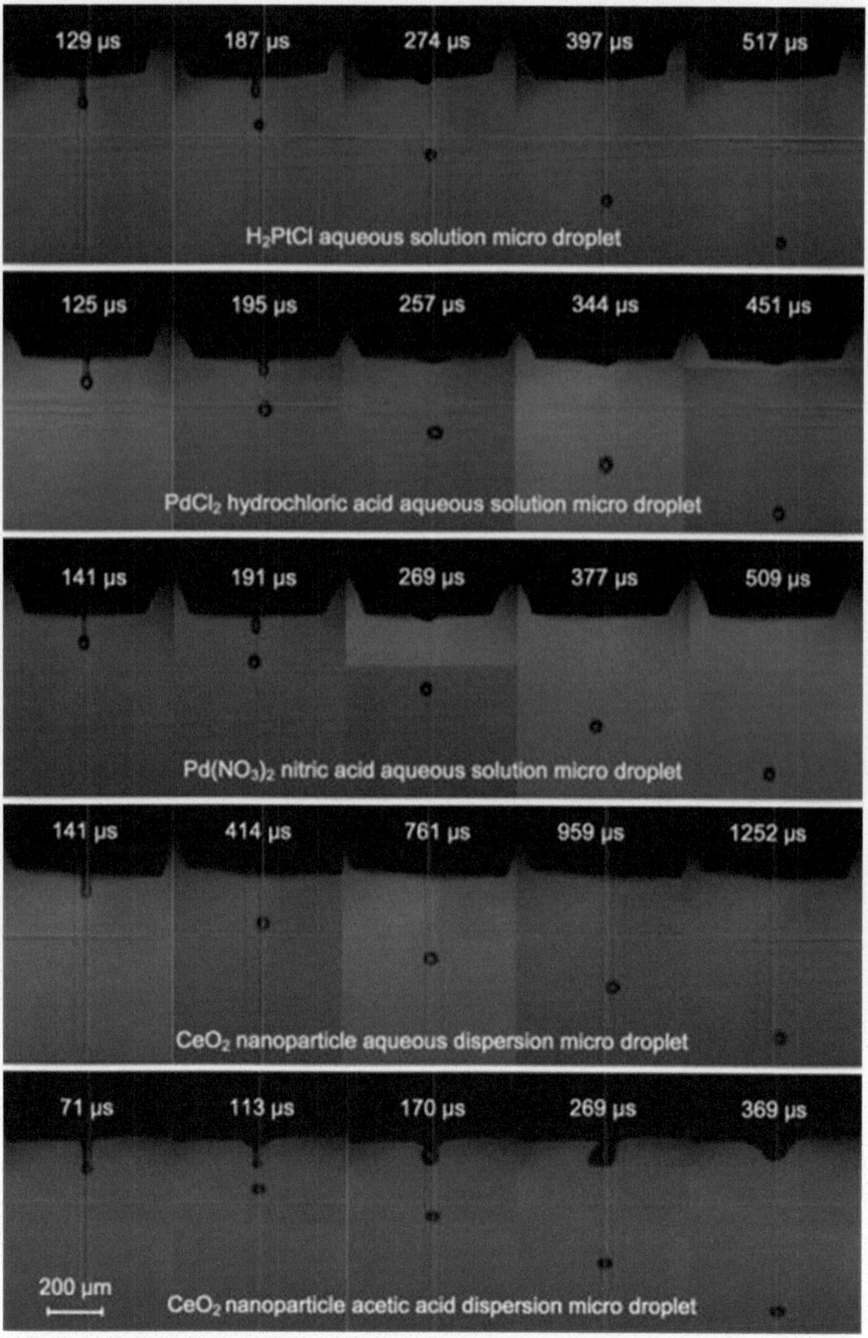

Fig. 3.4 In-flight trajectory of micro droplets with optimized parameters recorded by a synchronous CCD camera

Then, the loading capacity of a single component catalyst is expressed as

$$\omega_{\text{cat,total}} = \frac{m_{\text{cat}}}{m_{\text{support}} + m_{\text{cat}}} \times 100\% \tag{3.4}$$

where $\omega_{\text{cat,total}}$ is the mass fraction of single-component catalyst; m_{support} and m_{cat} represent the mass of carrier material and catalyst component, respectively; and m_{support} is defined by processed support materials. The mass of catalyst material can be calculated by Eq. (3.4) to determine the quantity of micro droplets loaded on the catalyst.

3.2.2.2 Device Fabrication

Due to the "layer-by-layer" and "core-shell" design, the sensors with proposed nanostructure catalyst can be directly in situ inkjet printed. To achieve advanced catalytic MEMS methane sensors, a low power consumption micro hotplate was fabricated by a standard lift-off process, reported by our previous work (Lu et al., 2016). The detailed in situ synthesis and fabrication of nano-catalyst model are illustrated in Fig. 3.5.

1. A layer of alumina paste (Senlont Electronic, China) with 10 μm thick was screen printed on the heating area of the quartz substrate, and a porous γ-Al_2O_3 frame with micropores was formed by calcining at 550 °C for 4 h (Fig. 3.5a).
2. Based on the microfluidic inkjet printing process, the CeO_2 NPs solution (with diameter of 25 nm, 10 wt% colloidal dispersions in aqueous solution, Sigma-Aldrich, USA) with a single droplet volume of 90 pL formed by a micro piezoelectric nozzle (MD-K-130, Microdrop, Germany, Fig. 3.1) was deposited into the porous alumina frame. The dispersing solvent and organic components in solution were removed by annealing in muffle at 450 °C for 20 min. The original

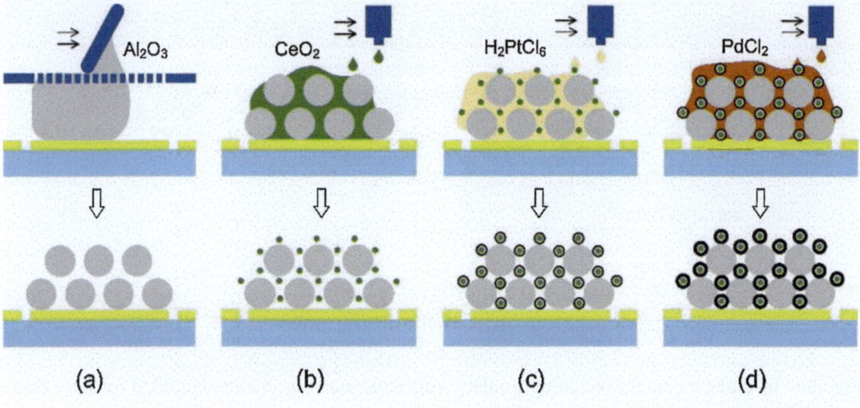

Fig. 3.5 Fabrication processes of the proposed CeO_2@Pd-Pt co-catalyst micro catalytic gas sensor based on layer-by-layer printing

γ-Al_2O_3 was modified by CeO_2 to form the morphology of dispersed NPs (Fig. 3.5b).

3. The droplets of H_2PtCl_6 (8 wt% aqueous solutions, Sigma-Aldrich) precursors were printed onto the CeO_2/γ-Al_2O_3 supporter. After high-temperature thermal decomposition, Pt monomers were obtained on the surface of CeO_2 NPs (Fig. 3.5c).
4. $PdCl_2$ (5 wt% in 10 wt% hydrochloric acid solutions, Sigma-Aldrich) as the precursor of the main catalyst of Pd, which was printed and decomposed at 500 °C for 2 h. Then, the uniform Pd-Pt alloy sphere clusters were formed covering with individual CeO_2 NPs (Fig. 3.5d).

The molar ratio between Pd and Pt solution was configured as 7:3 during the printing process, while the molar ratio between Pd-Pt and CeO_2 was adjusted as 10:1, 10:4, 10:8, and 10:32 for optimization (Fig. 3.6a–d). The loading content (mass ratio between loaded pure catalyst and carrier-supported catalyst) was configured as 20%. The catalytic MEMS methane sensor was bonded to the SOP footprint shell to complete the lead and packaging. The optical images of the MEMS gas sensors are shown in Fig. 3.6e, f.

3.2.2.3 Characterization and Results

The materials of γ-Al_2O_3, CeO_2/γ-Al_2O_3, CeO_2@Pd/γ-Al_2O_3, and $CeO_{2@}$Pd-Pt/γ--Al_2O_3 were characterized by SEM, respectively, and EDS analysis was carried out for the two core-shell materials (Figs. 3.7 and 3.8).

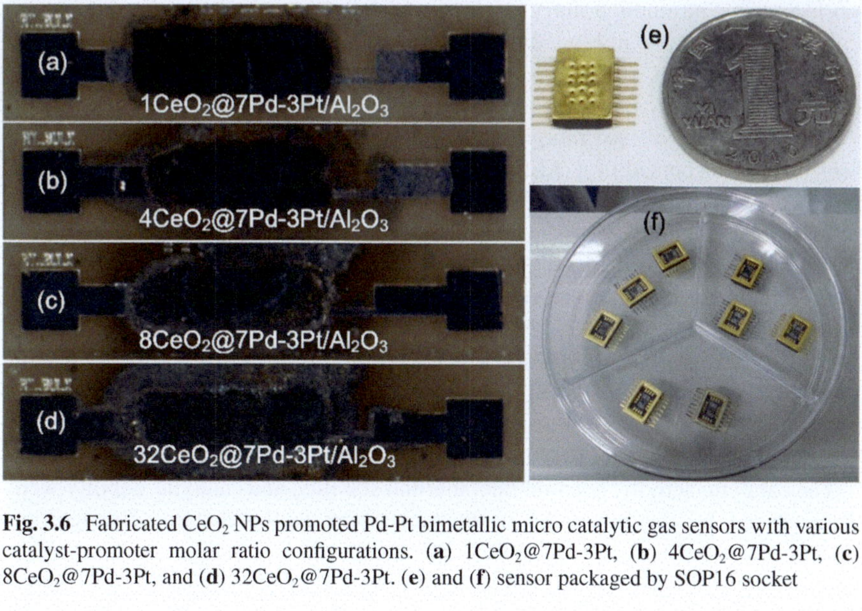

Fig. 3.6 Fabricated CeO_2 NPs promoted Pd-Pt bimetallic micro catalytic gas sensors with various catalyst-promoter molar ratio configurations. (**a**) 1CeO_2@7Pd-3Pt, (**b**) 4CeO_2@7Pd-3Pt, (**c**) 8CeO_2@7Pd-3Pt, and (**d**) 32CeO_2@7Pd-3Pt. (**e**) and (**f**) sensor packaged by SOP16 socket

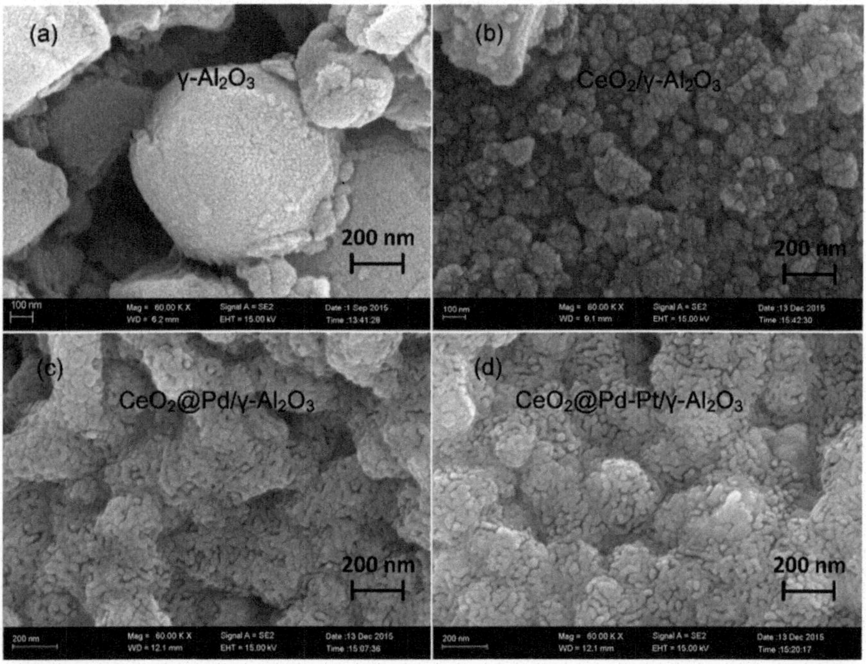

Fig. 3.7 SEM images of fabricated (**a**) γ-Al_2O_3 frame, (**b**) CeO_2/γ-Al_2O_3, (**c**) CeO_2@Pd/γ-Al_2O_3, and (**d**) CeO_2@Pd-Pt/γ-Al_2O_3

Highly dispersed catalyst was obtained by core-shell structure and optimizing the molar ratio between CeO_2 and Pd-Pt NPs. The mesoporous structure γ-Al_2O_3 was formed by the alumina slurry calcined at high temperature. The sub-micrometer-scale pores are shown in Fig. 3.7a. The CeO_2 NPs are distributed and filled in the gap of γ-Al_2O_3, with slight aggregation to form particles with dimensions from 60 to100 nm due to low thermal stability without Pd-Pt shells (Fig. 3.7b) (Cargnello et al., 2012). The surface morphology of the core-shell structure CeO_2@Pd and CeO_2@Pd-Pt dispersed by γ-Al_2O_3 is shown in Fig. 3.7c, d. In the final product of CeO_2@Pd-Pt, a particle shell with a thickness of 30–50 nm was constructed by the Pd-Pt alloy coated on CeO_2 nanoparticle, and a three-dimensional spherical structure of catalyst was formed with a diameter of 200 nm (Fig. 3.7d). From Fig. 3.7a–d, the incremental growth structure of the material is consistent with the layer-by-layer printing process.

Figure 3.8 represents the EDS characterization results of catalyst samples with four additives (platinum, palladium, cerium, aluminum). At the same time, the availability of multiple inkjet print method in doping-supported catalytic materials has been verified.

The catalytic activity, reaction rate, thermal stability, and catalytic stability were characterized by the bimetallic catalyst and core-shell structure. The ignition

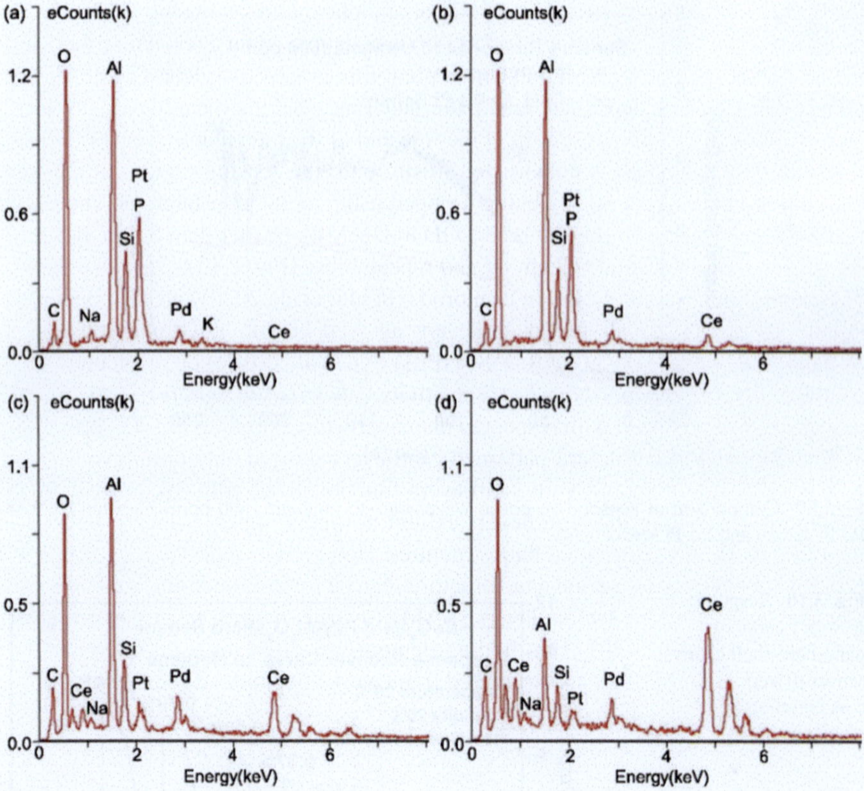

Fig. 3.8 EDS analysis of different molar ratio promoters (CeO_2) and catalysts (Pd-Pt) (**a**) 0.1, (**b**) 0.4, (**c**) 0.8, (**d**) 3.2

temperature, response time, sensitivity, long-term stability, and signal-to-noise ratio were improved evidently.

The sensing performance of the core-shell-structured catalyst with CeO_2 nanoparticle (cores) and Pd-Pt (shells) was investigated to 500 ppm methane between $4CeO_2$@$7Pd\text{-}3Pt/\gamma\text{-}Al_2O_3$ and $7Pd\text{-}3Pt/\gamma\text{-}Al_2O_3$ sensors. As shown in Fig. 3.9, the sensor response with CeO_2 promoter possesses much less signal fluctuation and better signal quality, compared with pure Pd-Pt bimetallic catalyst sensor. Pd catalyst is still not stable even at a high working temperature of 450 °C. The redox of PdO/Pd was suppressed by supplying sufficient oxygen of CeO_2 cores, which maintains the chemical phase stability of the Pd-Pt catalyst system. The released combustion heat to the sensing electrode of the sensor was also stabilized by CeO_2 at the same time. Additionally, the response time of the sensor is reduced from 50 s to 12 s with the existence of CeO_2-covered Pd-Pt. The significant increase of response time owes to the free oxygen provided by CeO_2, participating in the oxidation-reduction reaction of reactants near the active point. In this method, the conversion rate of each catalytic active point was improved (Liu et al., 2013), and

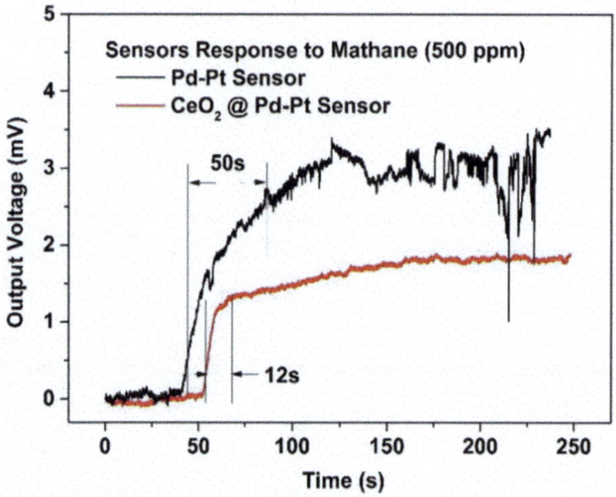

Fig. 3.9 Comparison of sensors' response waveforms to methane (500 ppm) between CeO_2@ Pd-Pt sensor and Pd-Pt sensor

Fig. 3.10 Response-recovery curve of nano-core-shell catalyst sensor to step concentration of methane air

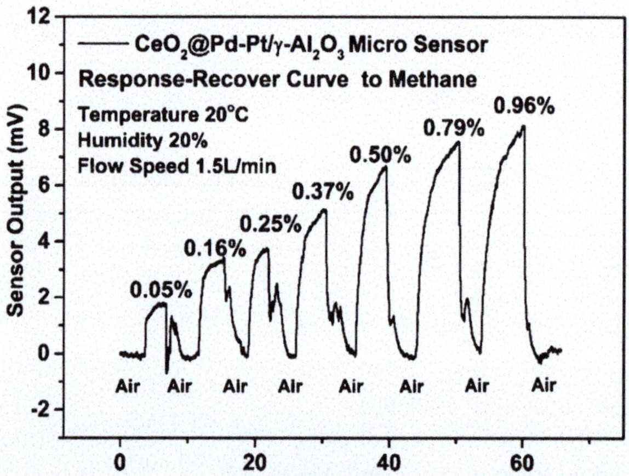

then the catalytic exothermic reaction rapidly reached the limit rate of mass transfer to obtain a stable state (Delgado, 2011).

Target methane (3% methane standard gas, 97% compressed air balance gas) and pure air were continuously and alternately introduced to measure the multi-concentration response recovery curve of the sensor. A multi-period response recovery waveform of the sensor was obtained by the alternative introduction of methane and air (Fig. 3.10). This experimental result presents high initial recovery stability of the sensor with a response time of 12 s and a recovery time of 10 s, respectively.

The response curves of the sensor with various configurations of catalyst-promoter molar ratio to methane ranging from 0 to 10,000 ppm are shown in

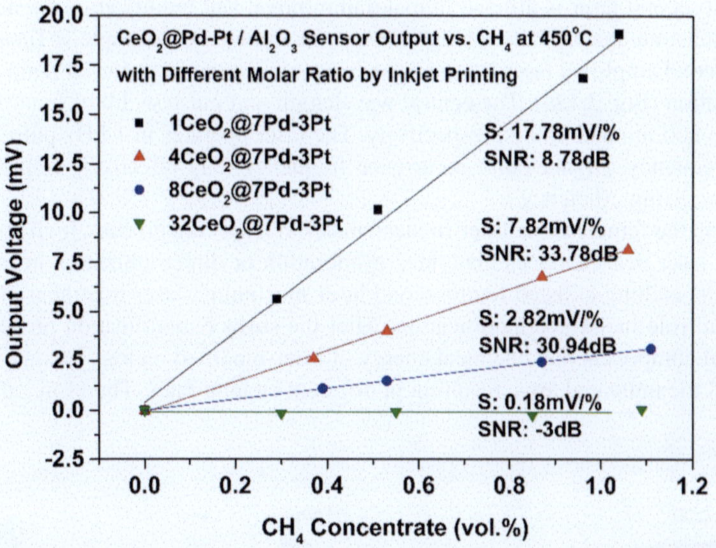

Fig. 3.11 Comparison of CeO_2@Pd-Pt sensors' sensitivity and SNR to methane (0 to 10,000 ppm) with various catalyst-promoter molar ratio configurations

Fig. 3.11. As the loaded CeO_2 promoter quantity increases from 1CeO_2@7Pd-3Pt to 4CeO_2@7Pd-3Pt, the signal-to-noise ratio (SNR) of the sensor increases significantly from 8.78 dB to 33.78 dB. The SNR decreases slightly to 30.94 dB as the promoter increases to 8CeO_2@7Pd-3Pt. This verifies the chemical stabilization effect of the CeO_2 cores to Pd-Pt shells. Meanwhile, the sensitivity of the sensor is in negative correlation to mole ratio. Thus, 4CeO_2@7Pd-3Pt is the optimized one compromising between SNR and sensitivity.

In this case, the mask-free inkjet printing enables the precise control of the functional material in components and morphology. This makes inkjet printing a competitive option in the fabrication of advanced MEMS gas sensors.

3.3 FsLDW for the Fabrication of Advanced MEMS Gas Sensor

3.3.1 Micro/Nano-Three-Dimensional Molding by FsLDW

FsLDW is an emerging technology to fabricate sensors in a mask-free manner. Various FsLDW technologies have been developed as important tools for subtractive (e.g., material removal, modification, etc.) and additive (e.g., 3D printing, etc.) fabrication of MEMS gas sensors.

Femtosecond laser is utilized in material removal and modification by sublimation of substrate material or etching-assisted machining. For example, a Ti/sapphire femtosecond amplified laser is used as an exposure source for material removal and modification (Fig. 3.12a). The central wavelength and pulse width of femtosecond laser are 800 nm and 150 fs, respectively. The laser operates in 1 kHz pulse recurrence frequency. Higher pulse recurrence frequency may effectively improve the laser processing efficiency.

During the femtosecond laser-induced material removing process, high power of focused laser beam induces materials' evaporation or direct ionization to remove. As for the etching-assisted femtosecond laser machining, laser treatment plays an important role in surface modification. After the surface modification process, the chemical component and chemical energy of laser-modified regions become different from the untreated area, resulting in different etching rates. Therefore, after the

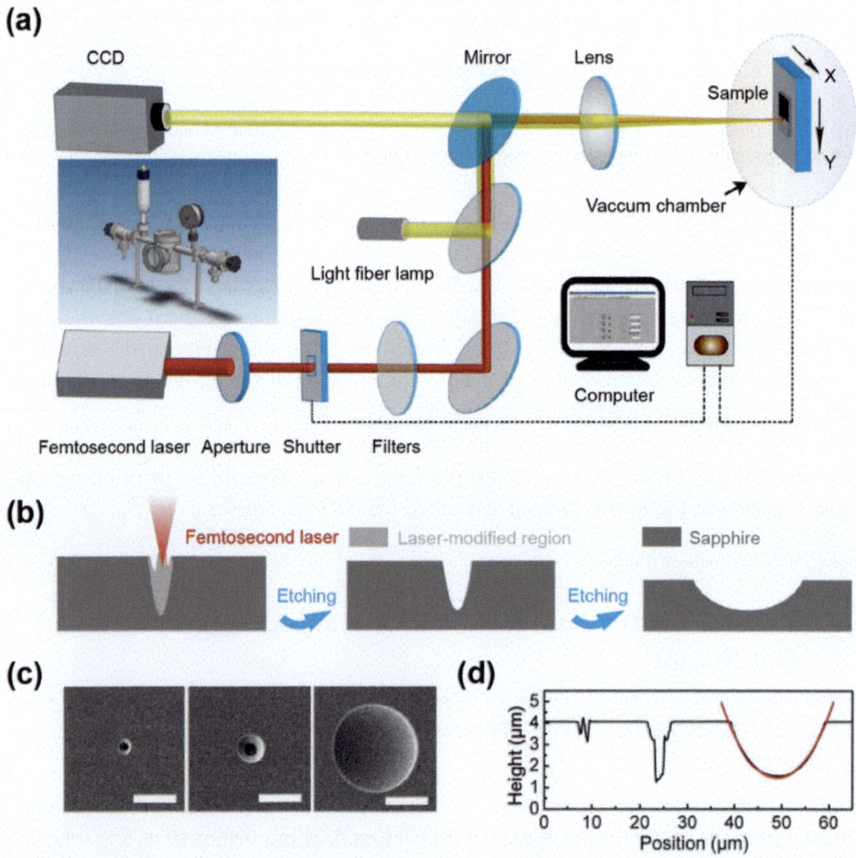

Fig. 3.12 (**a**) FsLDW system for material removal or modification (Yu et al., 2016). (**b–d**) Schematic diagrams of the femtosecond laser-assisted material removal process (Liu et al., 2019). Scale bar = 10 μm

etching process, three-dimensional (3D) structures are fabricated (Fig. 3.12b). Different physical parameters of these structures can be achieved by adjusting the laser power and etching conditions (Fig. 3.12c, d). Compared with laser-induced sublimation of substrate material, etching-assisted femtosecond laser machining is a more effective method to obtain smooth surfaces.

FsLDW has also been attractive for 3D printing, such as two-photon polymerization. In terms of 3D printing, a Ti/sapphire femtosecond oscillator (800 nm central wavelength, 120 fs pulse width, and 80 MHz repetition) is taken as an exposure source for two-photon polymerization (Fig. 3.13a). The main processes of 3D printing are demonstrated by the example of 3D nanofabrication based on SU-8 photoresist (Fig. 3.13b). The SU-8 photoresist is spin-coated on a clean substrate and prebaked for solvent evaporation. Then, femtosecond laser beams are focused into photoresists to prepare for the following three-dimensional scan. After the laser

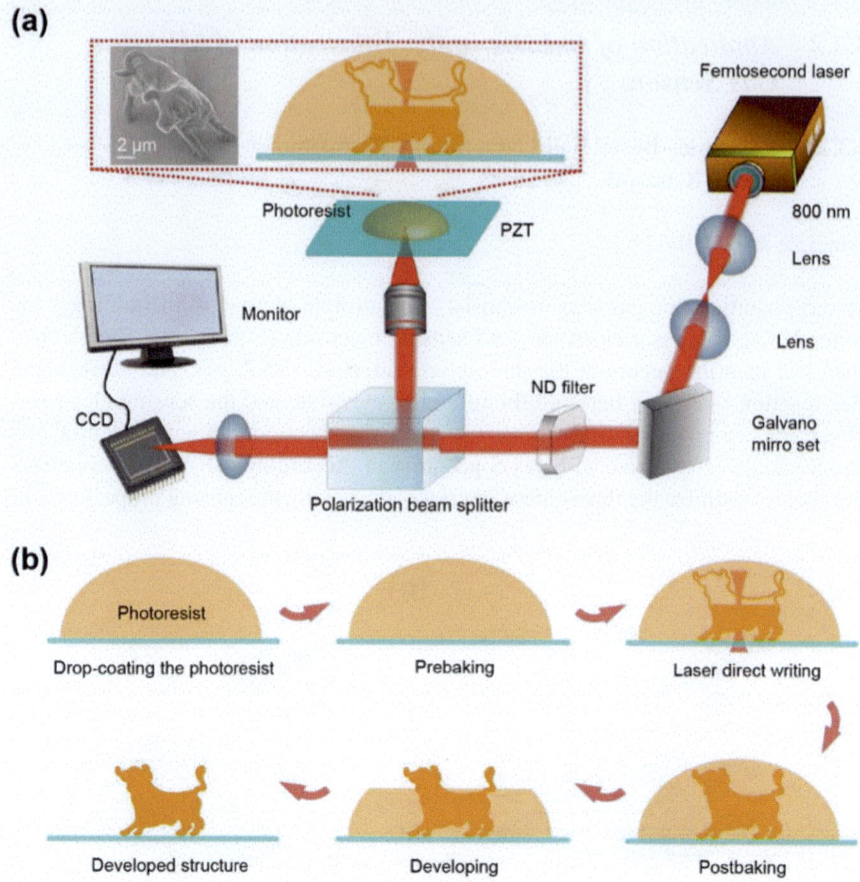

Fig. 3.13 (**a**) FsLDW system for 3D printing (Guo et al., 2014; Kawata et al., 2001). (**b**) Schematic diagrams of the femtosecond laser two-photon photopolymerization process

direct writing process, the scanned photoresist is post-baked. Finally, the unpolymerized photoresist is removed to leave the 3D structure on the substrate. The quality of the structure is related with laser power, scanning step, and exposure time. Similar principles apply to fabricate 3D micro/nanostructure based on polymethyl methacrylate and protein materials (Sun et al., 2012; Wei et al., 2018; Xiong et al., 2016).

With nonlinear femtosecond laser-matter interactions, femtosecond laser can be used to process various kinds of materials, such as SnO_2 (Guo et al., 2010), silver (Xu et al., 2010), reduced graphene oxides (Guo et al., 2014), lithium niobate (Tian et al., 2017), etc. The programmable processing ability enables FsLDW to fabricate complex patterns on the surface or inside of materials. These provide great 3D processing capability to sensor fabrication, integration, and interconnection in a mask-free and programmable manner.

3.3.2 Application of FsLDW in the Fabrication of MEMS Gas Sensors

3.3.2.1 Ceramics-Based Solid NO_2 Sensor Performance Improved by Laser Removal

Principle and Method

Mixed-potential-type gas sensors consist of electrolyte (e.g., yttrium-stabilized zirconia, YSZ), reference electrode, and sensing electrode (Fig. 3.14a). The electrochemical reaction occurs at the three-phase interface (YSZ, gas, and electrodes). Gas reaction rates vary between the reference electrode and the sensing electrode. This leads to a potential difference between the two electrodes. Such a potential difference is also correlative with gas concentration. Accordingly, FsLDW is an effective tool to optimize the three-phase interface, improving the sensing properties. For

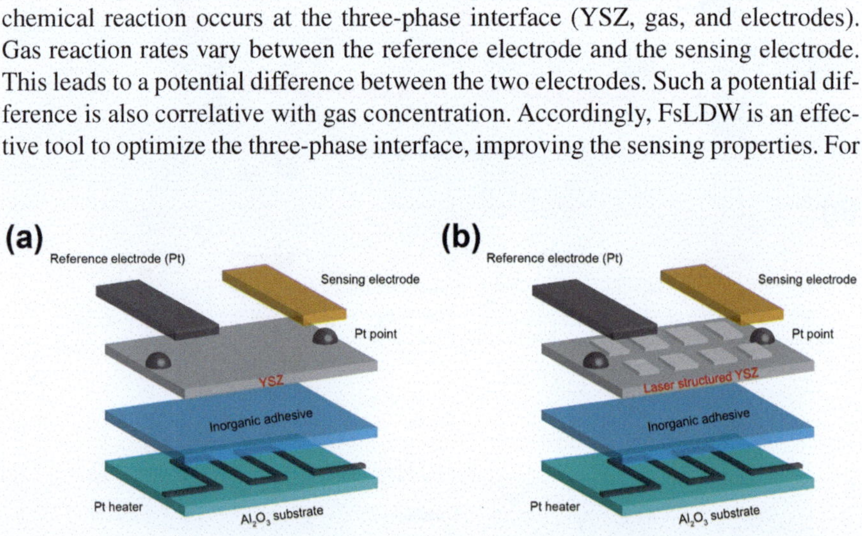

Fig. 3.14 Schematic of the mixed-potential-type NO_2 sensors based on (**a**) YSZ and (**b**) laser-structured YSZ

example, Guan et al. demonstrated high-performance NO_2 sensor with laser-structured YSZ (Fig. 3.14b) (Guan et al., 2014).

Device Fabrication

1. The laser-structured YSZ substrates ($T = 130$ μm and 240 μm) were fabricated by femtosecond amplified laser (Fig. 3.15).
2. The reference electrode was made of stripe-shaped Pt (0.5 mm × 2 mm). The sensing electrode (0.8 mm × 2 mm) was made of point-shaped Pt and $NiCr_2O_4$ (sensing materials).
3. The YSZ substrates containing reference electrode and sensing electrode were sintered (900 °C, 2 h). Bonding of heater was followed to facilitate the performance measuring.

Characterization and Results

As shown in Fig. 3.16, the NO_2 sensing performance can be improved by using the laser-structured YSZ. The response property of the sensor ($T = 130$ μm) is the best compared with those of all the tested sensors. The electric potential difference ($V_{NO_2} - V_{air}$) of the sensor ($T = 130$ μm) is 100 mV under 100 ppm NO_2. The response time is ~8 s. As the reaction speed is determined by the interface area, a larger three-phase interface of the sensor ($T = 130$ μm) is possibly the most important contributor to the evident improvement of response property.

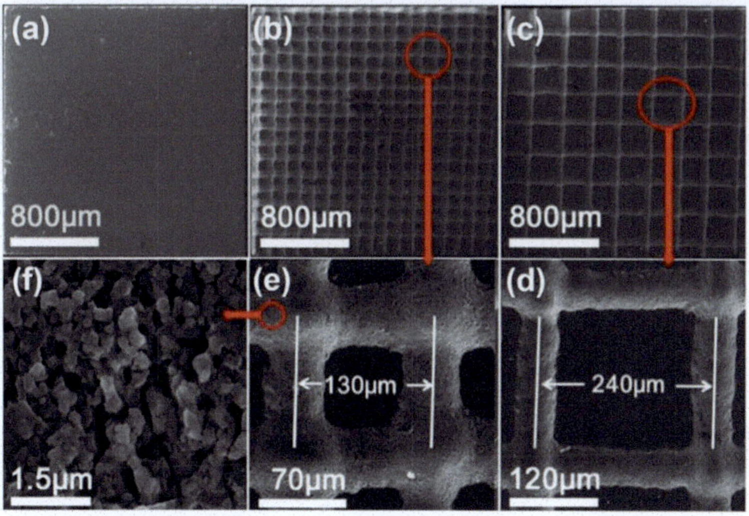

Fig. 3.15 SEM images of (**a**) the initial and (**b–f**) laser-structured YSZ surface (Guan et al., 2014)

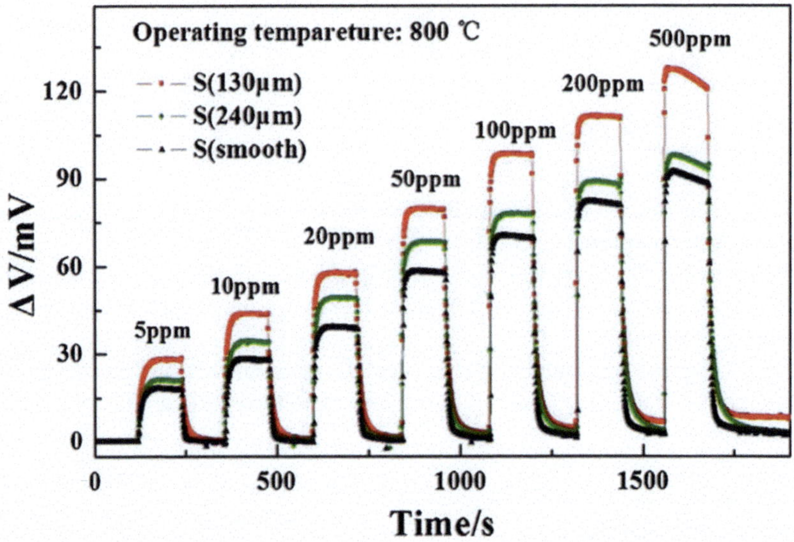

Fig. 3.16 NO_2 sensing characteristic based on the unprocessed and laser-structured YSZ substrate (Guan et al., 2014)

3.3.2.2 Graphene-Based Flexible Humidity Sensor Fabricated by Laser Modification

Principle and Method

Graphene oxide (GO) is a graphene derivative with plenty of oxygen-containing groups (OCGs). The existence of OCGs greatly increases the vaporized water molecule absorption of GO in high relative humidity (RH) via hydrogen bond. As water molecule absorption results in conductivity changes, GO is a promising material for humidity sensing (Fig. 3.17a). The sensing mechanism is based on the proton and ionic conductivity under different RHs. FsLDW is a remarkable way to realize the transformation from GO to reduced graphene oxides (RGO). During the reduction process, OCGs escape from GO nanosheets in the forms of gas (e.g., CO, CO_2, H_2O), leading to conductivity recovery. As a result, RGO has excellent conductivity to be the material of electrodes. Accordingly, a humidity sensor can be developed with GO (the sensing materials) and RGO (the electrodes). An et al. fabricated a humidity sensor by programmable FsLDW technology. In this humidity sensor, the RGO acts as electrodes and the unprocessed GO remains as the humidity sensing material (Fig. 3.17b) (An et al., 2017).

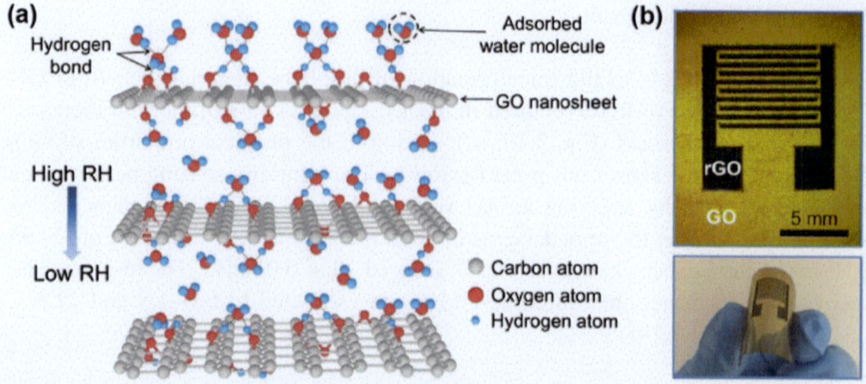

Fig. 3.17 (**a**) Schematic illustration of the interaction between water molecules and GO. (**b**) Optical images of RGO humidity sensors (An et al., 2017)

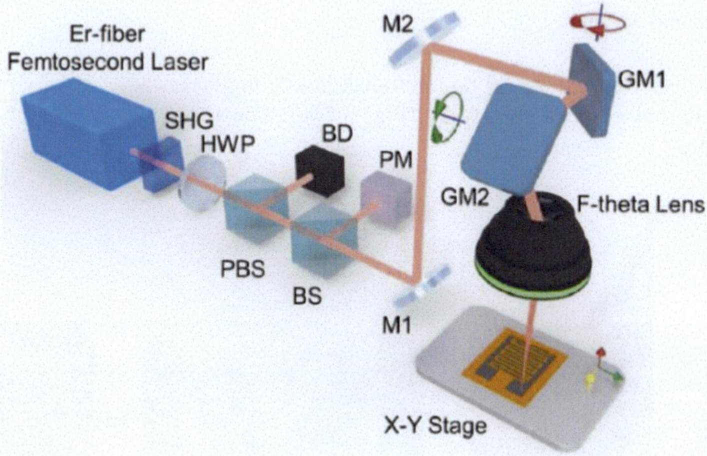

Fig. 3.18 FsLDW system for fabricating RGO humidity sensors (An et al., 2017)

Device Fabrication

1. O_2 plasma treatment (18 W, 1 min) was used on clean polyethylene terephthalate (PET) films to increase the hydrophilicity of PETs. This facilitates GO aqueous suspension to spread on the plasma-treated PET substrates. After drying, the GO film was placed on an X–Y translation stage.
2. GO film was reduced and patterned by programmable femtosecond laser. After laser scanning (focal length is ~100 mm; scanning speed is ~1 m/s), GO is transferred to the RGO film. The RGO electrode is made of ten interdigitated fingers with spacing of 300 μm. The width and length of each interdigitated finger are 300 μm and 8 mm, respectively (Fig. 3.18).

Characterization and Results

Laser reduction leads to the transformation of graphene material, i.e., from GO (colored in yellow) to RGO (colored in black). The thickness of the film increases in the reduction process (Fig. 3.19a–d). Owing to the physical properties of GO, RGO, and PET, the sensor has great flexibility. The humidity sensing performance was characterized by applying an AC voltage (1 V) at different frequencies. As shown in Fig. 3.19e, the impedance is in good linearity among the range of 15–95 RH%. The impedance changes from 10^7 Ω to 10^4 Ω at 0.02 kHz. The response and recovery performance, measured at 1 kHz, are estimated to be ~1.8 and 11.5 s, respectively (Fig. 3.19f).

3.3.2.3 PEG-DA-Based Humidity Sensor Fabricated by Two-Photon Polymerization

Principle and Method

As for whispering gallery mode microdisk lasers, the laser can be totally reflected at the internal edge of microdisk (Fig. 3.20a). The resonant wavelength is calculated by

$$2\pi NR = m\lambda \tag{3.5}$$

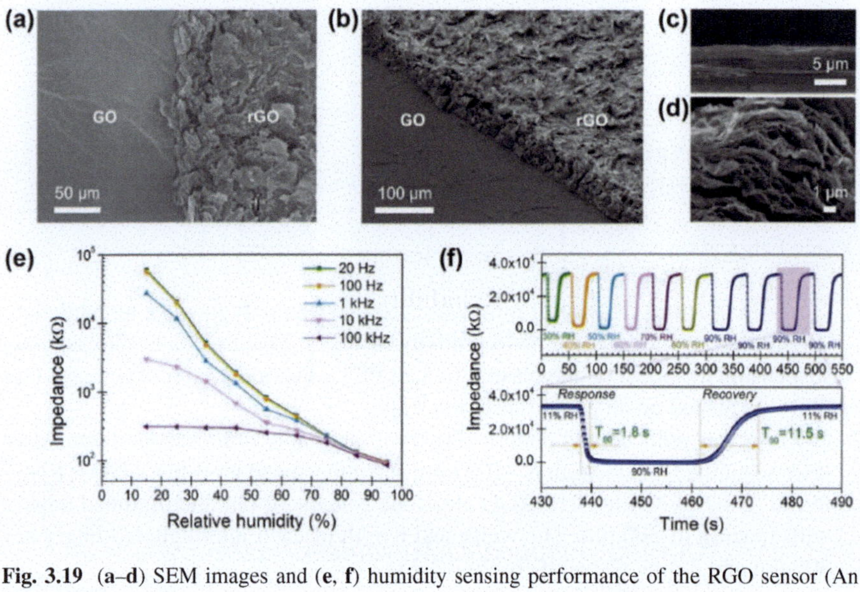

Fig. 3.19 (**a–d**) SEM images and (**e, f**) humidity sensing performance of the RGO sensor (An et al., 2017)

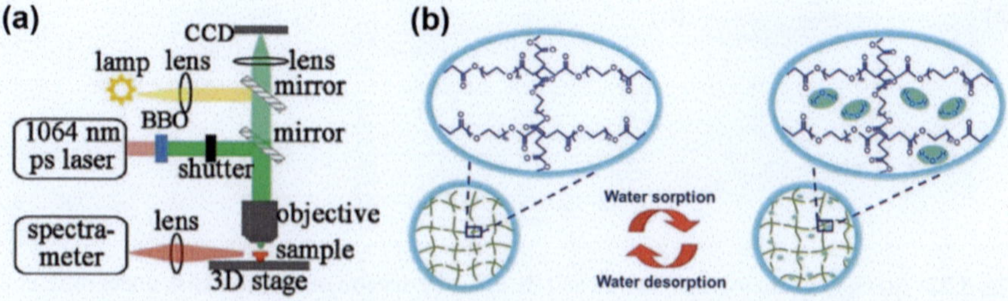

Fig. 3.20 Schematic illustration of (**a**) the measurement system (Huang et al., 2018) and (**b**) moisture-responsive PEG-DA (Lv et al., 2017)

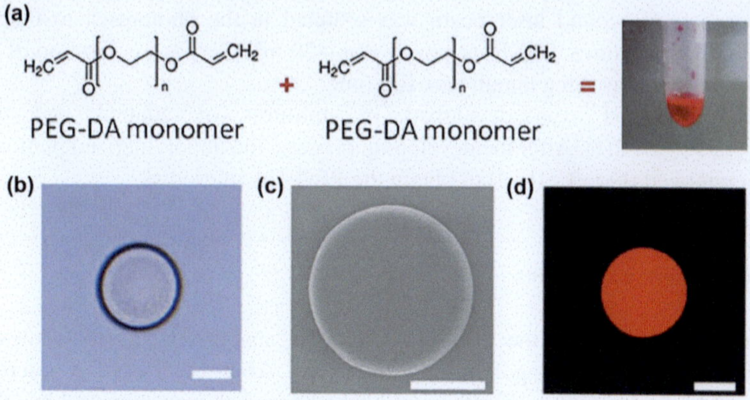

Fig. 3.21 (**a**) The synthesis of photoresist. (**b**) OM, (**c**) SEM, and (**d**) fluorescence microscope images of the rhodamine B-doped PEG-DA microdisk. All scale bars are 10 μm (Huang et al., 2018)

where m is an integer, R is the radius of the disk, and N is the refractive index of materials. Herein, the resonant wavelength (λ) is linearly related to the radius of the microdisk (R). As a result, the microdisk-laser-type gas sensor is able to be fabricated by using gas responsive materials. For example, poly(ethylene glycol) diacrylate (PEG-DA) hydrogel swells and shrinks by absorbing and releasing water molecules (Fig. 3.20b). Therefore, the radius of the disk changes under different relative humidity (RH), leading to resonant wavelength changes. Commonly, rhodamine B is doped into the PEG-DA monomer solutions, acting as a photoinitiator for two-photon polymerization and gain medium in microdisk lasers.

Device Fabrication

1. The photoresist was synthesized by mixing rhodamine B solution and PEG-DA monomer ($V_{\text{rhodamine B}}/V_{\text{PEG-DA}} \approx 3{:}10$). The mixture was shaken and dispersed by ultrasound treatment (Fig. 3.21a).

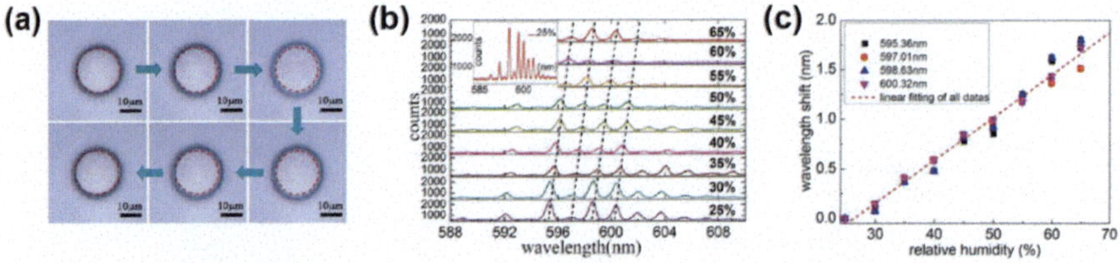

Fig. 3.22 (**a**) Optical micrograph images and (**b**, **c**) lasing spectra of the microdisk under different RH conditions (Huang et al., 2018)

2. The photoresist was spin-coated on a clean substrate. Femtosecond laser was focused by an oil immersion lens (NA = 1.35, 100×, Olympus). The highly focused femtosecond laser beam was scanned in the photoresist to fabricate microdisk structures. The laser power was ~20 mW. Exposure duration of each voxel was 500 μs. Step length was 100 nm.
3. The photoresist was developed in mixtures of ethanol and water ($V_{ethanol}/V_{water} \approx 2:1$). After the developing process, the unpolymerized photoresist was removed (Fig. 3.21b–d) to obtain the PEG-DA microdisk.

Characterization and Results

As shown in Fig. 3.22a, the PEG-DA microdisk absorbed water molecules and swelled as the surrounding humidity increased. After the PEG-DA microdisk expanded to the limitation, the diameter of the microdisk was increased by ~7.6% from the initial status. After the surrounding humidity decreased, the PEG-DA microdisk gradually shrunk to the initial size in 15 s. A red shift of the lasing spectrum was recorded when the RH increased from 25% to 65% (Fig. 3.22b). A wavelength variation (~1.72 nm) was observed (Fig. 3.22c), attributed to the microdisk diameter change.

3.4 Concluding Remarks

MEMS gas sensor is a miniaturized device for the identification of gas type and measurement of gas concentration, humidity, and temperature. The traditional standard MEMS fabrication has limitations for the future development of MEMS gas sensors. Recent researches highlight the potential employment of mask-free direct writing in next-generation MEMS gas sensors. In this chapter, several kinds of micromachining methods of non-silicon materials are exemplified in the leading-edge application of MEMS gas sensor, focusing on the preparation of surface nanomaterials and high-precision machining of different substrate materials at nano-level.

In the first section, we have proposed a method of in situ printing synthesis, based on a multicomponent catalyst, for the catalyst preparation of catalytic combustion MEMS gas sensor. The advantages of porosity, consistency, quantitative positioning accuracy, and component flexibility were verified by micromachining and characterization experiments.

In the second section, we have summarized the processing of sensitive materials and the interface between sensitive materials and target gas by FsLDW. Different types of materials, such as YSZ, GO, and PEG-DA, have been applied and femtosecond laser has been directly written to fabricate novel MEMS gas sensors. These cases send a tremendous positive signal that FsLDW would be a universal nanofabrication tool due to its high applicability to an unprecedented range of material.

Mask-free direct writing may bring revolutionary changes to micro/nanodevice fabrication technology in diversified scientific and industrial fields. The mask-free graphic and material growth process has the advantages of simple operation, high precision, and high consistency and customization. Further, due to the wide material applicability of direct writing processing, it is expected to replace the traditional lithography graphics fabrication in some special fields such as flexible sensing, biological detection, etc. This chapter has a prominent guiding significance in stimulating the inspiration of using mask-free direct writing to advance MEMS gas sensors.

Acknowledgments This work was supported by China's Postdoctoral Science Foundation (grant no. 2020M670037, 2020M670850), Foundation of Beijing Laboratory for Biomedical Detection Technology and Instrument (20201550018), National Key Research and Development Program of China, and National Natural Science Foundation of China (NSFC) under grant nos. 2017YFB1104300, 61905087, and 61935008.

References

Altmann, L., Sturm, H., Brauns, E., et al. (2012). Novel catalytic gas sensors based on functionalized nanoparticle layers. *Sensors and Actuators B: Chemical, 174*, 145–152.

An, J. N., Le, T. S. D., Huang, Y., et al. (2017). All-graphene-based highly flexible noncontact electronic skin. *ACS Applied Materials & Interfaces, 9*, 44593–44601.

Ando, T., & Fu, X. A. (2019). Materials: Silicon and beyond. *Sensors and Actuators A: Physical, 296*, 340–351.

Arlett, J. L., Myers, E. B., & Roukes, M. L. (2011). Comparative advantages of mechanical biosensors. *Nature Nanotechnology, 6*, 203–215.

Brauns, E., Morsbach, E., Schnurpfeil, G., et al. (2013). A miniaturized catalytic gas sensor for hydrogen detection based on stabilized nanoparticles as catalytic layer. *Sensors and Actuators B: Chemical, 187*, 420–425.

Briand, D., Colin, S., Gangadharaiah, A., et al. (2006). Micro-hotplates on polyimide for sensors and actuators. *Sensors and Actuators A: Physical, 132*, 317–324.

Cargnello, M., Jaen, J. J. D., Garrido, J. C. H., et al. (2012). Exceptional activity for methane combustion over modular Pd@CeO_2 subunits on functionalized Al_2O_3. *Science, 337*, 713–717.

Chang, W. Y., & Hsihe, Y. S. (2016). Multilayer microheater based on glass substrate using MEMS technology. *Microelectronic Engineering, 149*, 25–30.

Delgado, J. M. (2011). *Heat and mass transfer in porous media* (pp. 160–183). Springer.

DelRio, F. W., Cook, R. F., & Boyce, B. L. (2015). Fracture strength of micro- and nano-scale silicon components. *Applied Physics Reviews, 2*, 021303.

Farahani, R. D., Dube, M., & Therriault, D. (2016). Three-dimensional printing of multifunctional nanocomposites: Manufacturing techniques and applications. *Advanced Materials, 28*, 5794–5821.

Fischer, J., & Wegener, M. (2013). Three-dimensional optical laser lithography beyond the diffraction limit. *Laser & Photonics Reviews, 7*, 22–44.

Fischer, A. C., Forsberg, F., Lapisa, M., et al. (2015). Integrating MEMS and ICs. *Microsystems & Nanoengineering, 1*, 15005.

Guan, Y. Z., Li, C. H., Cheng, X. Y., et al. (2014). Highly sensitive mixed-potential-type NO_2 sensor with YSZ processed using femtosecond laser direct writing technology. *Sensors and Actuators B: Chemical, 198*, 110–113.

Guo, L., Xia, H., Fan, H. T., et al. (2010). Femtosecond laser direct patterning of sensing materials toward flexible integration of micronanosensors. *Optics Letters, 35*, 1695–1697.

Guo, L., Zhang, Y. L., Han, D. D., et al. (2014). Laser-mediated programmable N doping and simultaneous reduction of graphene oxides. *Advanced Optical Materials, 2*, 120–125.

Guo, X., Xue, Z., & Zhang, Y. (2019). Manufacturing of 3D multifunctional microelectronic devices: Challenges and opportunities. *NPG Asia Materials, 11*, 29.

Heule, M., Vuillemin, S., & Gauckler, L. J. (2003). Powder-based ceramic meso- and microscale fabrication processes. *Advanced Materials, 15*, 1237–1245.

Huang, Q. L., Xu, H. L., Li, M. T., et al. (2018). Stretchable PEG-DA hydrogel-based whispering-gallery-mode microlaser with humidity responsiveness. *Journal of Lightwave Technology, 36*, 819–824.

Jian, Y. Y., Hu, W. W., Zhao, Z. H., et al. (2020). Gas sensors based on chemi-resistive hybrid functional nanomaterials. *Nano-Micro Letters, 12*, 71.

Kawata, S., Sun, H. B., Tanaka, T., et al. (2001). Finer features for functional microdevices-micromachines can be created with higher resolution using two-photon absorption. *Nature, 412*, 697–698.

Lee, H. J., Choi, N., Yoon, E. S., et al. (2018). MEMS devices for drug delivery. *Advanced Drug Delivery Reviews, 128*, 132–147.

Liu, B. C., Yu, S. L., Wang, Q., et al. (2013). Hollow mesoporous ceria nanoreactors with enhanced activity and stability for catalytic application. *Chemical Communications, 49*, 3757–3759.

Liu, X. Q., Yang, S. N., Yu, L., et al. (2019). Rapid engraving of artificial compound eyes from curved sapphire substrate. *Advanced Functional Materials, 29*, 1900037.

Lu, W. S., Jing, G. S., Bian, X. M., et al. (2016). A quartz-based micro catalytic methane sensor by high resolution screen printing. *Journal of Micromechanics and Microengineering, 26*, 025021.

Lv, C., Xia, H., Shi, Q., et al. (2017). Sensitively humidity-driven actuator based on photopolymerizable PEG-DA films. *Advanced Materials Interfaces, 4*, 1601002.

Qiu, A. D., Li, P. L., Yang, Z. K., et al. (2019). A path beyond metal and silicon: Polymer/nanomaterial composites for stretchable strain sensors. *Advanced Functional Materials, 29*, 1806306.

Ren, Z. H., Chang, Y. H., Ma, Y. M., et al. (2020). Leveraging of MEMS technologies for optical metamaterials applications. *Advanced Optical Materials, 8*, 1900653.

Schroeder, V., Savagatrup, S., He, M., et al. (2019). Carbon nanotube chemical sensors. *Chemical Reviews, 119*, 599–663.

Seshadri, D. R., Li, R. T., Voos, J. E., et al. (2019). Wearable sensors for monitoring the physiological and biochemical profile of the athlete. *NPG Digital Medicine, 2*, 72.

Shin, W., Nishibori, M., Houlet, L. F., et al. (2009). Fabrication of thermoelectric gas sensors on micro-hotplates. *Sensors and Actuators B: Chemical, 139*, 340–345.

Sima, F., Sugioka, K., Vazquez, R. M., et al. (2018). Three-dimensional femtosecond laser processing for lab-on-a-chip applications. *Nanophotonics, 7*, 613–634.

Singh, M., Haverinen, H. M., Dhagat, P., et al. (2010). Inkjet printing-process and its applications. *Advanced materials, 22*(6), 673–685.

Sun, Y. L., Dong, W. F., Yang, R. Z., et al. (2012). Dynamically tunable protein microlenses. *Angewandte Chemie. International Edition, 51*, 1558–1562.
Tan, W. C., Wang, L., Feng, X. W., et al. (2019). Recent advances in black phosphorus-based electronic devices. *Advanced Electronic Materials, 5*, 1800666.
Tekin, E., Smith, P. J., & Schubert, U. S. (2008). Inkjet printing as a deposition and patterning tool for polymers and inorganic particles. *Soft Matter, 4*, 703–713.
Tian, Z. N., Hua, J. G., Hao, J., et al. (2017). Micro-buried spiral zone plate in a lithium niobate crystal. *Applied Physics Letters, 110*, 041102.
Ugrinic, M., deMello, A., & Tang, T. Y. D. (2019). Microfluidic tools for bottom-up synthetic cellularity. *Chem, 5*, 1727–1742.
Voiculescu, I., & Nordin, A. N. (2012). Acoustic wave based MEMS devices for biosensing applications. *Biosensors & Bioelectronics, 33*, 1–9.
Wei, D. Z., Wang, C. W., Wang, H. J., et al. (2018). Experimental demonstration of a three-dimensional lithium niobate nonlinear photonic crystal. *Nature Photonics, 12*, 596.
Wu, Z. G., Lin, X. K., Si, T. Y., et al. (2016). Recent progress on bioinspired self-propelled micro/nanomotors via controlled molecular self-assembly. *Small, 12*, 3080–3093.
Xiao, J. H., Wang, X. H., Luo, X., et al. (2015). In situ preparation of catalytic combustion films used as micro heat source by inkjet printing method. *Applied Surface Science, 327*, 400–405.
Xiong, W., Liu, Y., Jiang, L. J., et al. (2016). Laser-directed assembly of aligned carbon nanotubes in three dimensions for multifunctional device fabrication. *Advanced Materials, 28*, 2002–2009.
Xu, B. B., Xia, H., Niu, L. G., et al. (2010). Flexible nanowiring of metal on nonplanar substrates by femtosecond-laser-induced electroless plating. *Small, 6*, 1762–1766.
Yang, L., & Zhou, Z. M. (2014). Demonstration of reduced power consumption MEMS LEL sensor prepared by a novel digital microfluid technique. *IEEE Sensors Journal, 14*, 85–90.
Ye, R. Q., James, D. K., & Tour, J. M. (2019). Laser-induced graphene: From discovery to translation. *Advanced Materials, 31*, 1803621.
You, R., Liu, Y. Q., Hao, Y. L., et al. (2020). Laser fabrication of graphene-based flexible electronics. *Advanced Materials, 32*, 1901981.
Yu, X. Y., Lv, Z. H., Li, C. H., et al. (2016). The optical and electrical properties of co-doped black silicon textured by a femtosecond laser and its application to infrared light sensing. *IEEE Sensors Journal, 16*, 5227–5231.
Zheludev, N. I., & Kivshar, Y. S. (2012). From metamaterials to metadevices. *Nature Materials, 11*, 917–924.

Chapter 4
Composite Micro-Machining Technology on the Non-Silicon MEMS

Yunna Sun, Guifu Ding, Yan Wang, and Zhuoqing Yang

4.1 Introduction

In general, MEMS devices include microsensors, micro-actuators, micro-functional devices, and micro-power devices which are interconnected and integrated together through packaging technology. Meanwhile, MEMS devices mainly refer to the microscopic devices, particularly those with moving parts. Integrated circuit (IC) technology has stimulated the development of MEMS which inherits the advantages of IC and also possesses the characteristics and advantages of advanced cutting-edge technology. Therefore, MEMS devices are microminiaturized and diversified with high stability and performance. In addition, MEMS satisfies the demand for high integration and mass production at low cost. As a promising device, MEMS has been applied in various research fields, such as biomedicine, aerospace, Internet of Things (IOT), Internet of vehicles (IOV), mobile devices, environmental monitoring, etc. (Langley et al., 2012; Fernández et al., 2017). However, the application and development of traditional MEMS devices (silicon-based) have been limited and hindered by the single silicon substrate for the basic properties of silicon and fabrication process of IC technology (Mcdonald et al., 2000; Gong et al., 2014).

In order to break through the limitations of traditional MEMS technology, the basic substrate material can be broadened from silicon to non-silicon materials, such as polymers, glass, ceramics, paper, metals, and composite materials to produce a variety of high-quality components with microscale feature sizes and

Y. Sun (✉) · G. Ding · Y. Wang
National Key Laboratory of Micro/Nano Fabrication Technology, School of Electronic Information and Electrical Engineering, Shanghai Jiao Tong University, Shanghai, China
e-mail: Cecilia_Sun@sjtu.edu.cn

Z. Yang
National Key Laboratory of Science and Technology on Micro/Nano Fabrication, School of Electronic Information and Electrical Engineering, Shanghai Jiao Tong University, Shanghai, China

Z. Yang (ed.), *Advanced MEMS/NEMS Fabrication and Sensors*,
https://doi.org/10.1007/978-3-030-79749-2_4

nanoscale surface roughness. Therefore, NSMFT/NSNFT has received extensive attention and in-depth research from scholars, which provides new impetus for the further development of MEMS devices. NSMFT/NSNFT basically consists of a variety of high-performance materials, fabrication processes, device design and test schemes, and MEMS devices. The properties of materials are the foundation for obtaining a high-performance device, while the fabrication technology is a necessary tool for improving the performance of the devices. It is the objective of NSMFT/NSNFT to obtain a reliable MEMS device with high performance. The test method is a key stage for validating and analyzing the performance of MEMS, since the measurement of these properties and output performance are served as part of the development of a specific MEMS device. It should be noted that the performance of the device depends on all four parts.

Since the selection and fabrication of the basic substrate material are connected together with the fabrication process and structural design of the MEMS device, the influence of the substrate material and the design of the corresponding fabricated process will be analyzed together. Firstly, the properties of the material are essential to the performance of MEMS. Some materials with high thermal properties, mechanical properties, electrical properties, magnetic properties, or optical properties are promising to enhance the function of the device (Linsebigler et al., 1995; Singal et al., 2017). Meanwhile, the performance of the material may be affected by the fabrication process and conditions. The size and the arrangement density of the crystalline grain which affect the mechanical properties depend on the deposition conditions. The innovation of the device is embodied in two aspects of technology and design, which can be divided into new materials and novel structures. The structure and performance difference between the designed MEMS and the fabricated MEMS depends on the micro/nano-fabrication process. Thus, a device with good performance cannot be achieved only with the material and design but also related to the fabrication technology. Furthermore, the response to the test results provides a direction for the improvement of the device design and performance. With the development of NSMFT/NSNFT, the promising multidisciplinary cross-frontier research, MEMS, is further diversified, multifunctional, and high precision.

To state the development of MEMS based on the diversified NSMFT/NSNFT, there are three parts. The first section is about the materials and the fabrication technology. The second one is about the diversified MEMS devices, such as the inertial micro-actuator, vibration-based energy harvester, microchannel heat sink, and high-temperature thin film sensor. In this section, the abovementioned MEMS device will be analyzed from the structures, functional materials, and fabrication processes. In the last one, the test method will be analyzed, including the measurements of the properties of the basic materials and the output performance of MEMS device.

4.2 Materials and Fabrications

The typical fabrication process flow of silicon-based MEMS consists of bulk micro-machining, surface micro-machining, and CMOS MEMS technology. The key technology of bulk micro-machining is deep silicon etching and bonding technology. The core technology of surface micro-machining is the thin film fabricating technology. Among the three kinds of typical fabrication processes, CMOS MEMS technology is promising in the future. However, the single silicon substrate limits and impedes the widespread application of MEMS device. While selecting a variety of materials is dependent on the MEMS device performance, the quality of the material is decided by micro-machining. Just as shown in Fig. 4.1, the size and the arrangement density of the crystalline grain which affect the mechanical properties depend on the electroplated conditions. By optimizing the electroplated conditions, the silver with high quality can be obtained. Secondly, it is the design of MEMS devices.

4.2.1 Bulk Micro-Machining on Non-silicon Materials

Generally speaking, the mechanical, chemical, and thermal energy are used as the bulk micro-machining method on the non-silicon materials. As a conventional machining technology, the mechanical machining has been developed and applied for decades. By using the mechanical energy, the micro-turning, micro-grinding, and micro-milling enable the substrate to generate microstructures. However, the mechanical micro-machining application on hard materials is limited by its low stiffness and cutting tool strength. As one of the nonconventional manufacturing processes, the micro-ultrasonic machining (micro-USM) is capable of solving this

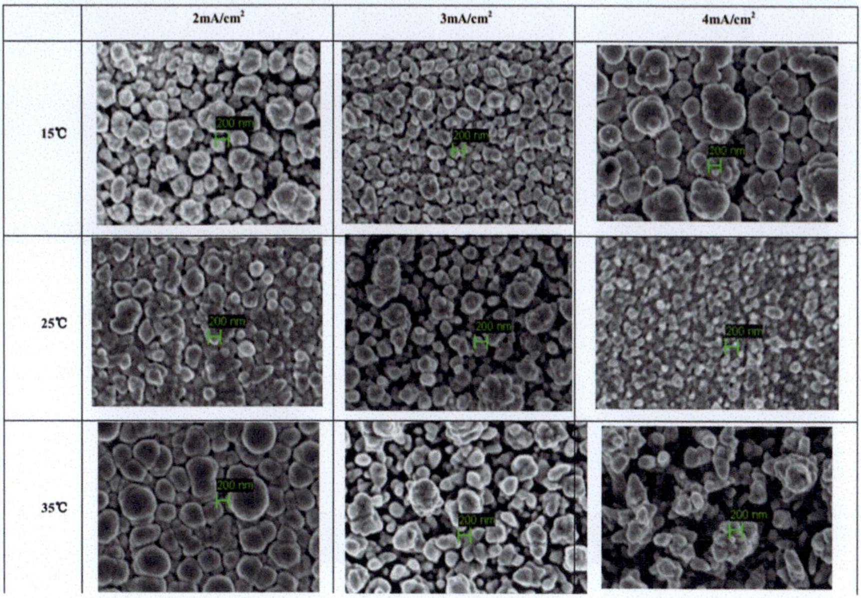

Fig. 4.1 The SEM images of the electroplating silver (Wang et al., 2016)

issue. The micro-USM method is beneficial for machining the hard and brittle non-conductive materials without thermal damage. With the help of micro-USM, free surface and high accuracy can be obtained, just as shown in Fig. 4.2a that a 1/8 tapered spherical ball has been achieved.

In terms of the chemical energy, the wet etching (a type of chemical machining) can achieve the machining process based on the special mask materials. The higher etchant resistance of the mask is the basis for obtaining a deeper etching microstructure. The polymers and rubbers, which have stable chemical performance at the etching solution, easy pattern, and inexpensive removal after etching, are often used as the mask materials to protect the substrate surface. By adjusting the concentration of etching solution and etching temperature, the etching rate and the roughness of the etching sidewall can be controlled. The chemical energy-based machining methods maintain good machining rate on materials with smooth surface. With the reaction going on, the etching rate and etching profile are limited. Electrochemical machining (ECM), especially electrochemical micro-machining (EMM), can figure out this issue (Bhattacharyya et al., 2002). During ECM and EMM stage, there is no additional thermal or mechanical stresses on the substrate, and no tool wear occurs. Furthermore, ECM and EMM possess a wider range of materials, better precision, higher machining rate, and controlled material removal.

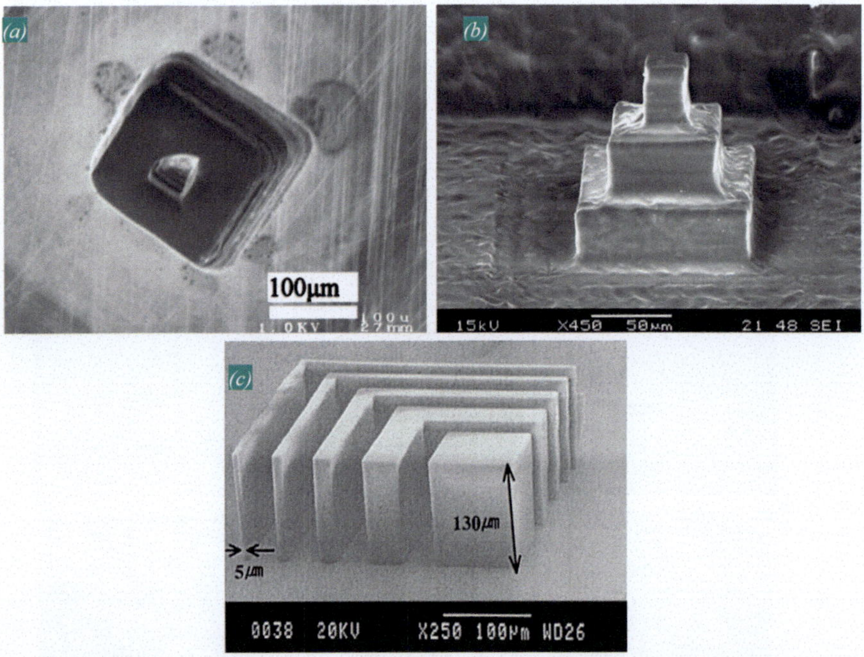

Fig. 4.2 (**a**) Micro-cavity by micro-USM (Yu et al., 2004), (**b**) micro-pyramid by ECDM (Xuan et al., 2009), and (**c**) microstructure by UV-LIGA (Chang & Kim, 2000)

Laser beam machining (LBM) and electrical discharge machining (EDM) are based on thermal energy. With LBM, microstructures can be machined on ceramics, polymers, and metals. It should be pointed out that the unavoidable thermal stress and the high cost of the equipment limit its wide application. EDM brings a relatively higher material removal rate. However, the tool wear rate is extremely high due to the skin effect and area effect induced by EDM (Liu et al., 2016). Electrochemical discharge machining (ECDM) has attracted interest of many scholars by combining ECM and EDM. ECDM possesses good dimension tolerances inherited from EDM, and also has good surface integrity of machined parts (Xuan et al., 2009). The material removal is related to the chemical etching rate, tool electrode, voltage, electrolyte, wettability, and the dimension of microstructures. With the layer-by-layer machining and pulsed voltage, 3D microstructures have been fabricated with high precision shown in Fig. 4.2b.

4.2.2 *Surface Micro-Machining on Non-silicon Materials*

Different from the bulk machining method mentioned above, surface micro-machining on non-silicon materials refers to the use of surface sacrificial layers, patterns, and deposition technology on the surface of non-silicon substrates to manufacture multifilms or microstructures. Surface micro-machining technology saves removing or etching the substrate materials. Similar to the silicon-based surface micro-machining technology, the non-silicon-based surface micro-machining is to make use of LIGA, sacrificial layer technology, and electroplating technology to prepare functional structure on non-silicon substrate. A variety of active and movable microstructures, such as micro bridge, cantilever beam and cantilever block, etc., as well as the sensitive components and actuators, such as resonant microsensor, the resonant micro-gyroscope, micro-accelerometer, micro-motor, all kinds of brake, etc., can be obtained by removing the sacrificial layer.

Surface micro-machining process mainly composites by sacrificial layers and structure layers which is fabricated by the deposition process which can be separated to physical deposition and chemical deposition. In terms of the material used as the sacrificial layers, it shall be fabricated with high precision and removed easily. The positive or negative resists, water-soluble polymer, SiO_2, glass, Cu, Ni, Ag, Ti, and so on often serve as the sacrificial layer. As far as the microstructure with high aspect ratio is concerned, it is mainly fabricated with the conductive material including metal (Cu, Ni, Ag, Au, Sn, and so on), conductive nano-material (CNT, graphene, graphdiyne), and composite conductive material by the electroplating and electrophoresis technology. Given the compatibility, high precision, and simplicity of electroplating, it has been widely used in MEMS. The grain size and properties of the electroplated layer depend on the additive, external forced convection, microstructure dimension, electric field intensity distribution, solution temperature, seed layer quality, and so on (Zhang et al., 2015; Huang et al., 2016). In order to enhance the electrical, thermal, and mechanical properties of the deposited metal, the composite electroplating has attracted the interest of many scholars. The electrical resistivity of electroplated graphene-Cu film is decreased greatly, and the thermal conductivity is improved compared with Cu film (Huang et al., 2016). It has been found that the composite electroplated Ni-W alloy film possesses excellent wearing resistance and corrosion resistance than that of the Ni film (Wang et al., 2010). Moreover, the Ni-W alloy film still has good properties after heat treatment. The composite Cu-diamond film possesses a higher thermal conductivity (Wu et al., 2019) and depressed coefficient of thermal expansion compared with Cu film.

By using surface micro-machining technology, an extremely finer microstructure with higher aspect ratio can be fabricated. Furthermore, the fabricated micro-device has better performance along with low cost and reduced materials. A SU-8 microstructure with 5 μm thickness and 130 μm depth has been made by UV-LIGA, as shown in Fig. 4.2c. UV-LIGA technology also has been used to fabricate the folded waveguide (Xie et al., 2015). With multistep SU-8 UV-LIGA technology and a thin TiO_2 film formed on the Ti substrate surface, the folded waveguide was fabricated with high precision, high performance, and high reliability. A non-silicon-based micro-channel heat sink has been designed and fabricated through the surface micro-machining process including patterning the SU-8 mask sacrificial layer, electroplating the Cu substrate integrated with pin fin, and bonding the top and bottom non-silicon substrate together as depicted in Fig. 4.3. In this design, high thermal conductivity Cu has been used as the pin fin integrated with the bottom substrate of the composite heat sink via UV-LIGA technology, and AlN is used as the top substrate (Zhao et al., 2014). With SU-8 mask sacrificial layer, the height of the pin fin can be fabricated among 0.2 mm–1.5 mm. With the higher conductivity Cu and optimized pin fin height, the maximum dissipation heat flux of the composite heterogeneous heat sink can reach 209 W/cm^2 when the chip maximum temperature is under 85 °C at a flow rate of 90 mL/min. However, the surface micro-machining technology can only fabricate 2D structures.

To sum up, the advantages and disadvantages of the bulk micro-machining technologies and surface micro-machining technologies are listed in Table 4.1. These machining methods are capable of satisfying different needs of MEMS devices fabricated by NSMFT/NSNFT. The non-silicon-based MEMS device will be more functional and stable to broaden the application of MEMS devices, whereas a complicated MEMS device may not be fabricated only by one kind of micro-machining technology.

4.2.3 Composite Micro-Machining on Non-silicon Materials

Some complicated MEMS devices are fabricated based on bulk micro-machining together with surface micro-machining. The diversity of MEMS devices is broadened by composite micro-machining method. The performance of the devices is also improved through composite micro-machining.

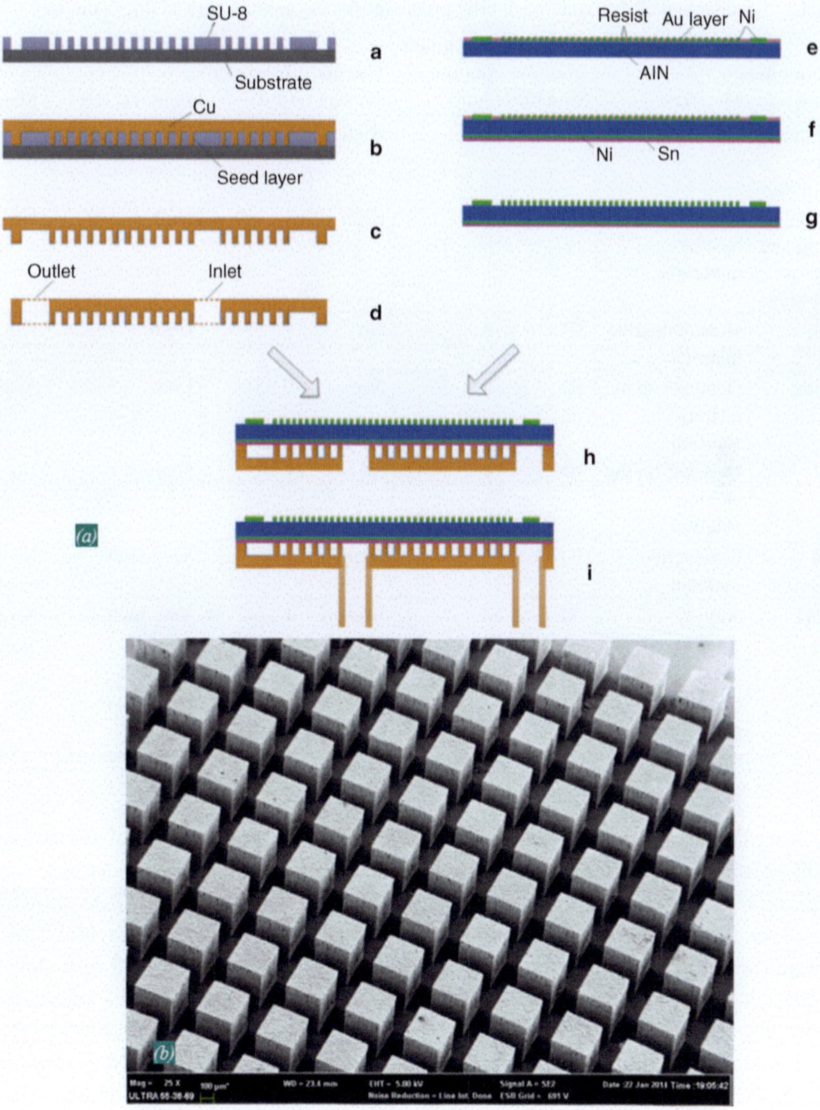

Fig. 4.3 (**a**) The fabrication process of the heat sink and (**b**) the top view SEM image of an array of pin fin (Zhao et al., 2014)

As a promising and advantageous package method, the through via of the glass interposer was formed by the focused electrical discharging method, and the redistribution layer was fabricated by the surface sacrificial layer technology. The composite polymer interposer used micro-machining technology to prepare the composite polymer and filled the via before forming the composite polymer substrate. In terms of this diversified technology and the isolation substrate, it decreases fabrication cost, improves yield, and guarantees the electrical reliability (Sun et al., 2018; Liu et al., 2017).

Table 4.1 Comparison of different machining processes for the manufacture of micro-devices

Machining method	Materials	Structure	Minimum feature (μm)	Thermal damage	Tolerance (μm)	Removal rates	Mask
Mechanical micro-machining	Any	3D	10	High	1	High	No
Diamond milling and micro-machining	Not suitable for ferrous materials	3D	10	High	1	High	No
USM	Nonconductive material	3D	5	No	1	High	No
Etching	Depends on the etching solution	3D	1	No	0.05	Low	Yes
ECM	Depends on the etching solution	3D	1	No	1	High	No
EDM	Conductive material	3D	10	High	1	Very high	No
ECDM	Any	3D	1	High	1	Very high	No
Laser	Any	3D	1	Very high	0.5	Low	No
LIGA	Cu, Ni, permalloy, ceramic	2D	0.01	No	0.02	Depends on process electroformable	Yes

A complicated axial flux electromagnetic micro-motor composed of the rotor and the stator has been manufactured with EDM and laser technology as well as lithographic and electroplating processes, respectively (Yang et al., 2001). As shown in Fig. 4.4a, the SmCo permanent magnetic alloy has been manufactured with EDM to obtain the rotor with diameter of 1.0–2.0 mm and thickness of 0.2–0.3 mm. And the center hole with diameter of 0.4–0.6 mm was drilled by laser. With a series of lithographic and electroplating processes (Fig. 4.4b), the thickness of each winding was 15 μm and the minimum spacing between them was 1 μm. With this design and fabrication technology, the micro-motor has been successfully demonstrated in micro-pumps, endoscopes, and micro-helicopters.

Thus, the composite micro-machining method is promising in the future for designing and fabricating MEMS devices with high performance and reliability.

4.2.4 Diversified Materials

NSMFT/NSNFT broadens the substrate to non-silicon materials, such as polymer, glass, ceramic, paper, and metal. Thus, MEMS device based on the NSMFT/NSNFT is more diversified in performance, structure, and fabrication to satisfy different

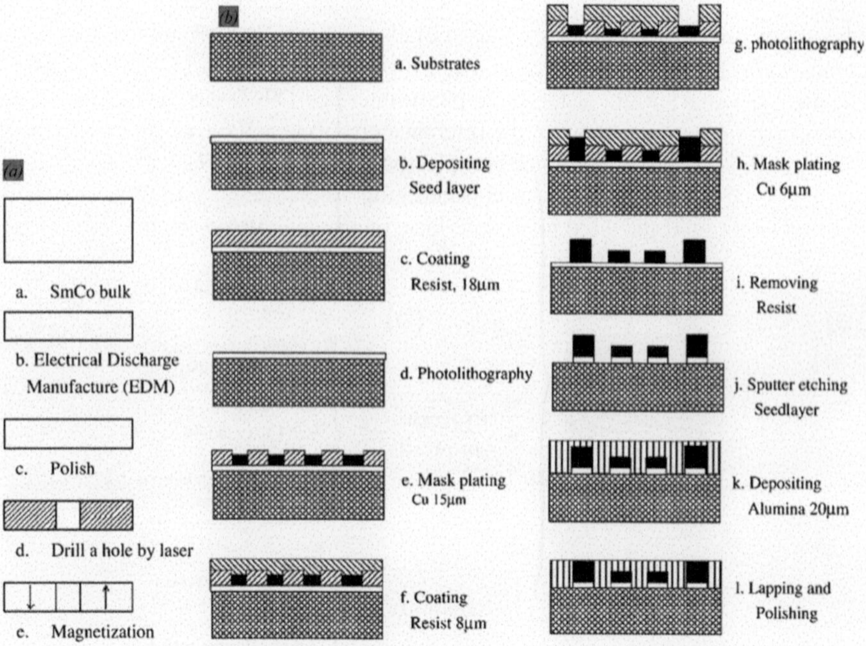

Fig. 4.4 Fabrication processes for the (**a**) rotor and (**b**) stator windings (Yang et al., 2001)

application environments, as shown in Fig. 4.5. Firstly, the diversified performances include electrical, mechanical, thermal, optical, and biological properties and compatibility. Secondly, the diversified structures can be achieved since the non-silicon-based MEMS fabrication technology is meaningful to obtain movable and dangling MEMS device with a larger aspect ratio. Finally, the fabrication process is diversified for the fabrication process flow of NSMFT/NSNFT consisting of bulk micro-machining, surface micro-machining, and composite micro-machining. In detail, the basic fabrication processes of the NSMFT/NSNFT include electroplating, electrophoretic deposition, sputtering, UV-LIGA, quasi-LIGA, bonding, slicing, etching, polishing, etc., which have great influence on the properties of the materials and performance of the MEMS device.

With the help of NSMFT/NSNFT, the conductive films and pillars can be fabricated by electroplating (such as Cu, Ni, Ag, Au, Sn, Pt, Ti, and Ta), and also can be prepared by chemical vapor deposition (CVD), composite plating, and chemical plating (such as CNT, graphene, and metal composites). The insulating film such as Al_2O_3 and SiO_2 can be obtained by CVD, ion beam sputtering, thermal oxidation, and sintering, and the polymer and composite polymer serving as insulating film can be formed by coating and filling under vacuum. The diversified non-silicon substrate accelerates the novel emergence of various fabrication methods. The new fabrication methods bring the materials with high performance to MEMS devices. As an effective and compatible technology, NSMFT/NSNFT broadens the application of MEMS devices.

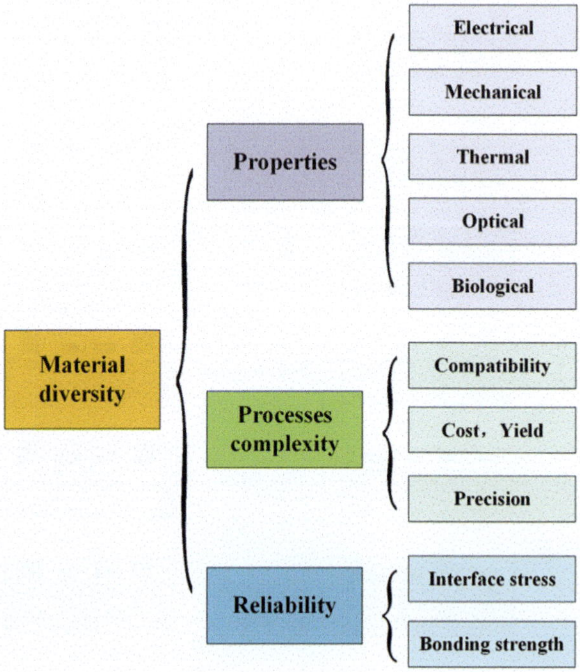

Fig. 4.5 Influences of material diversity

The design scheme and material selection together with fabrication process of the MEMS device are the basic factors to obtain a MEMS device with high performance and high reliability. Since the main indexes of the microsensors are the sensitivity, measurement range, phase shift, and frequency range, in order to obtain a high sensitive pressure sensor, a sandwiching ultrathin gold nanowire-impregnated tissue paper between two thin polydimethylsiloxane sheets has been proposed (Gong et al., 2014). Furthermore, this sensor has a wide measurement range that is sensitive to pressing, bending, torsional forces, and acoustic vibrations. Polymer composite possessed composite has been widely applied in the advantage packaging (Sun et al., 2018), since the polymer extremely high thermal conductivity, reliable electrical performance, and composite micro-machining. Figure 4.6 shows the fabricated polyimide composite with SiC-whisker and AlN nanoparticles. Sun et al. (2017) designed a nonlinear electromagnetic energy harvester based on the large deformation of the thin Ni-based guided-beam structure shown in Fig. 4.7a. The good bandwidth performance of the proposed energy harvester is a result of the merits of both the mechanical properties and guided-beam structure. In addition, the nonlinearity of the harvester was controlled by adjusting the thickness of the central platform.

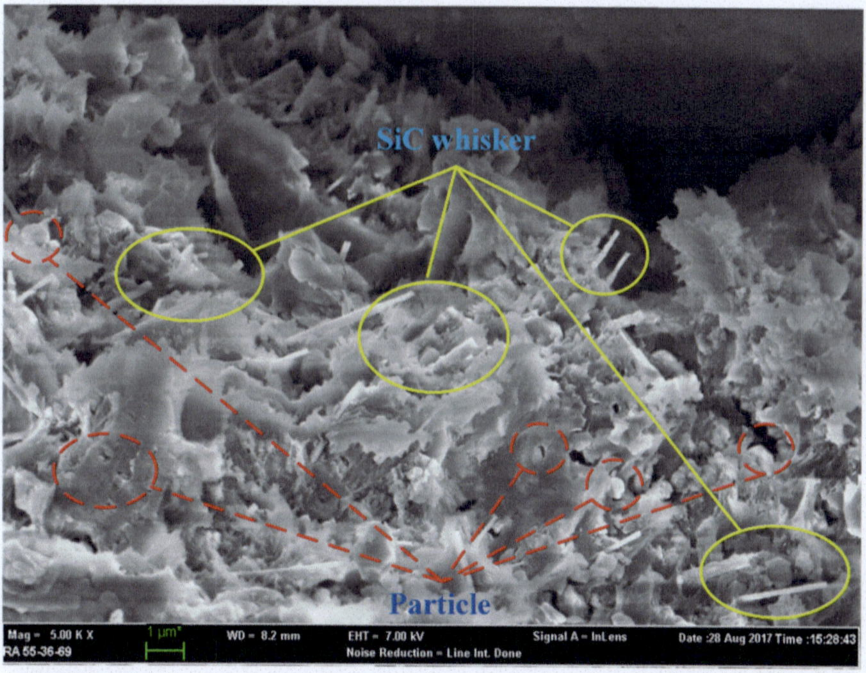

Fig. 4.6 SEM image of the composite polyimide with SiC-whisker and AlN nanoparticles (Sun et al., 2018)

In order to satisfy the different applications, the design scheme, material selection, and fabrication process are different. A spark micro-actuator based on NSMFT/NSNFT was proposed and fabricated to meet the needs of pulsed power application. As shown in Fig. 4.7b, an in-plane spark gap actuator with dual-trigger electrode for pulsed power application has been designed and fabricated on a glass wafer with the surface micro-machining technology (Liu et al., 2013). This proposed actuator consisted of two main electrodes and a dual-trigger electrode which is trigged at a high peak current and a smaller trigger voltage. Based on NSMFT/NSNFT, this actuator has a minimized dimension and reduced trigger voltage with low fabrication cost. However, the design and fabrication of inertial micro-actuator focus on threshold acceleration, response time, contact time, and sensing range.

4.3 Non-silicon MEMS Device

As a promising multidisciplinary cross-frontier research field based on NSMFT/NSNFT, non-silicon-based MEMS, particularly moving devices, possess many advantages such as high performance, high reliability, and low cost. MEMS are made up of components ranging in size from 1 μm and 100 μm, while MEMS

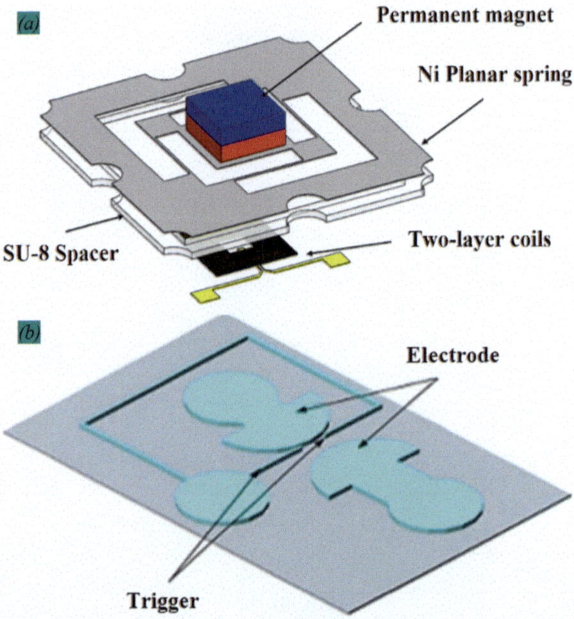

Fig. 4.7 (**a**) 3D schematic of the proposed energy harvester (Sun et al., 2017). (**b**) Schematic representation of the spark gap actuator (Liu et al., 2013)

devices generally range in size from 20 μm to 1 mm. The structure diversity will bring performance diversity, but also induce the process complexity and challenge the device reliability, as shown in Fig. 4.8. The thermal mismatch stress and mechanical stress induced by the composite structure and fabrication processes are serious thermal reliability problems for devices, especially for the moveable MEMS. MEMS is mainly composed of microsensors, micro-actuator, and micro-power with packaging technology. In the following parts, the inertial micro-actuator, vibration-based energy harvester, and high-temperature sensor will be reviewed as the typical examples of micro-actuator, microsensors, and micro-power, respectively. In addition, non-silicon-based heat sink will also be reviewed as the special devices of functional MEMS devices. In terms of the NSMFT/NSNFT, the inertial micro-actuator presents the MEMS with complicated moveable structure, the vibration-based energy harvester represents the MEMS with diversified structures, the heat sink stands for the MEMS with diversified functional materials, and the high-temperature sensor delegates the MEMS with diversified fabrication processes.

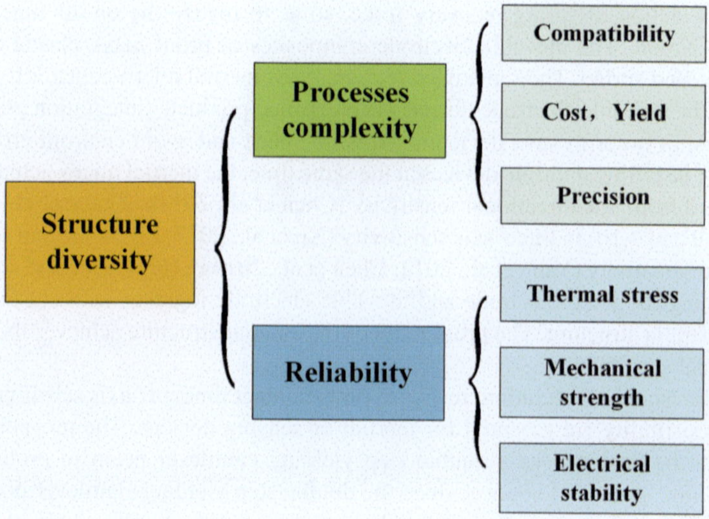

Fig. 4.8 Influences of structure diversity

4.3.1 *MEMS with Complicated Movable Structures: Inertial Micro-Actuator*

As an important component of MEMS, the micro-actuator is developing rapidly. Electrostatic, magnetostatic, piezoelectric, and electrothermal mechanisms are the main actuation trigger mechanisms commonly used in micro-actuators. Micro-actuators including ohmic contact actuator, capacitive actuator, inertial micro-actuator, and RF actuator have been applied in many industries. As a typical inertial sensor, the inertial micro-actuator has been reported since 1972 (Frobenius et al., 1972). This micro-cantilever electromechanical inertial micro-actuator has been developed rapidly. Due to their good electrical conductivity, good mechanical properties, and toughness, metallic materials have been used as structural materials for MEMS micro-actuator to improve working performance.

Based on surface micro-electroplating technology, the actuator can be fabricated as a multilayer suspended movable structure. Since the inertial micro-actuator possesses sensing and actuating functions, the inertial micro-actuator can replace the complex detection system which is composed of accelerometer, control unit, and actuation unit. In addition, it makes up for the shortcomings of traditional accelerometer, such as the spurious trigger caused by electromagnetic interference, the complex signal processing circuit, and so on. Furthermore, it is a passive device without any power consumption and that has small size, fast response speed, and low manufacturing cost and is lightweight.

The working principle of the inertial micro-actuator is that the movable electrode moves toward the fixed electrode until collision occurs, that there is external acceleration beyond the threshold, and finally that it returns to the equilibrium position under the action of spring recovery force, so as to realize the on-off function of external circuit. The movable electrode composites of proof mass, elastic contact structure, and spring. The sensitive direction of the inertial micro-actuator is dependent on the movable electrode design. As electronic products' integration gradually improves, in order to save device installation space and avoid centroid error produced by installing multiple devices at the same time, the inertial micro-actuator has developed from one directional sensitivity (Chen et al., 2014b; Zhang et al., 2016), (Figs. 4.9 and 4.10) to three-axis sensitivity (Xu et al., 2017; Fig. 4.11) and omnidirectional sensitivity (Yang et al., 2012; Chen et al., 2014a), (Figs. 4.12 and 4.13), by redesigning the fixed electrode and movable electrode together with elastic beam and constraint structure. The proof mass of the annular structure achieves the omnidirectional sensitivity.

The threshold acceleration, response time, contact time, off-axis sensitivity, and shock resistibility are essential for inertial or sensing devices. The movable electrode can be designed as a double step yielding cantilever beam to prolong the contact time and avoid bounce, since the double step yielding cantilever possesses large stiffness. To reduce the contact bouncing and damage effectively, the fixed electrodes in horizontal and vertical direction can be respectively fabricated as cantilever and multi-hole crossbeam with Ni by low-temperature photoresist modeled electroplating technology. Off-axis sensitivity and shock resistibility were guaranteed by the constraint sleeve and reverse stop-block structures (Zhang et al., 2016). The response time and contact time are related to the stiffness of the movable electrode and the fix electrode, the gap distance between the movable electrode and the fix electrode, the effective mass of the system, the overload acceleration, and the acceleration pulse width. Firstly, the contact time can be prolonged effectively by the flexible structure for offering sufficient elastic deformation during the contact of the electrodes. A complaint cantilever beam has been proposed as the stationary electrodes of the micro-actuator (Fig. 4.9). Instead of rigid structure, fabricating an elastic cantilever Ni beam on the movable electrodes has shortened the response time and prolonged the contact time as shown in Fig. 4.10. With the flexible design

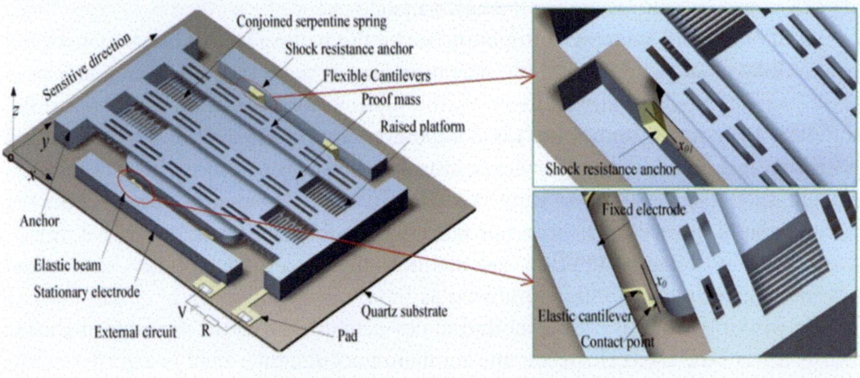

Fig. 4.9 Sketch of the inertial micro-actuator device (Chen et al., 2014b)

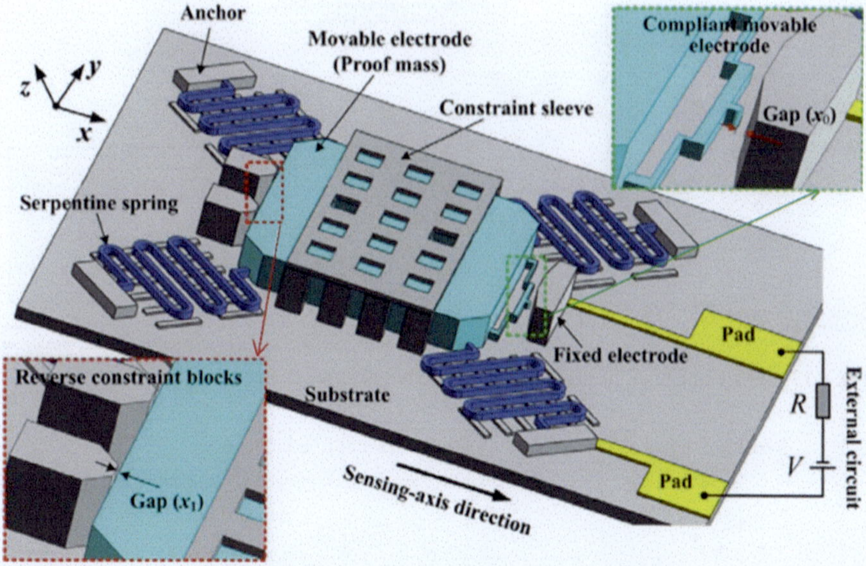

Fig. 4.10 Sketch of the designed inertial micro-actuator (Zhang et al., 2016)

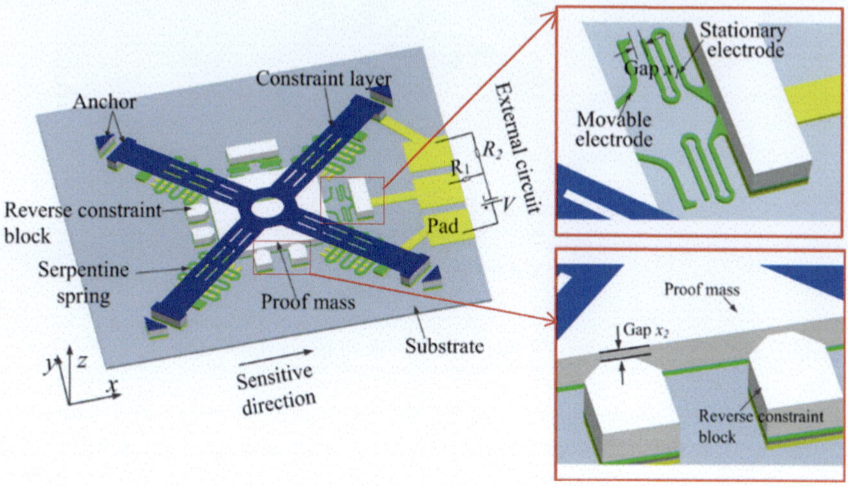

Fig. 4.11 The schematic diagram and structure parameters of the designed inertial micro-actuator (Xu et al., 2017)

on the electrodes, the synchronous follow-up inertial micro-actuator achieved a continuous contact (Fig. 4.11). In addition, the elastic materials and CNTs with high electrical conductivity and mechanical strength were also used to enhance the electromechanical contact performance. The polymer (Fig. 4.13) and CNTs have been introduced to fabricate a micro-actuator to achieve soft contact (Yang et al., 2012;

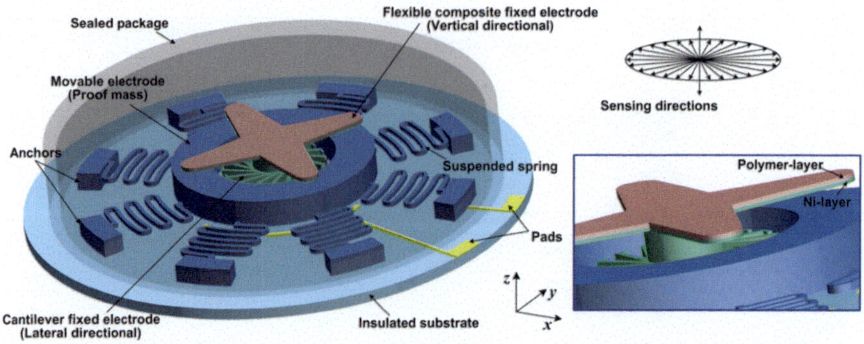

Fig. 4.12 3D schematic of the proposed inertial micro-actuator device (Yang et al., 2012)

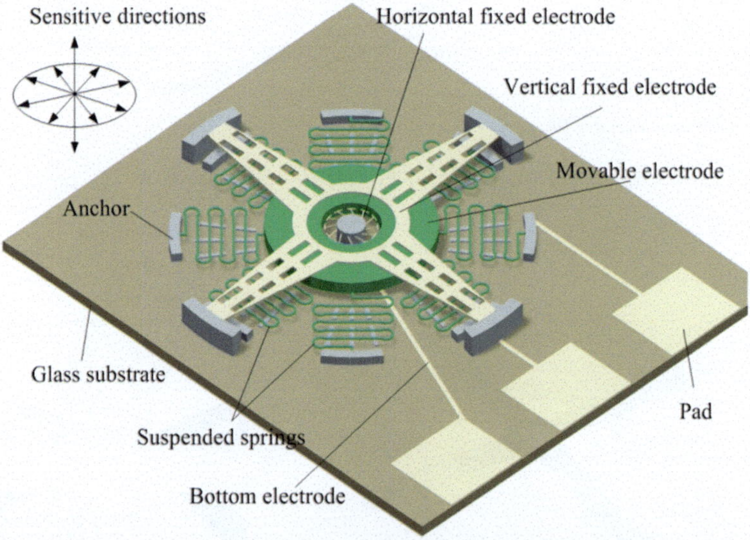

Fig. 4.13 The structure diagram of the designed threshold sensor (Chen et al., 2014a)

Chen et al., 2016). Furthermore, a novel MEMS inertial micro-actuator with electrostatic force assistance and multistep pull-in behavior has been proposed (Li et al., 2017). In this design, the flexibly connected multi-plane design reduced the pull-in voltage, improved the contact stability of the inertial actuator, and prolonged the contact time. According to the value of the threshold acceleration, the inertial micro-actuator can be divided into low threshold inertial micro-actuator, high threshold inertial micro-actuator, and adjustable threshold inertial micro-actuator. In the current reports about inertial micro-actuator, the threshold acceleration is higher than 30 g. A novel segmental circle spring with lower stiffness compared with traditional semicircle spring was proposed to design a low-g inertial micro-actuator (Fig. 4.14).

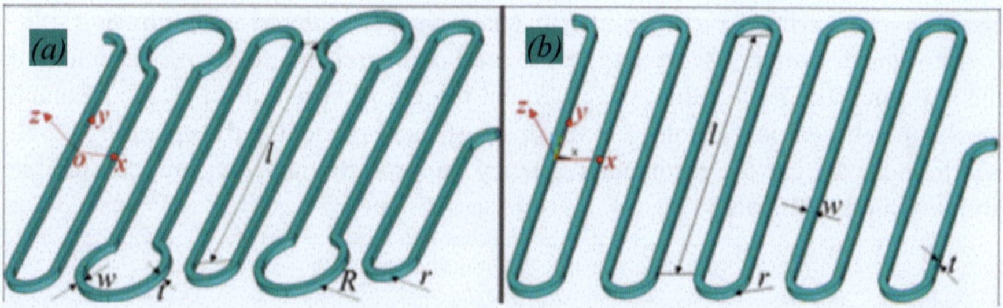

Fig. 4.14 (**a**) Segmental circle spring, (**b**) conventional semicircle spring (Chen et al., 2016)

With this design, the test results of the fabricated inertial micro-actuator indicate that the threshold acceleration was about 25 g.

In terms of performance, adjusting the threshold and reducing the interaxial interference of the micro-actuator are the main directions for further study. In order to further improve the performance of the micro-actuator, the micro-mechanical vibration conduction theory and micro-mechanical contact collision theory shall be explored in detail. In addition, the overall structure size still has room for further reduction. With the decrement of micro-actuator volume, the influence of size effect and material characteristics will become more and more significant. Thus, some of the microcosmic forces, such as van der Waals force, electrostatic forces, etc. need to be studied emphatically to be applied in optimal design of the movable electrode and fixed electrode together with elastic beam and constraint structure.

4.3.2 *MEMS with Diversified Structures: Vibration-Based Energy Harvester*

With the miniaturization of MEMS devices and the rapid development of packaging methods, the size of the MEMS is determined by the energy supply unit. Hence, the energy supply device has become one of the bottlenecks in the development and application of MEMS. Since the power consumption of MEMS (especially wearable electronics, embedded wireless sensors, and low power electronics) continues to decrease with the miniaturization of MEMS devices, battery, as one of the most common power choice, does not keep pace with the development of power requirements. Therefore, micro-power generation is introduced to replace the role of battery in traditional power supply field. Micro-power generator aims to convert a variety of waste and unused energy such as temperature gradient, solar, wind, acoustic noise, vibrations, etc. into electric energy, and provide continuous power to permit an autonomous operation of electronic devices.

In view of the advantages of vibration energy in terms of power density, endurance to harsh environment, long lifetime, and easy to find and transform, vibration-based energy harvesters (VEHs) transformed by the piezoelectric, electromagnetic, electrostatic, or triboelectric mechanisms have been considered as a promising solution to replace battery. According to the method of improving the performance of the harvester, the main future development directions which most of the resonant mechanism based on VEH need to make efforts are broadening the operating bandwidth and reducing the resonant frequency along with enhancing the compatible ability with the MEMS.

4.3.2.1 Broadening the Operating Bandwidth

The conventional linear one-degree-of-freedom (1-DOF) resonant oscillator based on VEH is only effective in the operating bandwidth around its sole resonant frequency; otherwise, the converting efficiency will drop dramatically. However, the majority of practical vibration sources is commonly disturbed in a wide frequency range or under random excitations. In order to broaden the bandwidth of VEH, many studies have been carried out on frequency tuning devices, multimodal structures, and nonlinear techniques.

The tuning device can be divided into manual tuning and self-tuning. Based on magnetic levitation, Mann (Mann & Sims, 2009) produced an oscillator with tunable resonance. VEH can also be adjusted by tuning the screw manually. L. Gu and C. Livemore focused on rotation motion and proposed a passive self-tuning piezoelectric energy harvester which utilizes the centrifugal force emerging during rotation to adjust the stiffness (Gu & Livermore, 2010). Nevertheless, self-tuning VEH always requires extra device to tune the resonant frequency, which means the extra space and energy are required to tune the frequency. Since the frequency up-conversion (FUC) based on VEH can convert low-frequency excitation to a higher frequency range, an impact-driven energy harvester based on FUC has been designed and fabricated with Ni and Si (Lee et al., 2007).

Multiple bending modes with continuous beam and an array of cantilevers are the basic designs of the multimodal harvester. Hybrid structures mainly include dual-mass cantilever beam, dual-beam system, and dynamic amplifier system. With L-shaped beam configurations, the bandwidth of the VEH has been widened (Erturk et al., 2009). A frequency sweeper employed a series of cantilevers with varying lengths, and resonance frequencies were proposed to broaden the bandwidth as illustrated in Fig. 4.15. Such 2-DOF VEHs with multifrequencies provide a reliable broadened bandwidth without any extra tuning efforts or auxiliary components.

The bandwidth can be broadened when the nonlinearity is sufficiently strong. The nonlinear techniques mainly include monostable and bistable mechanisms. An electromagnetic self-powered low-level VEH with electroplated Ni resonator exhibited monostable nonlinear characteristic when experiencing large displacement (Wang et al., 2009). Wang and Dai reported a broadband electromagnetic VEH based on MEMS (Fig. 4.16); the nonlinearity was introduced by the interaction of stretching and bending of the electroplated Ni cantilever. The bistable nonlinearity can be achieved by permanent magnet (Mann & Sims, 2009). To further improve the conversion efficiency, A. Cammarano et al. utilized the high permeability material

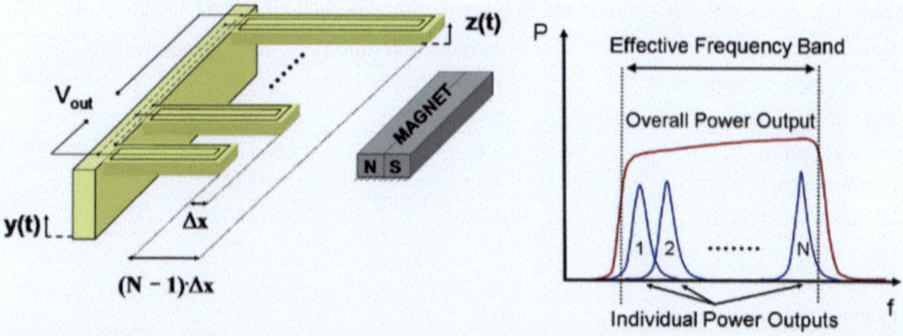

Fig. 4.15 Frequency sweeper (Sari et al., 2008)

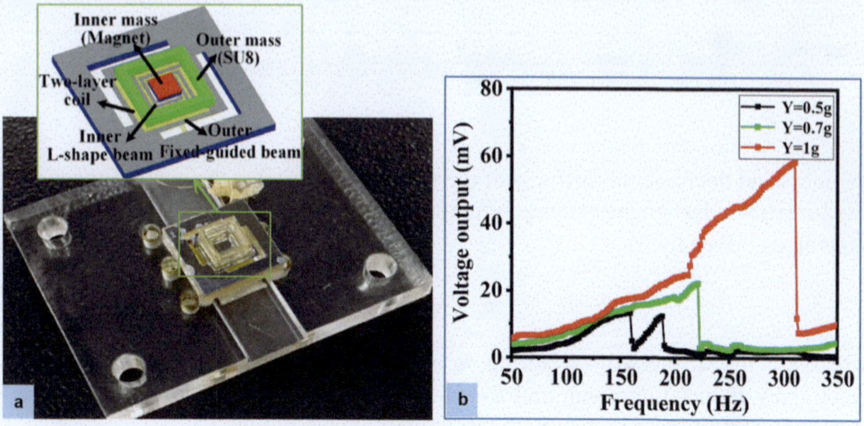

Fig. 4.16 (**a**) Energy harvester consisting of a 2-DOF planar spring and a micro two-layer coil, (**b**) experiment results of voltage against frequency-up sweep (Wang et al., 2020)

to induce the flux path to go through the coil (Cammarano et al., 2010). The flux experienced direction changes during vibration, which created strong coupling between mechanical and electrical domains. Due to the interaction between the permanent and soft magnet, the bistable compliance characteristic was brought into the system.

4.3.2.2 Reducing the Resonant Frequency

Low-level VEH can enhance the energy conversion efficiency of a large number of the low-frequency and low-amplitude vibrations in the environment shown in Table 4.2. Basically, the harvester volume is usually large, if the natural frequency of the harvester is very low. That is, it is a great challenge that a desired low-level VEH with effective energy output has a miniaturized size and is compatible with MEMS. The main methods to reduce the resonant frequency of VEH are to change

Table 4.2 Acceleration magnitude and frequency of fundamental vibration (Roundy et al., 2003)

Vibration source	Acceleration (m/s^2)	Frequency peak (Hz)
Car engine compartment	12	200
Base of three-axis machine tool	10	70
Blender casing	6.4	121
Clothes dryer	3.5	121
Car instrument panel	3	13
Door frame just after door closes	3	125
Small microwave oven	2.5	121
CD on notebook computer	0.6	75
HVAC vents in office building	0.2–1.5	60
Windows next to a busy road	0.7	100
Second story floor of busy office	0.2	100
Walking	9.8	5
Heartbeat	2.9	51
Person nervously tapping their heel	3	1

the mass and decrease the stiffness of the spring owing to that the maximum energy can be extracted when the excitation frequency matches the natural frequency of the system, ω_n, given by

$$\omega_n = \sqrt{k / m} \tag{4.1}$$

where k is the stiffness of the spring, m is the seismic mass, and c_e and c_m are the electrically induced damping and the mechanical damping, respectively. A schematic of a spring-mass system with vibration is shown in Fig. 4.17.

By using low stiffness materials or optimized spring microstructures, the spring with lower stiffness can be achieved. A polymer-based helical spring array (Fig. 4.18) was fabricated by projection micro-stereo-lithography. This VEH has a relatively smaller feature size and broader operation frequency range at a lower resonant frequency due to the 3D polymeric springs with lower stiffness. As shown in Fig. 4.19, SU-8 membrane with AlN pads is fabricated to convert mechanical vibration of the membrane to voltage. The nonlinear stiffness was achieved due to the stretching strain in the membrane that provided a wider bandwidth (~146 Hz) than conventional linear oscillators at lower frequency. In order to change the mass of VEH, the additional masses or the vertical tubes have been designed to enhance the mass of the micro-system, and the system resonant frequency was reduced to improve the efficiency of the energy conversion (Fig. 4.20). It is capable of producing useful power from a vibration level of 60 mg, delivering 46 μW to a resistive load of 4 kΩ when the device is shaken at its resonant frequency of 52 Hz and the power density is about 307 μW/m^3. With the help of NSMFT, the 3D structural VEH with high performance and small feature size can be obtained. The feature size should be minimized and the efficiency of VEH should also be improved in the future.

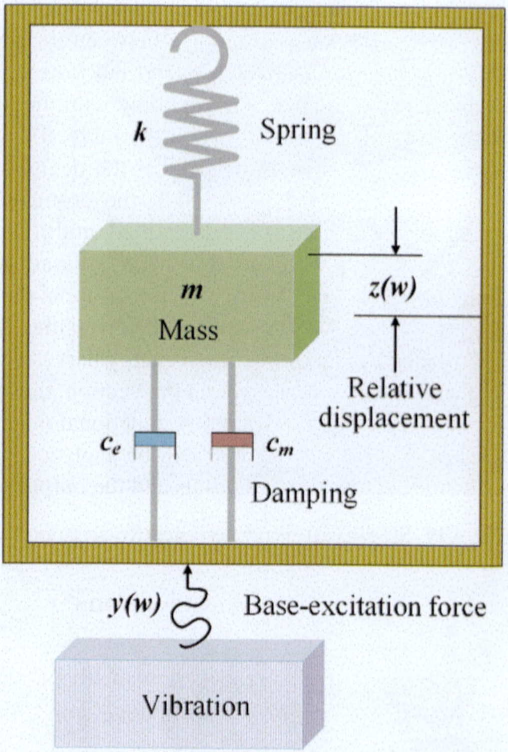

Fig. 4.17 Schematic of a spring-mass system with vibration

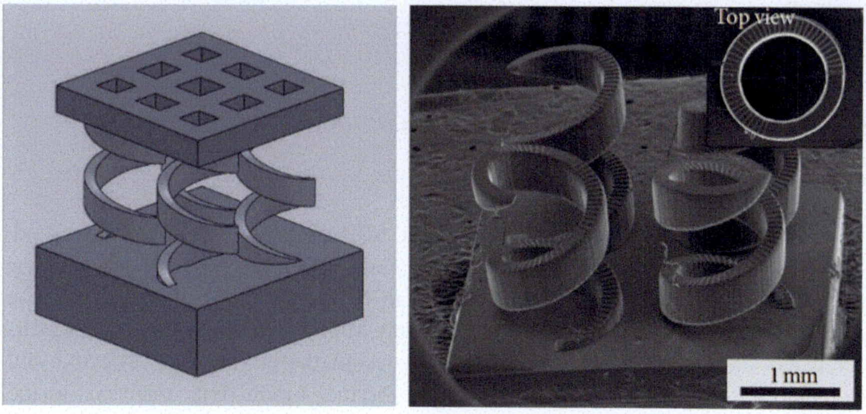

Fig. 4.18 The 3D polymeric helical spring structure (Baker et al., 2012)

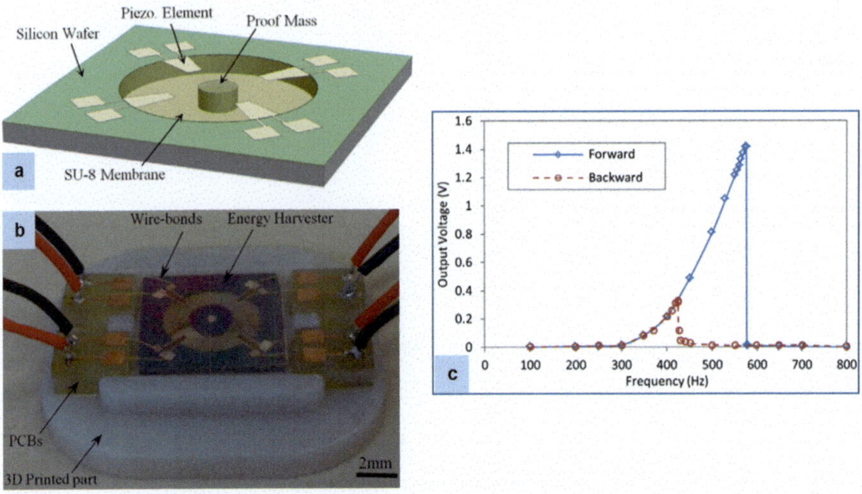

Fig. 4.19 SU-8-based energy harvester (Rezaeisaray et al., 2015)

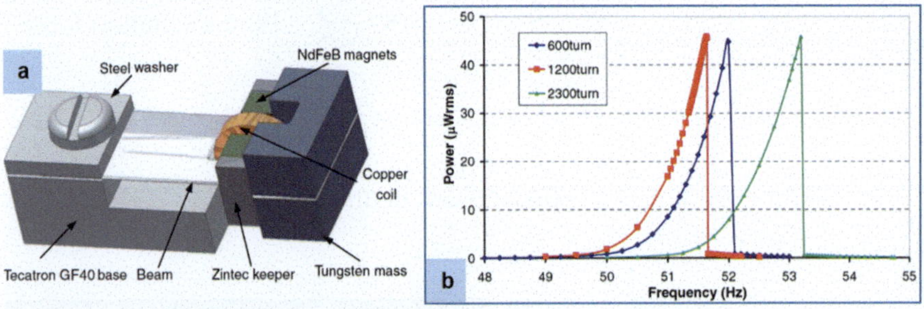

Fig. 4.20 Micro cantilever generator (Beeby et al., 2007)

4.3.3 MEMS with Diversified Functional Materials: Microchannel Heat Sink

In terms of thermal management of integrated MEMS system, the effective method, heat sink, is usually applied to keep hot spots under the junction temperature (85 °C). Since Tuckerman et al. (Tuckerman & Pease, 1981) first proposed the microchannel heat sink (MCHS) in 1981, many MCHS have been studied from different aspects. In order to achieve an optimal design, there exists some effective and rational directions, for example, using high thermal conductivity materials as the architecture and optimizing the microchannel, inlets and outlets, and fluid flower of the MCHS. A MCHS with a high thermal conductivity metal-ceramics composite structure (Fig. 4.21) has been designed and fabricated through UV-LIGA technology to enhance the heat dissipation capacity of the heat sink. The MCHS consists of the chip (or simulative heat source), top baseplate connected with the chip microchannel, bottom baseplate, and the inlet and outlet, as shown in Fig. 4.21.

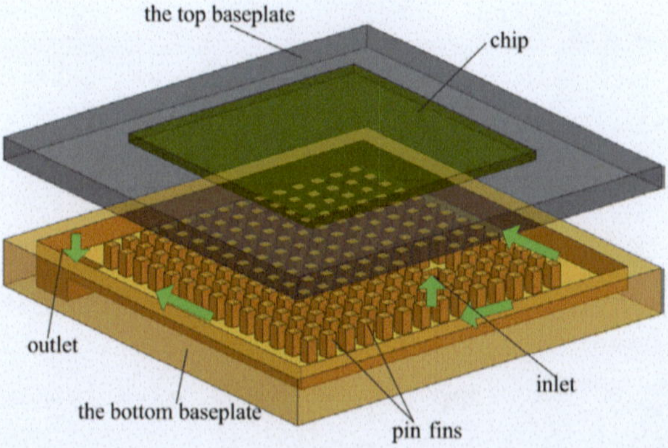

Fig. 4.21 Sketch of the microchannel heat sink (Zhao et al., 2014)

The high thermal conductivity ceramics (HTCCs) are with good insulation performance; thus, HTCC, including diamond, CNT, graphene, SiC, AlN, and their composite materials, is promising for improving the heat dissipation capacity of the MCHS. By using HTCC as the top plate of MCHS which is connected to the chip, the heat of the hot spot moves rapidly to the surrounding area. In addition, the electric performance of the chip can be guaranteed. With the compatible process, the microstructures can be obtained on the bottom surface of the HTCC to achieve an interconnection of near node. In this way, the thermal resistance is reduced greatly, which occupies a large proportion of the total resistance.

The high thermal conductivity materials including HTCC and metals with graphic characteristic are the candidates of the microchannel. The diversified and multidimensional microchannel with high precision is the key part for the heat dissipation capacity. The structure, pitch, and dimension of the microchannel have influence on the effective heat exchange area, and the status of fluid flow is related to the heat dissipation efficiency. Given the mature fabrication technology of Si, it can also be used as the microchannel (Erp et al., 2020). The manifold MCHS is designed and monolithically integrated with Si to avoid the bonding step (Yang et al., 2016). studied the influence of electroplated Cu pin fin with different shapes, and the cross sections of the pin fins were triangle, square, pentagon, hexagon, and circle geometries as illustrated in Fig. 4.22. The pin fins with hexagonal and circular cross sections showed better cooling performance; for the hexagonal cross section, there are some disturbed effects with an angle in its back which is not obvious in circular case. In addition, three types of electroplated Cu pin fins (rhombus, hydrofoil, and sine (Fig. 4.23)) with symmetry geometries are compared in detail in Yang et al. (2017). The samples were made by NSMFT/NSNFT, and sine cross section pin fins showed the best performance among the three cases. In order to obtain an optimized heat dissipation performance of MCHS, the height, width, and pitch of the microchannel shall be focused (Chein & Chen, 2009).

The bonding surface with low stress, low thermal resistance, and high bonding strength is urgent for the MCHS. Firstly, the interface thermal resistance induced by

Fig. 4.22 Inner views of the bottom baseplate with different pin fins (Yang et al., 2016)

Fig. 4.23 Inner view and overview of heat sinks: (**a**) pin fins with rhombus geometry, (**b**) pin fins with hydrofoil geometry, and (**c**) pin fins with sine geometry (Yang et al., 2017)

the bonding process has great influence on the heat dissipation capacity. In addition, the high stress induced by the mechanical and thermal loads during the bonding process will cause warpage, cracks, and defects on the interface. Thus, the thermal resistance will be enlarged, and thermal mechanical reliability and bonding strength will be weakened and reduced. Finally, the bonding strength is a direct challenge to the fluid speed, which determines the upflow limit of the fluid, and so to the maximal heat dissipation capacity. During analysis on the temperature distribution and temperature gradient, the pressure drop of the MCHS shall also be focused for the higher speed of the fluid and large scale of microchannels which will enhance the capacity of the MCHS accompanied with a higher pressure drop which will reduce the mechanical reliability of the MCHS.

Since the fluid temperature in the inlet region is the lowest, the design of the inlet will affect the heat dissipation capacity. The influence of inlet/outlet arrangement on fluid flow and heat transfer has been analyzed (Chein & Chen, 2009). With the same geometric dimensions of the heat sink and channel, six types of inlet/outlet arrangement have been designed. The heat sinks (I-, N-, D-, and S-type) with horizontal fluid supply and collection has more serious velocity misdistribution than heat sinks (U- and V-type) with vertical fluid supply and collection. Thus, temperature nonuniformity is more serious in the heat sinks with horizontal fluid supply and collection. It was found that the V-type heat sink has the best performance among the heat sinks studied. The location of the inlet has been analyzed and optimized to alleviate the impact of hot spots as shown in Fig. 4.24. Compared with designs with other inlet locations, the maximum temperature of the chip is reduced by up to 26.7% during the position ratio (the ratio of the distance between the inlet and the center to half the diagonal length) of 0.1.

Using high thermal conductive fluid, nanofluids or multi-phase flow as the fluid is an effective way to enhance heat dissipation efficiency. Since the thermal conductivity of solids is higher than that of most fluids, nanofluids have intrigued many research groups' considerable interest in heat transfer. Al_2O_3, SiO_2, CrO_2, Fe_2O_3, SiC, graphene oxide, and diamond nanoparticles all could enhance the overall thermal conductivity of MCHS. As shown in (Jang & Choi, 2005), heat dissipation performance of a MCHS with water-based diamond nanofluids (1 vol.%, 2 nm) at the fixed pumping power of 2.25 W was enhanced by about 10% compared with that

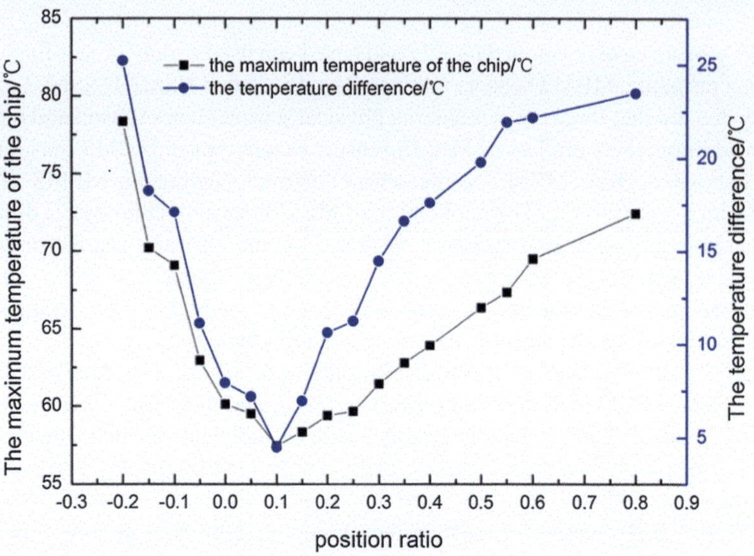

Fig. 4.24 Effect of inlet position ratio on the heat dissipation when the power is 72 W and flow rate is 60 mL/min (Zhao et al., 2014)

of a MCHS with water. This is because the thermal resistance and temperature difference between the heated microchannel wall and the coolant are reduced by the diamond nanofluids. Pool boiling, flow boiling, jet, and sprays based on the merits of phase change have been designed as cooling schemes to enhance the heat dissipation capacity. As one of the most important phase changes, boiling consists of pool boiling and flow boiling. Boiling with intrinsically small thickness offers extremely large convective heat transfer coefficients and relatively small temperature rise compared with the single-phase MCHS. Thus, boiling-cooling possesses a high ability to dissipate high heat fluxes because boiling can effectively depress overall thermal resistance between the heat source and heat sink, reduce the fluid flow rate and pressure drop, and provide more uniform temperature distribution along the flow direction. Nevertheless, it is difficult to minimize the flow instabilities during boiling when improving the critical heat flux to maximize heat dissipation. In order to solve this issue, various structures, such as inlet restrictors, artificial nucleation sites, vapor venting membranes, pin fins, and nanowire-coated surfaces, can be integrated into MCHS.

High heat dissipation capacity, high reliability, and integration with MEMS are the main development in future. The mechanism of the MCHS within nano–/microscale shall be focused and analyzed more to effectively improve the performance of the heat sink.

4.3.4 MEMS with Diversified Fabrication Processes: High-Temperature Sensor

MEMS sensors have developed rapidly and have been used widely. According to the working principle, MEMS sensors fabricated by diversified NSMFT/NSNFT can be simply divided into three types including physical sensor, chemosensor, and biosensor. In addition, there are lots of MEMS sensors except typical MEMS sensor mentioned above, such as MEMS tactile sensor for micro judgment, MEMS sensor, MEMS image sensor, etc. The application of MEMS sensor technology is different in different application environments. With the continuous improved requirements of measurement for power engineering, especially in dynamic rotation, limited measuring space, and other extreme measuring environment, high-temperature sensors with high precision have to be designed and fabricated.

High-temperature sensors for monitoring the state of device surface in extreme environments are divided into line type, patch type, and thin film. With the advantages of low-dimensional material characteristics, small volume, high surface cladding, fast response, low heat capacity, and small time constant, high-temperature thin film sensor (HTTFS) has developed rapidly as an alternative technique possessing effective and compatible surface measurement. Thin film sensors based on NSMFT/NSNFT are promising to meet the urgent needs in aeronautics, hypersonic flight, and aerospace research. HTTFS mainly consists of the component substrate to be tested, transition layer, insulating layer, sensitive body, protective layer, and lead wire. The transition layer and insulating layer are usually prepared by plasma spraying, CVD (LSCVD, PECVD, and MOCVD), PVD (magnetron sputtering deposition, IBSD, and DIBSD), and sol-gel method.

The component substrate to be tested is superalloy; thus, MCrAlY (Fe, Co, Ni, and their combinations are the candidates of M) is serving as the transition layer and BN, Al_2O_3, SiN, and MgO are the candidates of the insulating layer. After the transition layer is deposited in vacuum environment, heat treatment will be carried out in Ar atmosphere, so that Al element in the alloy film is diffused to the surface of the transition layer. And then a dense oxide layer is formed on the surface of the transition layer by thermal oxidation. The thermal oxidation time and temperature have great influence on microstructure and properties of oxide layer. After that, the insulating layer will be prepared by deposition. In this way, adhesion between the transition layer and the insulating layer will be guaranteed. In order to enhance the reliability of the insulating layer, the insulating layer with multi-layers is promising.

The sensitive body includes ceramic-based materials (TaN, TaoN, TiN, TiAlN, ZrN, AlN, BTi_2, $TiSi_2$, $TaSi_2$, WSi_2, ITO, and so on), metal-based materials (karma alloy, constantan alloy, iron-chromium-aluminum alloy, platinum-based alloy, W-Re alloy, and so on). There exist three ways to form the sensitive body. The first one is chemical etching which has the advantages of fast graphic process and high precision, and also has a certain potential corrosion to the substrate due to the residual etching solution. Firstly, the sensitive film is sputtered on the insulating layer, and then the photoresist is in situ spin coated and exposed for development. After that, the thin film sensor is fabricated by etching with etching solution and removing the residual photoresist. The second way is sputtering with hard mask. In this way, the sensitive body can be prepared directly. But the line width is limited; generally, the line width is only 500 μm. The third one is lift-off. The main processes include in situ spin coating photoresist (photoresist is used as sacrificial layer), patterning, and sputtering sensitive films, and finally, the patterned films can be obtained by removing the photoresist. Together with UV-LIGA, its process has the characteristics of high precision, small linewidth, and wide range of application, which is a very common film patterning process at present.

The materials for the insulating layer are often used as the protecting layer. And the prepared process is similar, except for some higher requirements for preparation process to avoid damages to the preferred layers. Nevertheless, annealing process and multi-layers are needed to ensure the density and uniformity of protective film. During high-temperature service stage, thin film sensors are easily oxidized which will reduce the reliability of the sensors or lead to failure (Zhang et al., 2017). One of the most effective ways to improve the oxidation resistance is to coat a passivation layer, such as Al_2O_3 or ZrO_2, on the surface of thin film sensors, as depicted in Fig. 4.25. In terms of the stability and sensitivity of sensors, the annealing treatment, doping, and multi-composite films are often used to deal with the sensors before application.

The noble metals with high-temperature resistance and good conductivity are often selected for the lead wire. The lead wire is welded with conductive paste to

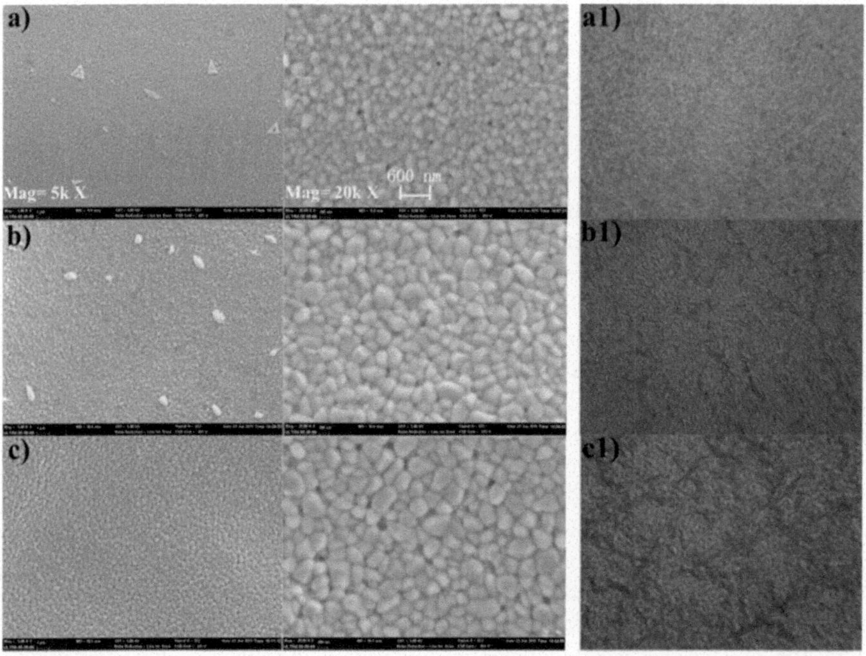

Fig. 4.25 SEM images of aluminum oxidized in air at 1100 °C for different time: (**a**) 5 h, (**b**) 10 h, and (**c**) 20 h, and (**a1**), (**b1**), and (**c1**) were the corresponding optical microscope images (Niu et al., 2016)

connect to the film and then sintered at a high temperature, such as platinum paste, which contains organic matter. After high-temperature sintering, organic matter in platinum paste is volatilized and burned. After cooling to room temperature, metallic platinum is obtained.

Since the surface stress, heat transfer, and temperature gradients are high, which have serious effects on the aerodynamics of the device, thin film sensors are needed to monitor these parameters (stress, heat flux, and temperature) in the engine systems to evaluate advanced materials and components. Thus, there exist three main HTTFSs, namely, thin film strain gauges (TFSGs), thin film thermopile array heat flux sensors (TFTPs), and thin film thermocouples (TFTCs).

Non-intrusive TFSGs are ideally suitable for in situ strain monitoring the components because TFSGs are fabricated directly on the surface of the components. In addition, TFSGs have minimal impact on thermal, strain, and vibration patterns and possess high robustness stability during high "g" loading associated with high-speed revolution. Generally, metals and related alloys, such as PdCr, Pt, and NiCr, are chosen as piezoresistive materials for TFSGs at high temperature (below 800 °C), while the semiconductor ceramics such as ITO, AlN, and SiC are usually used for ultrahigh temperature strain measurement (above 800 °C) (Kazi et al., 2006; Rasia et al., 2010). With Al_2O_3/Al bilayer coatings as protective layer, a remarkable improvement in piezoresistive response and stability has been made, which resulted from the excellent structure stability and the oxidation resistance provided by in situ-formed Al_2O_3 interlayer combined with sputtered Al_2O_3 overcoat (Yang et al., 2018). According to Fourier's law, TFTPs are utilized to measure the temperature difference over the thermal resistance of known thermal conductivity and heat flux (Zribi et al., 2016). The basic measurement principle of the heat flux sensors is the same as that of TFTCs which is called Seebeck effect. Due to temperature nonuniformity of TFTPs and TFTCs, the detected temperature range is relatively small, and the measurement error is extremely high. The high-temperature stable Pt-PtRh and Pt-ITO with larger electrical conductivity difference are used as the main materials of TFTPs and TFTCs to enhance the temperature range and reduce errors.

4.4 Test Methods for the Non-silicon MEMS Device

The measurement of the material properties and device performance is one of the key parts of NSMFT/NSNFT. Some reliable and reasonable measurement methods have to be designed and set up to validate and calibrate the performance and design of MEMS. The performance of the materials is basically divided into mechanical performance, thermal performance, and electrical performance. The output performance of MEMS devices and the basic structural feature and properties have to be measured and characterized.

4.4.1 Material Properties

The mechanical, thermal, and electrical properties of microscale materials are distinct with those of the bulk materials. The characterization and measurement of microscale material can provide precise data for the modeling design and analysis of the MEMS device. Young's modulus, fracture strength, yield strength, Poisson's ratio, thermal conductivity, coefficient of thermal expansion, electrical conductivity, permittivity, etc. have to be tested and characterized to analyze the reliability of material and MEMS. During the stage of measuring the basic properties of material, there exist several challenges: preparing a reliable and simple test specimen, setting up a simple test platform with high precision, and analyzing the test data based on a reasonable theory (Sun et al., 2018; Yi & Kim, 1999).

Since the microstructure of MEMS devices is complicated and minimized, a simplified microstructure test specimen should be designed to match well with the test methods and test system. One way to prepare test specimen is fabricating a sample with similar dimensions and structure of the device, just as the test specimen shown in Fig. 4.26. In this way, this test results may match well with the actual value of the device. Another way is designing and fabricating a new structure that depends on the device and test system, as shown in Fig. 4.27. Because the complicated

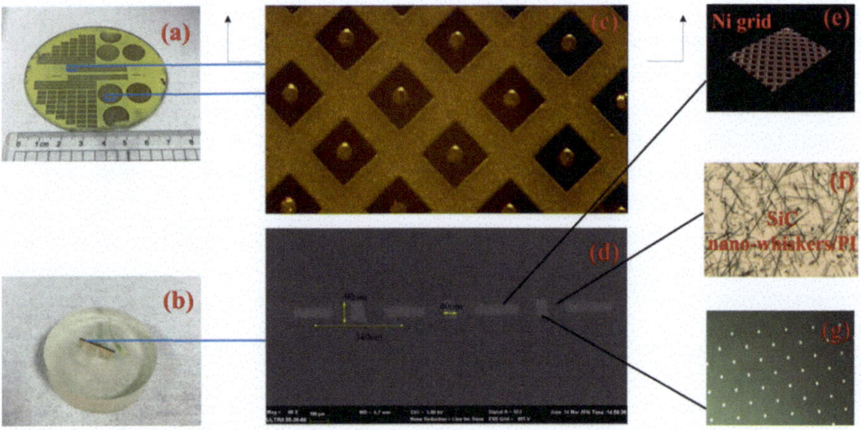

Fig. 4.26 The micrographs of the complex reinforced polymer interposer (Liu et al., 2017)

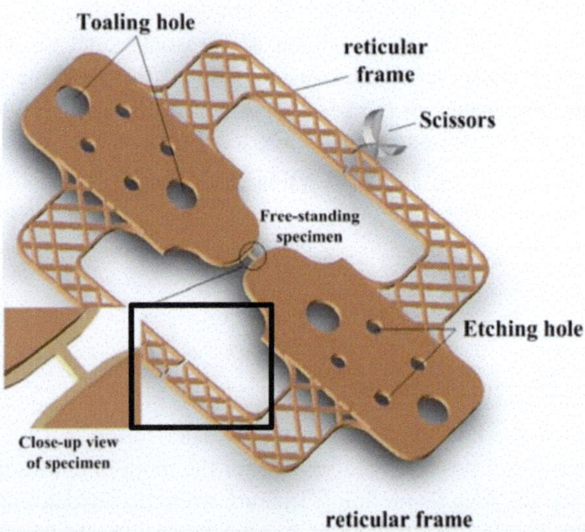

Fig. 4.27 Schematic view of the test sample for the Cu specimen and supporting frame (Li et al., 2011)

structure cannot be tested directly, a simplified structure with the basic structural feature of the device should be designed to satisfy the test needs of the test system.

Without a cost-effective and reasonable test system and method, the measured properties may be inaccurate. The main methods used to test mechanical properties are beam-bending, compression method, tension test, intrinsic frequency, interferometry, and micro-friction (Yi & Kim, 1999; Gu et al., 2014). Thermal properties including thermal conductivity, specific heat capacity, thermal diffusion, and coefficient of thermal extension are also key to analyzing the reliability of device. The steady-state method and the laser flash method have been used to test the thermal conductivity parallel and perpendicular to the surface, respectively (Ogata et al., 2015). The interface bonding strength and internal stress have also been measured for analyzing the reliability of materials and fabrication processes (Sun et al., 2018). As the basic properties of materials, electrical properties have developed many years ago. With the rapid development of IC, the test system and measurement methods for the electrical properties have become very mature.

4.4.2 Output Performance and Structural Feature of MEMS

The morphology features, reliability, and the output performance will be tested, characterized, and analyzed after the MEMS device is fabricated. In the measurement of the output performance, a complete test system basically consists of the platform with specimens, input signal, monitor, sensor, actuator, and data collection. The interference of external signal should be avoided during the design of a test system.

With SEM, TEM, FIB, 3D profiler, stylus profiler, AFM, micro-Raman, X-ray, scanning acoustic microscopy, laser ultrasonics, and magneto-optical imaging, the structural feature and surface morphology of the device are tested with high precision. These data are useful for analyzing the output performance, evaluating reliability, and further optimizing the design of the device. In order to characterize other unconventional characteristic of the device, a test system should be set up. A micro-force and micro-displacement system was designed to test a nonlinear force-displacement behavior (mechanical performance) of the clamped-guided beams as shown in Fig. 4.28. The nonlinear behavior of the clamped-guided beams has great influence on reducing pull-in voltage, extending travel range, and simplifying configurations of the micro-actuator. With this test system composed of optic microscope, micro-force sensor probe, piezoelectric motion platform with position resolution of 5 nm, and force acquisition system, the measurement range is

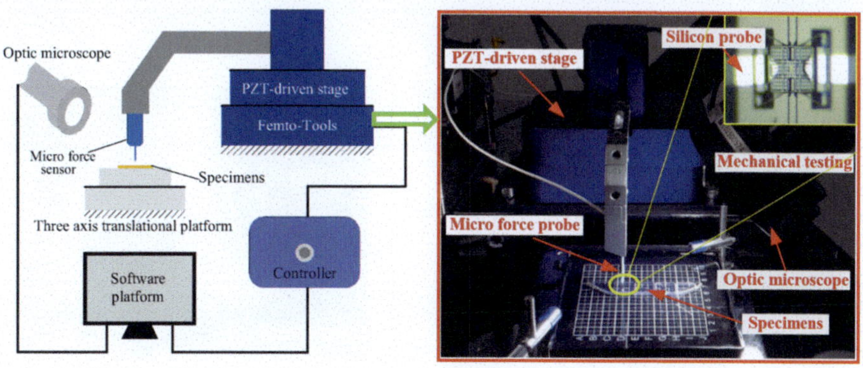

Fig. 4.28 Spring force testing system by Femto-tools stage (Xiang et al., 2018)

0–1000 μN and the resolution is 50 nN. Furthermore, the test data matched well with the simulation results.

The basic performance and output performance of the microsensor, micro-actuator, harvester, antenna, heat sink, micro-needle, and flapping wing are listed in Table 4.3. To obtain the precise measurement results, a stable, reliable, and reasonable test system is necessary. As shown in Fig. 4.29, a micro-displacement test system consists of high-precision laser displacement sensor, optic microscope, moving platform, metal probe, signal generator, voltage amplifier, oscilloscope, and vibration absorbed platform. With this test system, the static and dynamic output performance of the micro-actuator, the voltage-displacement, can be measured under room temperature and atmospheric pressure. Since test conditions are similar to the application conditions, the high-precision sensor and vibration absorbed platform are introduced to ensure measurement accuracy.

Table 4.3 Basic performances of the MEMS devices

MEMS	Performance index			
Microsensor	Sensitivity	Measurement range	Phase shift	Frequency range
Micro-actuator	Threshold	Response time	Sensitivity	Shock resistibility
Harvester	Output	Resonant frequency	Operating bandwidth	–
Antenna	Gain	Bandwidth	Radiation efficiency	Directional diagram
Heat sink	Heat dissipation capacity	Anti-pressure	–	–
Micro-needle	Mechanical strength	Flow velocity	–	–
Flapping wing	Flapping and rotation angle	Frequency	Lift	–

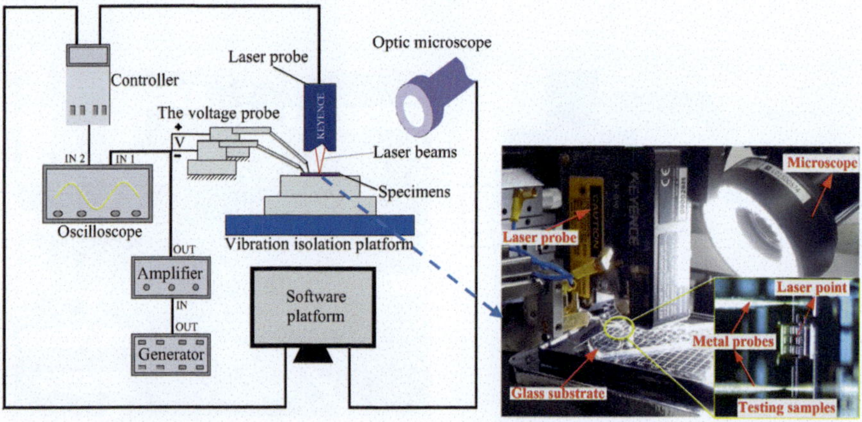

Fig. 4.29 Dynamic performance measurements of the electrostatic actuators (Xiang et al., 2018)

4.5 Conclusions

This chapter mainly introduces the research progress of NSMFT/NSNFT, hoping to provide valuable guidance and reference for researchers in related research fields. NSMFT/NSNFT consists of a variety of materials, fabrication technology, and MEMS device design and performance measurement. As the basic part of NSMFT/NSNFT, the selection and preparation of materials with good properties are extremely important. After obtaining the basic materials, the fabrication processes have been carried out to achieve the performance index of the MEMS device. Next, the test system with high precision must be used to measure the performance of the device and validate the feasibility and rationality of material selection, fabrication processes, and design scheme.

Under the background of the rapid development of NSMFT/NSNFT, miniature MEMS devices with high performance are available. The development of the micro-machining technologies, new materials with good properties, and test systems for the device and materials has been introduced and summarized simply. The development of some typical MEMS devices, micro-actuator, VEH, MCHS, and high-temperature sensor has been introduced as the typical sample to introduce the key technology of NSMFT/NSNFT. Different devices have different design thoughtfulness, project, and aim. However, the miniaturization, multi-functionalization, and compatibility with other devices are future development of the devices.

A number of methods have been proposed to improve the performance of MEMS devices. These include:

1. Optimizing the design of the device by finite element method and theoretical analysis.
2. Preparing the basic materials with good properties.
3. Selecting a reliable and feasible fabrication process.
4. Testing the performance and key parameters of the device to validate the design.
5. Setting up the reliable and reasonable test system to measure the special characteristics and performance of the device or materials.

Acknowledgments The authors would like to thank supports from the National Key Research and Development Program of China (2020YFB2008503), the National Natural Science Foundation of China (No. 61974088), the Advanced Research Ministry of Education Joint Foundation (6141A02022424), and Professional Technical Service Platform for Non-Silicon Micro-Nano Integrated Manufacturing (20DZ2291300).

References

Baker, E., Reissman, T., Zhou, F., Wang, C., Lynch, K., & Sun, C. (2012). Microstereolithography of three-dimensional polymeric springs for vibration energy harvesting. *Smart Materials Research, 2012*(2012-5-21), 2012(5 b).

Beeby, S. P., Torah, R. N., Tudor, M. J., Glynnejones, P., O'Donnell, T., Saha, C. R., et al. (2007). A micro electromagnetic generator for vibration energy harvesting. *Journal of Micromechanics and Microengineering, 17*(7), 1257–1265.

Bhattacharyya, B., Mitra, S., & Boro, A. K. (2002). Electrochemical machining: New possibilities for micromachining. *Robotics and Computer-Integrated Manufacturing, 18*(3–4), 283–289.
Cammarano, A., Burrow, S. G., & Barton, D. A. W. (2010). An energy harvester with Bistable compliance characteristics. In *ASME 2010 international design engineering technical conferences and computers and information in engineering conference* (pp. 725–732).
Chang, H. K., & Kim, Y. K. (2000). UV-LIGA process for high aspect ratio structure using stress barrier and C-shaped etch hole. *Sensors and Actuators A: Physical, 84*(3), 342–350.
Chein, R., & Chen, J. (2009). Numerical study of the inlet/outlet arrangement effect on microchannel heat sink performance. *International Journal of Thermal Sciences, 48*(8), 1627–1638.
Chen, W., Ding, G., Wang, Y., Wang, H., Zhao, X., Yang, C., et al. (2014a). An all-metal passive threshold sensor for omni-directional vibration monitoring application. In *IEEE Sensors* (pp. 428–431).
Chen, W., Wang, Y., Wang, Y., Zhu, B., Ding, G., Wang, H., et al. (2014b). A laterally-driven micromachined inertial switch with a compliant cantilever beam as the stationary electrode for prolonging contact time. *Journal of Micromechanics and Microengineering, 24*(6), 065020.
Chen, W., Yang, Z., Wang, Y., Ding, G., Wang, H., & Zhao, X. (2016). Fabrication and characterization of a low-g inertial microswitch with flexible contact point and limit-block constraints. *IEEE/ASME Transactions on Mechatronics, 21*(2), 963–972.
Erp, R. V., Soleimanzadeh, R., Nela, L., Kampitsis, G., & Matioli, E. (2020). Co-designing electronics with microfluidics for more sustainable cooling. *Nature, 585*(7824), 211–216.
Erturk, A., Renno, J. M., & Inman, D. J. (2009). Modeling of piezoelectric energy harvesting from an L-shaped beam-mass structure with an application to uavs. *Journal of Intelligent Material Systems and Structures, 20*(5), 529–544.
Fernández, L. J., Wiegerink, R. J., Flokstra, J., Sesé, J., Ansen, J. H. V., & Elwenspoek, M. (2017). A capacitive RF power sensor based on MEMS technology. *Journal of Micromechanics and Microengineering, 16*(7), 1099.
Frobenius, W. D., Zeitman, S. A., White, M. H., & O'Sullivan, D. D. (1972). Microminiature ganged threshold accelerometers compatible with integrated circuit technology. *IEEE Transactions on Electron Devices, 19*(1), 37–40.
Gong, S., Schwalb, W., Wang, Y., Chen, Y., Tang, Y., Si, J., et al. (2014). A wearable and highly sensitive pressure sensor with ultrathin gold nanowires. *Nature Communications, 5*(2), 3132.
Gu, L., & Livermore, C. (2010). Passive self-tuning energy harvester for extracting energy from rotational motion. *Applied Physics Letters, 97*(8), R175.
Gu, T., Cheng, P., Wang, S., Wang, H., Dai, X., Wang, H., et al. (2014). Mechanical property evaluation of tsv-cu micropillar by compression method. *Electronic Materials Letters, 10*(4), 851–855.
Huang, G., Wang, H., Cheng, P., Wang, H., Sun, B., Sun, S., Zhang, C., Chen, M., & Ding, G. (2016). Preparation and characterization of the graphene-cu composite film by electrodeposition process. *Microelectronic Engineering, 157*, 7–12.
Jang, S. P., & Choi, S. U. S. (2005). Cooling performance of a microchannel heat sink with nanofluids. *Applied Thermal Engineering, 17*(17), 2457–2463.
Kazi, I. H., Wild, P. M., Moore, T. N., & Sayer, M. (2006). Characterization of sputtered nichrome (Ni–Cr 80/20 wt.%) films for strain gauge applications. *Thin Solid Films, 515*(4), 2602–2606.
Langley, D., Coutu, R. A., Jr., & Collins, P. J. (2012). Using inductance as a tuning parameter for RF meta-atoms. *Nano-Micro Letters, 4*(2), 103–109.
Lee, D. G., Carman, G. P., Murphy, D., & Schulenburg, C. (2007). Novel micro vibration energy harvesting device using frequency up conversion. In *IEEE, International Solid-State Sensors, Actuators and Microsystems Conference* (pp. 871–874).
Li, J., Wang, H., Wang, H., Zhang, Z., Cheng, P., & Ding, G. (2011). Design and fabrication of cu-tsv free-standing specimen for uniaxial micro-tensile test. In *Advanced packaging materials international symposium on* (pp. 318–321).
Li, J., Sun, Y., Wang, Y., Yang, Z., Ding, G., Wang, H., et al. (2017). A MEMS inertial switch with electrostatic force assistance and multi-step pull-in for eliminating bounce and prolonging

contact time. In *International conference on solid-state sensors, actuators and microsystems* (pp. 183–186).

Linsebigler, A. L., Lu, G., & Yates, J. T. (1995). Photocatalysis on TiO2 surfaces: Principles, mechanisms, and selected results. *Chemical Reviews, 95*(3), 735–758.

Liu, C., Yang, Z., Ding, G., Zhou, Z., Liu, Q., Huang, Y., & Zheng, Y. (2013). Design, simulation, and characterization of a low-cost in-plane spark gap microswitch with dual-trigger electrode for pulsed power applications. *IEEE Transactions on Industrial Electronics, 60*(8), 3240–3247.

Liu, Q., Zhang, Q., Zhu, G., Wang, K., Zhang, J., & Dong, C. (2016). Effect of electrode size on the performances of micro-EDM. *Advanced Manufacturing Processes, 31*(4), 391–396.

Liu, Y., Sun, Y., Wang, Y., Ding, G., Sun, B., & Zhao, X. (2017). A complex reinforced polymer interposer with ordered Ni grid and SiC nano-whiskers polyimide composite based on micromachining technology. *Electronic Materials Letters, 13*(1), 29–36.

Mann, B. P., & Sims, N. D. (2009). Energy harvesting from the nonlinear oscillations of magnetic levitation. *Journal of Sound and Vibration, 319*(1), 515–530.

Mcdonald, J. C., Duffy, D. C., Anderson, J. R., Chiu, D. T., Wu, H., Schueller, O. J., et al. (2000). Fabrication of microfluidic systems in poly(dimethylsiloxane). *Electrophoresis, 21*(1), 27–40.

Niu, D., Wang, Q., Zhang, C., Cheng, P., Wang, Y., Zhang, J., Ding, G., Xie, K., Shao, J., Liu, Y., & Duan, L. (2016). Preparation, characterization and application of high-temperature Al_2O_3 insulating film. *Surface & Coatings Technology, 291*, 318–324.

Ogata, N., Jimenez, G., Kawai, H., & Ogihara, T. (2015). Structure and thermal/mechanical properties of poly(l-lactide)-clay blend. *Journal of Polymer Science Part B Polymer Physics, 35*(2), 389–396.

Rasia, L. A., Mansano, R. D., Damiani, L. R., & Viana, C. E. (2010). Piezoresistive response of ito films deposited at room temperature by magnetron sputtering. *Journal of Materials Science, 45*(15), 4224–4228.

Rezaeisaray, M., El Gowini, M., Sameoto, D., Raboud, D., & Moussa, W. (2015). Wide-bandwidth piezoelectric energy harvester with polymeric structure. *Journal of Micromechanics and Microengineering, 25*(1), 015018.

Roundy, S., Wright, P. K., & Rabaey, J. (2003). *A study of low level vibrations as a power source for wireless sensor nodes*. Elsevier Science Publishers B. V.

Sari, I., Balkan, T., & Kulah, H. (2008). An electromagnetic micro power generator for wideband environmental vibrations. *Sensors and Actuators A: Physical, 145–146*(1), 405–413.

Singal, S., Srivastava, A. K., & Rajesh, R. (2017). Electrochemical impedance analysis of biofunctionalized conducting polymer-modified graphene-CNTs nanocomposite for protein detection. *Nano-Micro Letters, 9*(1), 7.

Sun, S., Dai, X., Sun, Y., Xiang, X., Ding, G., & Zhao, X. (2017). Mems-based wide-bandwidth electromagnetic energy harvester with electroplated nickel structure. *Journal of Micromechanics and Microengineering, 27*(11), 115007.

Sun, Y., Luo, J., Yang, Z., Wang, Y., Ding, G., & Wang, Z. (2018). Development of a polyimide/SiC-whisker/nano-particles composite with high thermal conductivity and low coefficient of thermal expansion as dielectric layer for interposer application. In *IEEE Electronic Components & Technology Conference*.

Tuckerman, D. B., & Pease, R. F. W. (1981). High-performance heat sinking for vlsi. *IEEE Electron Device Letters, 2*(5), 126–129.

Wang, P., Dai, X., Zhao, X., & Niu, L. (2009). Electromagnetic self-powered low-level vibration energy scavenger with microelectroplated nickel resonator. *Electronics Letters, 45*(16), 832–833.

Wang, H., Liu, R., Cheng, J., Cao, Y., Ding, G., & Zhao, X. (2010). Electrodepositing amorphous Ni-W alloys for MEMS. *Microelectronic Engineering, 87*(10), 1901–1906.

Wang, B., Wang, H., Cheng, P., Yang, Z., & Ding, G. (2016). Optimizing the electroplating condition of an acid low cyanide silver bath and its application in micromachining. In *IEEE, International Nanoelectronics Conference (INEC), 2016* (pp. 1–2).

Wang, K., Shao, X., Dai, X., Yang, Z., & Zhao, X. (2020). A planar architecture with dual-nonlinearity for bandwidth extension of a multi-modal energy harvester. In *2020 IEEE 33rd international conference on micro electro mechanical systems (MEMS)*.

Wu, Y., Sun, Y., Luo, J., Cheng, P., Wang, Y., Wang, H., et al. (2019). Microstructure of cu-diamond composites with near-perfect interfaces prepared via electroplating and its thermal properties. *Materials Characterization, 150*, 199–206.

Xiang, X., Dai, X., Sun, S., Ding, G., & Zhao, X. (2018). An inherent nonlinear electrostatic actuator without any additional auxiliary to reduce pull-in voltage and extend travel range. *Journal of Micromechanics and Microengineering*.

Xie, F., Ding, G., Zhao, X., & Cheng, P. (2015). Design, fabrication and measurement of a novel 140ghz folded waveguide based on su-8 uv-liga technology. *Journal of Micromechanics and Microengineering, 25*(8), 085010.

Xu, Q., Yang, Z., Sun, Y., Lai, L., Jin, Z., Ding, G., Zhao, X., Yao, J., & Wang, J. (2017). Shock-resistibility of MEMS based inertial microswitch under reverse directional ultra-high g acceleration for IoT applications. *Scientific Reports, 7*, 45512.

Xuan, D., Bo, H., & Chong, N. (2009). Micro-structuring of glass with features less than 100 μm by electrochemical discharge machining. *Precision Engineering, 33*(4), 459–465.

Yang, C., Zhao, X., Ding, G., Zhang, C., & Cai, B. (2001). An axial flux electromagnetic micromotor. *Journal of Micromechanics and Microengineering, 11*(2), 113–117.

Yang, Z., Zhu, B., Chen, W., Ding, G., Wang, H., & Zhao, X. (2012). Fabrication and characterization of a multidirectional-sensitive contact-enhanced inertial microswitch with a electrophoretic flexible composite fixed electrode. *Journal of Micromechanics and Microengineering, 22*(22), 45006–45015.

Yang, D., Wang, Y., Ding, G., Jin, Z., Zhao, J., & Wang, G. (2016). Numerical and experimental analysis of cooling performance of single-phase array microchannel heat sinks with different pin-fin configurations. *Applied Thermal Engineering, 112*, 1547–1556.

Yang, D., Jin, Z., Wang, Y., Ding, G., & Wang, G. (2017). Heat removal capacity of laminar coolant flow in a micro channel heat sink with different pin fins. *International Journal of Heat and Mass Transfer, 113*, 366–372.

Yang, S., Li, H., Lin, X., Yao, J., Yang, Z., Zhang, C., Wang, H., & Ding, G. (2018). Effect of Al_2O_3/Al bilayer protective coatings on the high-temperature stability of PdCr thin film strain gages. *Journal of Alloys and Compounds, 759*, 1–7.

Yi, T., & Kim, C. J. (1999). Measurement of mechanical properties for MEMS materials. *Measurement Science & Technology, 10*(8), 706–716.

Yu, Z., Rajurkar, K. P., & Tandon, A. (2004). Study of 3D micro-ultrasonic machining. *Journal of Manufacturing Science and Engineering, 126*(4), 727–732.

Zhang, Y., Ding, G., Wang, H., & Cheng, P. (2015). Effect of external factors on copper filling in 3D integrated Through-Silicon-Vias (TSVs). *Journal of the Electrochemical Society, 162*(9), D427–D434.

Zhang, Q., Yang, Z., Xu, Q., Wang, Y., Ding, G., & Zhao, X. (2016). Design and fabrication of a laterally-driven inertial micro-switch with multi-directional constraint structures for lowering off-axis sensitivity. *Journal of Micromechanics and Microengineering, 26*(5), 055008.

Zhang, Y., Cheng, P., Yu, K., Zhao, X., & Ding, G. (2017). Ito film prepared by ion beam sputtering and its application in high-temperature thermocouple. *Vacuum*, 146.

Zhao, J., Wang, Y., Ding, G., Sun, Y., & Wang, G. (2014). Design, fabrication and measurement of a microchannel heat sink with a pin-fin array and optimal inlet position for alleviating the hot spot effect. *Journal of Micromechanics and Microengineering, 24*(11), 115013–115021.

Zribi, A., Barthès, M., Bégot, S., Lanzetta, F., Rauch, J. Y., & Moutarlier, V. (2016). Design, fabrication and characterization of thin film resistances for heat flux sensing application. *Sensors and Actuators A: Physical, 245*, 26–39.

Chapter 5
Nano-in-Nano Integration Technology for Advanced Fabrication of Functional Nanofluidic Devices

Jinbin Yang and Yan Xu

5.1 Introduction

Nanofluidics is the study of nanoscale fluid behaviors, physical phenomena, liquid and interfacial reactions, and substances confined in nanostructures (e.g., nanochannels) (Xu, 2018; Mawatari et al., 2014; Mukhopadhyay, 2006; Holt, 2009; Prakash et al., 2008). Because the evaluation scale reaches the nanometer scale, which is close to the thickness of the electric double layer, Debye length, and the scales of many molecules, nanofluidics is emerging with numerous novel characteristics, phenomena, and applications that are uncommon in conventional bulk solution and population level studies (Sparreboom et al., 2009; Bocquet & Tabeling, 2014; Rauscher & Dietrich, 2008; Cao et al., 2018; Poddar et al., 2016; Perry et al., 2010; Guo et al., 2015; Hatsuki et al., 2013). Therefore, nanofluidic studies are attractive and promising in multidisciplinary fields (Haywood et al., 2015; Harms et al., 2015a; Öz et al., 2019; Kim et al., 2011; Lin et al., 2018; Napoli et al., 2010; Abgrall & Nguyen, 2008; Zhang et al., 2009; Liu et al., 2005; Sui et al., 2016). Generally, among all nanofluidic studies, manufactured device-based nanofluidic studies, which utilize devices/chips with well-designed nanostructures for evaluation, play an increasingly important role (Duan et al., 2013; Peng & Li, 2016; Kovarik & Jacobson, 2009; Mijatovic et al., 2005; Syed et al., 2014; Arayanarakool et al.,

J. Yang
Department of Chemical Engineering, Graduate School of Engineering, Osaka Prefecture University, Osaka, Japan

Y. Xu (✉)
Department of Chemical Engineering, Graduate School of Engineering, Osaka Prefecture University, Osaka, Japan

Japan Science and Technology Agency (JST), PRESTO, Kawaguchi, Saitama, Japan

NanoSquare Research Institute, Osaka Prefecture University, Osaka, Japan
e-mail: xu@chemeng.osakafu-u.ac.jp

Z. Yang (ed.), *Advanced MEMS/NEMS Fabrication and Sensors*,
https://doi.org/10.1007/978-3-030-79749-2_5

2011; Mao & Han, 2005; Kim et al., 2010; Harms et al., 2015c). Two main reasons permit the significance of device-based nanofluidics. First, the manufactured devices have nanostructures, such as nanochannels and nanowells, which are designable, well-defined, and regular. These designed nanostructures ensure the regularity of the evaluation and lower the randomness and uncertainty of the processes. Second, the substrates of devices are usually glass or polymers that are transparent and chemically/physically stable. These devices allow many different evaluations inside nanospaces and make observation and detection possible (Wang et al., 2015, 2019b; Mokkapati et al., 2011; He et al., 2011; Branton et al., 2008; Shirai et al., 2014). Consequently, manufactured nanofluidic devices have potential to clarify the mechanisms in the microworld, and the nanofluidic evaluations are no longer a "black box."

Progress has been made in the past; however, nanofluidic studies are currently facing a crucial limitation, that is, the functionalization of nanofluidic devices (Zand et al., 2013; Mawatari et al., 2018; Friedrich et al., 2017; Louër et al., 2013; Marie & Kristensen, 2012; Frykholm et al., 2014; Piruska et al., 2010b). Limited by the fabrication technology, most nanofluidic devices are focused on utilizing normal and bare nanochannels. Bare nanochannels are those without modification or without any other component. Admittedly, even bare nanochannels show high potential in many research fields owing to their confining capability for nanoscale targets. However, we should also be aware that scientific research is emerging with higher requirements for the device's function. Simple restrictions by the electrostatic interactions of nanochannel walls are currently insufficient. Functions such as single-molecule manipulation, nanofluid control, or sensing in nanospaces are difficult to achieve. As a result, the application of nanofluidic devices is limited.

To fabricate functional nanofluidic devices for extending the application range, we proposed a novel concept, Nano-in-Nano Integration, over the past few years. As shown in Fig. 5.1, Nano-in-Nano Integration refers to the construction of nanocomponents inside the nanochannel. Compared to conventional bare nanochannels, the smaller internal component provides an increasing number of possibilities for nanofluidic devices. Functional nanofluidic devices with Nano-in-Nano components are

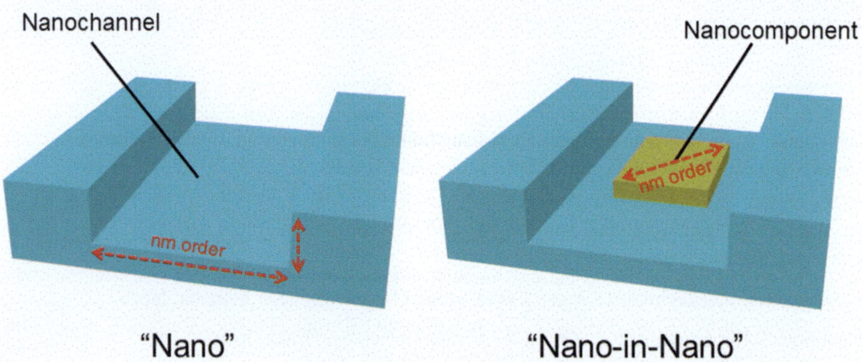

Fig. 5.1 Concept drawing of nanochannel and Nano-in-Nano Integration

promising for achieving various complex functions that have been previously unavailable.

Nano-in-Nano Integration is a key element in functional nanofluidic devices. First, the modification of the nanochannels is easier, including site-specific modification. Previously, bare nanochannels were limited by the substrate material; thus, there are not many options for this modification. Most of the modifications in bare nanochannels are nonspecific modifications, which may result in a lower signal-to-noise ratio. Second, precise nanoscale manipulation was possible. Nanochannels without functions permit basic handling processes for the samples. To achieve precise nanoscale manipulation, the introduction of internal functional components, such as valves, is necessary. Finally, the throughput of nanochannels has the potential for increase. Functional nanoarrays can be integrated into nanochannels, allowing regular, multiple, and parallel processes inside the nanospace to obtain substantial information.

However, bare nanochannels are still the majority in nanofluidic studies. The fabrication of Nano-in-Nano Integration is a hurdle that interrupts the development of functional nanofluidic devices. In this chapter, we introduce Nano-in-Nano Integration technology, including the fabrication and demonstrated applications in nanofluidics. We hope that this chapter will be helpful in the understanding of Nano-in-Nano Integration and contribute to the advanced fabrication of functional nanofluidic devices.

5.2 Fabrication of Nano-in-Nano Integration

Owing to the development of nanofabrication technologies, an increasing number of possibilities for nanofluidic devices have been emerging in different areas (Duan et al., 2013; Mijatovic et al., 2005; Park et al., 2018; Wang et al., 2019a; Menard & Ramsey, 2011; Wu et al., 2017; Perry & Kandlikar, 2006; Kim et al., 2015; Kudr et al., 2015; Kamai & Xu, 2021). Based on technologies such as electron beam (EB) lithography, focus ion beam (FIB), and nanoimprinting, well-defined nanochannels can be fabricated on various substrate surfaces, including silica glass and polydimethylsiloxane (PDMS). The highly confined nanochannels have size features at the nanometer scale in one or more dimensions (length, width, depth, or all three), which are promising for handling nanoscale targets. Although ultra-small nanochannels can be fabricated, Nano-in-Nano fabrication remains challenging. One reason for this is that at the nanoscale, many conventional modification and fabrication technologies for constructing functional components are in vain. For example, UV-based photolithography is difficult to apply to nanoscale fabrication because of the diffraction limit. Although there are advanced technologies such as EBL, which utilizes an electron beam with an ultra-short wavelength rather than conventional UV light, to precisely "put" the ultra-small nanocomponent inside the nanochannel is still not easy. Hence, the second reason that makes Nano-in-Nano fabrication challenging is the alignment precision.

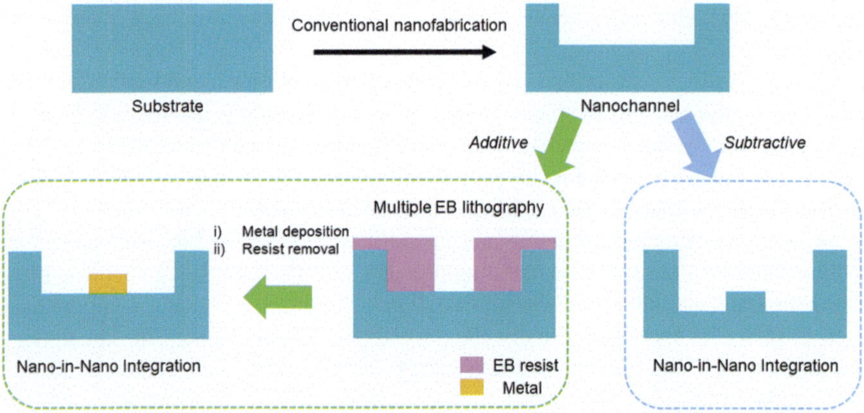

Fig. 5.2 Schematic drawing of the typical processes of Nano-in-Nano fabrication

Based on previous nanofabrication technologies, we successfully fabricated Nano-in-Nano Integration using our unique Nano-in-Nano fabrication methods. As shown in Fig. 5.2, Nano-in-Nano fabrication can be divided into two types. After the conventional nanofabrication processes, a nanochannel was obtained. To further obtain the Nano-in-Nano structure, one method is to perform milling/manufacturing in the obtained nanochannel, and then removing the unwanted areas; the remaining part is the Nano-in-Nano structure. This is a subtractive method, which is simple, direct, and fast. However, it is worth noting that this milling method only permits fabrication using a single material. Another method of Nano-in-Nano fabrication is to perform multiple EB lithography. Using EB resist masks, further processing at the desired position of the nanochannel is possible. This method facilitates the introduction of new materials, which indicates that it is an additive method.

5.2.1 *Focus Ion Beam Milling for Mono-Material Nano-in-Nano Integration*

The FIB milling method is a typical subtractive method for Nano-in-Nano fabrication. As a fabrication technology that is widely used in semiconductor applications, FIB technology utilizes focused ion beams for the removal of materials at the nanoscale level. However, in the field of nanofluidic studies, the fabrication of nanostructures by FIB is difficult because of the nonconductivity of substrate materials. To achieve FIB milling on nonconducting surfaces such as silica glass surfaces, a layer of gold was pre-deposited and utilized to serve as a conducting medium. Gallium ion (Ga^+) beams are used for high-rate gold layer milling to obtain various gold patterns. The obtained gold patterns can be further used as masks to maintain charge conductance to achieve glass milling (Fig. 5.3) (Xu & Matsumoto, 2015). After one- or multistep milling processes on the glass surface, Nano-in-Nano

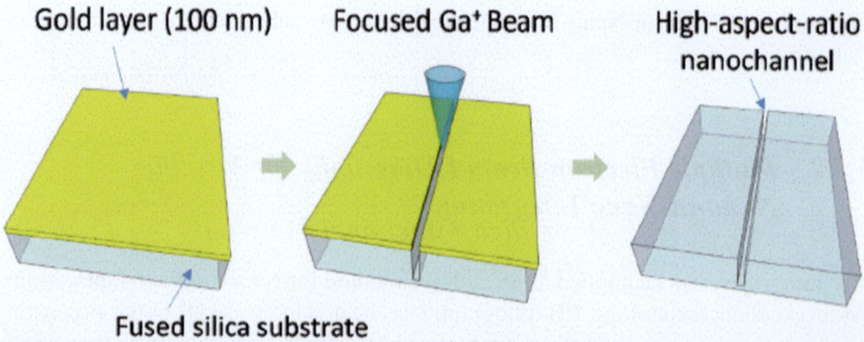

Fig. 5.3 Schematic drawing of FIB milling for nanochannel fabrication on glass surface. (Reproduced from Xu and Matsumoto (2015) with permission from the Royal Society of Chemistry)

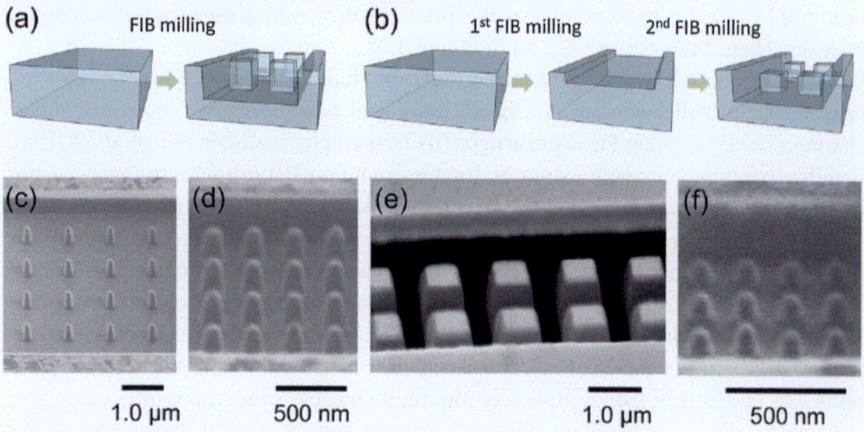

Fig. 5.4 Schematic drawing of FIB milling-based Nano-in-Nano fabrication: one-step milling (**a**) and multistep milling (**b**). SIM images for FIB-based Nano-in-Nano structures: low-density nanoarray (**c**) and high-density nanoarray (**d**) by one-step milling, nanoarrays in a planar nanochannel (**e**) and arrays in a square nanochannel (**f**) by multistep milling. (Reproduced from Xu and Matsumoto (2015) with permission from the Royal Society of Chemistry)

structures are obtained, and the gold marks can be removed by chemical methods (Fig. 5.4). From the scanning ion microscope (SIM) images, patterns with different densities, heights, and aspect ratios were successfully fabricated inside the nanochannel (Fig. 5.4c–f). It should be noted that the as-obtained patterns have high aspect ratios. This is a feature of the FIB milling method for Nano-in-Nano Integration. However, because of the subtractive manufacturing mechanism, the FIB milling method is limited in fabricating Nano-in-Nano Integration with composite materials. The addition of other materials is difficult when using FIB milling.

In conclusion, although the FIB milling method has advantages such as simple and direct processes and can obtain high-aspect-ratio patterns, the FIB milling-based Nano-in-Nano Integration still cannot realize increasingly complex functions.

5.2.2 *Multiple Electron Beam Lithography for Metallic Nano-in-Nano Integration*

EB lithography can be applied as an additive method for Nano-in-Nano fabrication. As an excellent technology, EB lithography has been widely used for high-precision and high-regularity nanofabrication on glass surfaces. It is revolutionary that small patterns can be easily fabricated without the effect of the diffraction limit owing to the ultra-short wavelength of the electron beams. However, when patterning at the nanoscale, a significantly small deviation may result in irreparable problems, such as unconnected channels or dislocation patterns. Here, the alignment system of EB lithography permits high-precision placement control, which ensures the success of Nano-in-Nano Integration.

Our work on EB lithography-based Nano-in-Nano fabrication has been reported in the past. Metallic nanoarrays, nanovalves, and nanoelectrodes inside the nanochannels were fabricated using multiple EB lithography methods (Xu et al., 2015a). Briefly, bare nanochannels were obtained by standard EB lithography-based nanofabrication processes, which is the first EB lithography. The key step in EB-based Nano-in-Nano fabrication is the placement control step of the alignment system. A pair of pre-fabricated gold reference marks help to finish the alignment, and then the second or even third EB lithography can be applied subsequently to fabricate Nano-in-Nano or Nano-in-Nano-in-Nano structures (Fig. 5.5). The reference marks can be detected using a backscatter detector of the EB machine, and the relative positions can be determined. In this way, the multistep EB lithography processes will not cause a deviation problem. Figure 5.6 shows high-precision placement control after multiple EB lithography. A gold nanopattern was fabricated in a cross nanochannel, and the actual position information in the *X* and *Y* axes was measured. As

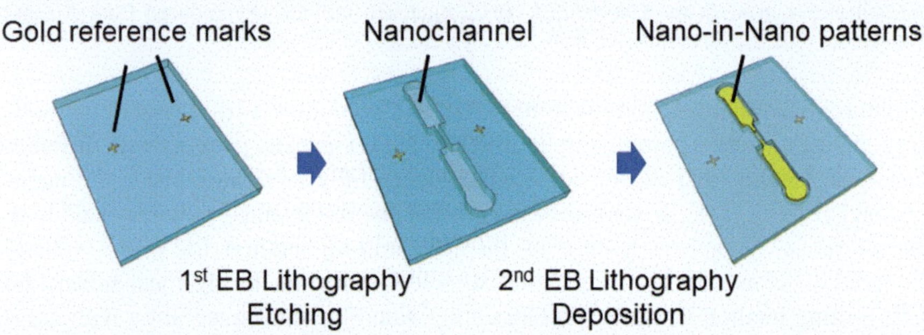

Fig. 5.5 Schematic drawing of gold reference marks for the multiple EB lithography-based Nano-in-Nano fabrication. (Adapted with permission from Xu and Xu (2015). Copyright 2015, Wiley-VCH)

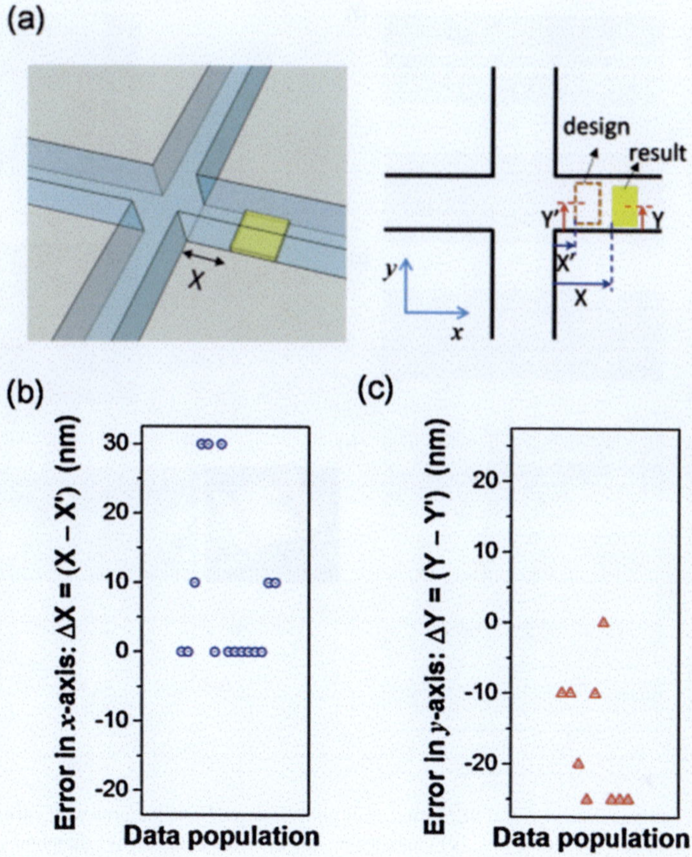

Fig. 5.6 (**a**) Schematic drawing of gold pattern in cross nanochannel and the deviation measurement for the position. Graphs of deviation in (**b**) *X*-axis and (**c**) *Y*-axis. (Reproduced from Xu et al. (2015a) with permission from the Royal Society of Chemistry)

a result, only very small deviations were observed, indicating that the gold nanopatterns were precisely fabricated at the desired positions (Fig. 5.6b, c).

Owing to the precise placement control, gold nanopatterns with different shapes, sizes, and amounts were successfully fabricated inside the nanochannels. Various nanopatterns, including nanodots, nanowires, nanostripes, gradient patterns, and diamond, square, triangle, and heart-shaped nanopatterns are shown in the scanning electron microscopy (SEM) images (Fig. 5.7a–e). From these images, it is clear that all the patterns are located inside the nanochannels, indicating the excellent quality of Nano-in-Nano fabrication. Moreover, the fabrication also has a high efficiency. Nanodot and nanosegment arrays consisting of dozens or even hundreds of nanopatterns were fabricated easily, as shown in Fig. 5.7f, g. Nested Nano-in-Nano-in-nano structures are also possible by adding an alignment and EB lithography process (Fig. 5.7h, i). Gold nanopatterns were fabricated precisely inside the nanowells,

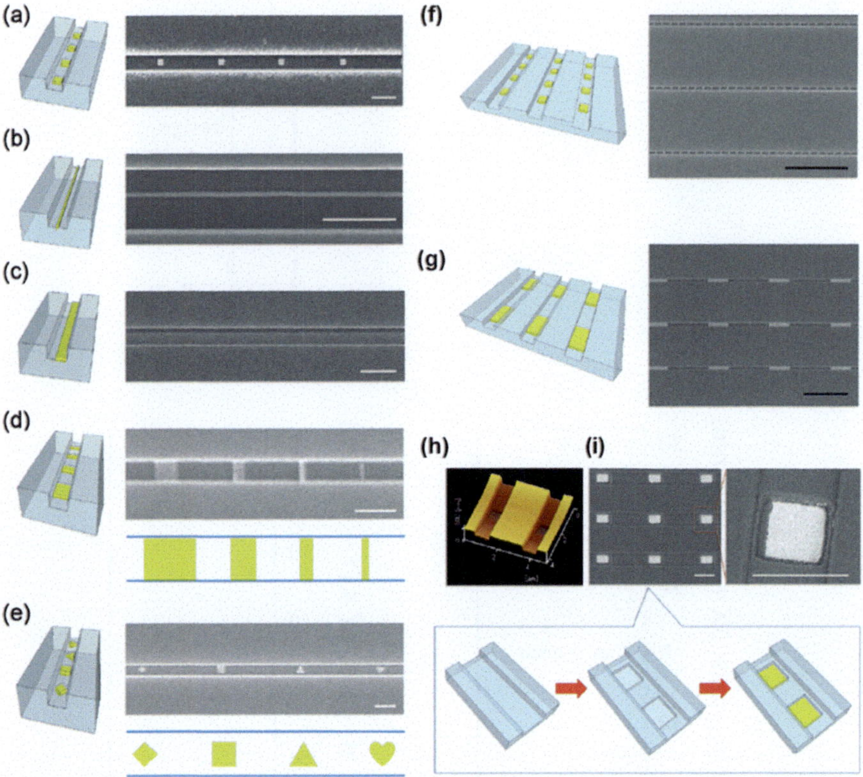

Fig. 5.7 Schematic drawings and SEM images of gold Nano-in-Nano patterns with various structures: (**a**) nanodots; (**b**) nanowires; (**c**) nanostripes; (**d**) gradient patterns; (**e**) diamond-, square-, triangle-, and heart-shaped patterns; (**f**) nanodot array; (**g**) nanosegment array; and (**i**) Nano-in-Nano-in-Nano structure. (**h**) AFM image of Nano-in-Nano-in-Nano structures (white scale bar: 1 μm; black scale bar: 10 μm). (Reproduced from Xu et al. (2015a) with permission from the Royal Society of Chemistry)

which were also precisely located in the nanochannels. These components with two kinds of elements (nanowell and gold nanodot) have the potential to provide more functions in applications.

As a short summary of this part, multiple EB lithography is promising for fabricating composite-material Nano-in-Nano structures. The newly introduced materials permit more functions in nanofluidic devices. Integrated components, such as gold nanopatterns, can be functionalized and serve as valves or electrodes. With these functional components, the obtained nanofluidic devices are promising for multidisciplinary applications. For the proof-of-concept, we introduce three applications that we have recently made.

5.3 Nano-in-Nano Integration for Functional Nanofluidic Devices

To demonstrate the Nano-in-Nano technology in the fabrication of functional nanofluidic devices, three devices with Nano-in-Nano structures functioning as arrays, valves, and electrodes are introduced.

5.3.1 Nano-in-Nano Integration for High-Throughput Molecule Capture Arrays

The ability to restrict small molecules using nanochannels with similar size features is an advantage of nanofluidic devices (Shirai et al., 2014; Mawatari et al., 2018; Paik et al., 2012; Shirai et al., 2018; Faez et al., 2015). The motions of nanoscale targets are confined by electrostatic interactions, ensuring a stable state for the following reaction or detection. Additionally, owing to the nanoscale level of the channel depth, the background from the solution is also decreased. In summary, nanofluidics is considered a suitable platform for molecular level analysis. However, when reaching single-molecule level analysis, the lower concentration and complex sample conditions are still the hurdles. Further restriction of the targets inside the nanochannels and reduction of effects from other interfering substances are necessary. The chemical/physical modification of nanochannels is an ideal way to solve such problems, which is also a conventional strategy in fields such as microfluidics. However, owing to the highly closed environment of nanofluidic channels, the modification is still challenging in the nanofluidic field. Although several successful modifications of nanochannels, such as silver-coated nanochannels or lipid-modified nanochannels, have been reported in the past, such modifications are still uniform and nonspecific in bare nanochannels (Kim et al., 2017; Lin et al., 2017). Their functions may not completely satisfy the development of current nanofluidic studies. Additionally, undesired absorption may result in incorrect conclusions. When the target molecules have a relatively low signal intensity, it is difficult to distinguish the target from the background (low signal-to-noise ratio) under such conditions. Moreover, uniform modification may also lead to a lower throughput, which is a problem in current complex and numerous bioanalyses.

To achieve site-specific modification in nanochannels, Nano-in-Nano Integration is necessary. The multiple EB lithography-based Nano-in-Nano Integration enables the selective, partial modification of nanochannels. Nanocomponents such as gold nanopatterns are easily functional. For example, self-assembled monolayers (SAMs) are widely used for gold modification. Molecules are immobilized and form a layer on the gold surface based on Au-S interactions, Si-O interactions, or other interactions. SAM technology is mature; however, it is applied in bulk-level solutions. As an early exploration, we first made it achievable in a nanofluidic system. The target molecules were selectively immobilized on the surfaces of gold Nano-in-Nano arrays via self-assembly, indicating successful molecule capture in nanofluidic devices.

As a proof-of-concept study, HOOC–$(CH_2)_{10}$–S–S–$(CH_2)_{10}$–CONH–fluorescein in ethanol solution was used (Xu et al., 2015a). The disulfide bonds of these molecules can interact with the gold surface selectively and form SAMs on the nanopattern surfaces. Simultaneously, only weak nonspecific absorption occurred in the glass areas without gold patterns. Thus, the fluorescence was detectable only on the array surfaces, revealing the molecule capture capability of the nanofluidic device (Fig. 5.8a). Figure 5.8b shows an image of a Nano-in-Nano array with many gold nanopatterns, which ensured the high throughput of the system. After the

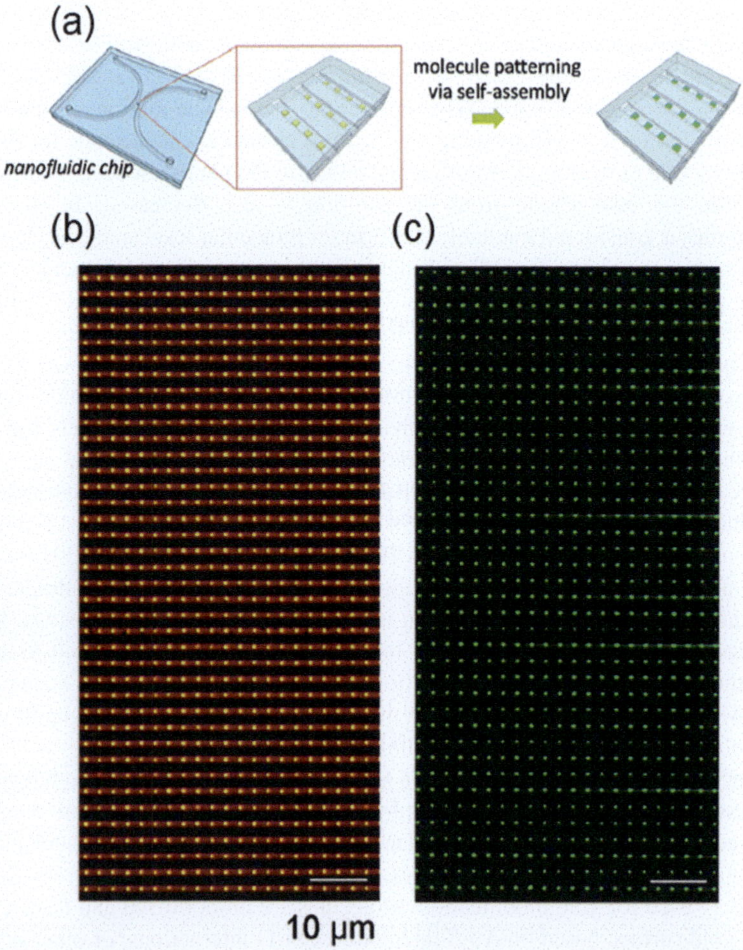

Fig. 5.8 (**a**) Schematic drawings of molecule capture using Nano-in-Nano arrays. (**b**) Bright-field image of gold nanopattern arrays. (**c**) Fluorescence image of the as-formed SAM molecule arrays. (Reproduced from Xu et al. (2015a) with permission from the Royal Society of Chemistry)

nanofluidic processes, including sample introduction, reaction, and rinsing, fluorescent arrays were observed, as shown in Fig. 5.8c. The fluorescent image shows an obvious difference between the arrays and backgrounds, revealing a high signal-to-noise ratio.

The demonstration of SAM arrays in nanochannels proved that Nano-in-Nano Integration can play an important role in nanoscale modification and function as molecule capture arrays. We expect that this functional nanofluidic device will be a powerful tool for the capture of biomolecules such as DNA and proteins, which is helpful in bioanalysis.

5.3.2 Nano-in-Nano Integration for Active Nanovalve

Because of the highly closed and ultra-small size features of nanofluidic channels, the manipulation inside nanochannels remains a challenge (Ouyang et al., 2010; Kneller et al., 2016; Piruska et al., 2010a; Harms et al., 2015b). Unlike conventional fluidic experiments, the liquids inside nanochannels are difficult to "touch" directly because of a lack of proper tools for fabrication or functioning at the nanoscale. The valve is a powerful, widely used component in fluid control areas, which is reliable in fluidic experiments to achieve sample introduction, flow direction control, flow rate control, and reaction control (Au et al., 2011; Lee et al., 2018). It is obvious that the valve function is important in a fluidic system. Regular manipulation in nanochannels is also necessary for nanofluidic-based synthesis, recognition, or analysis processes. Hence, nanofluidic devices also require valves. However, the fabrication of valves in nanofluidic systems is difficult, resulting in a weak fluid control capability in current nanofluidic devices. In the past, indirect methods have been used to control the liquid flow inside nanochannels. These indirect methods cannot satisfy the requirement of high-precision manipulation in nanofluidic areas, especially in emerging single-molecule studies.

Owing to the development of nanofabrication, achievements have been made regarding nanovalves in nanochannels, such as Laplace nanovalve. Femtoliter droplet handling was achieved using a Laplace nanovalve in the nanochannel (Mawatari et al., 2012). The embedded nanopillar structure in the nanochannel was constructed as a wetting boundary for the trapping of liquids and resisted the Laplace pressures at the liquid surface to function as a valve. In this way, the opening and closing of the nanochannel could be controlled by adjusting the introduction pressure. Furthermore, a femtoliter-level droplet was generated using this valve for the first time. Although there are achievements, one problem is that, currently, most nanovalves are passive valves, which means that active control in nanofluidic systems is still challenging. Additionally, some nanovalves cannot maintain their function after use, which means they are single-used valves and have limitations in practical applications. To advance the application of functional nanofluidic devices for fluid manipulation, reusable, active nanovalves in nanochannels for the regular control of ultra-small volumes of liquid are necessary.

Based on Nano-in-Nano Integration technology, it is possible to fabricate gold nanopatterns inside nanochannels. These gold patterns have the potential to serve as active valves after suitable modification, which can be used in nanoscale fluid manipulation. For example, some polymers have special sensitivity, such as thermosensitivity and photosensitivity, which can respond to external stimuli such as heat and light. When coated with such polymers, gold nanopatterns can be used as nanovalves that operate actively. The polymers change their shapes once activated. This change may result in the blocking of nanochannels, demonstrating the capability of active manipulation in nanospaces.

As a proof-of-concept study, we proposed a polymer-based nanovalve fabricated using poly (N-isopropylacrylamide) (PNIPAM) terminated with a trithiocarbonate (TTC) group at one end (TTC-t-PNIPAM) (Xu et al., 2016). This is a specific thermoresponsive polymer, which exhibits swelling/shrinking behavior under temperature changes. The nanochannels can then be blocked/opened actively as desired. Briefly, to fabricate the active nanovalves, gold nanopatterns were first fabricated at the expected positions for the valves. Subsequently, the as-synthesized polymers, TTC-t-PNIPAM, were coated onto the gold surface and formed a polymer brush using typical nanofluidic processes. The mechanism can be explained as follows: when the temperature is below the lower critical solution temperature (LCST), the polymers are swollen, resulting in a closed state of the nanochannels. When the temperature is higher than the LCST, the thermoresponsive polymers show a phase transition and shrink, resulting in the open state of the nanochannels (Fig. 5.9a). A

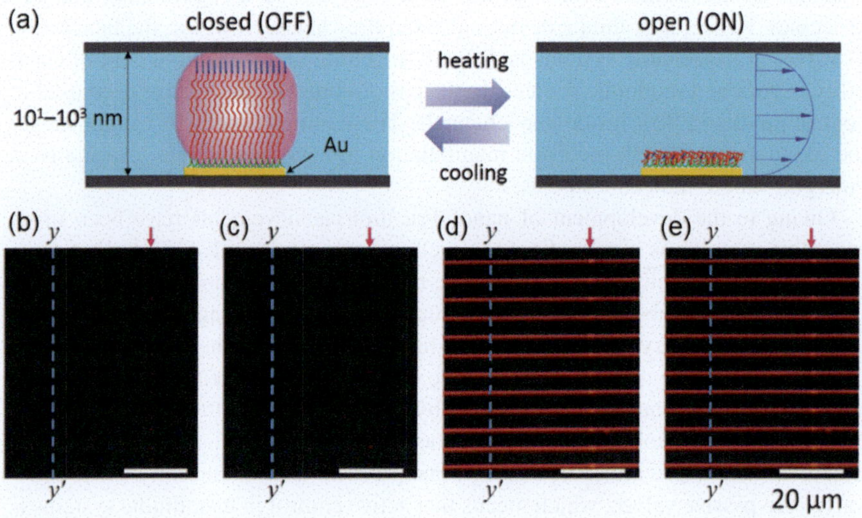

Fig. 5.9 (**a**) Schematic drawing of the nanovalve for active manipulation. (**b**, **c**) Fluorescence images of nanochannels when introducing sulforhodamine B solution under different conditions: (**b**) 10–20 s and (**c**) 20 min after introduction without heating ("closed" state), (**d**) 10–20 s, and (**e**) 17.5 min after heating at 45 °C ("open" state). (Adapted with permission from Xu et al. (2016). Copyright 2016, Wiley-VCH)

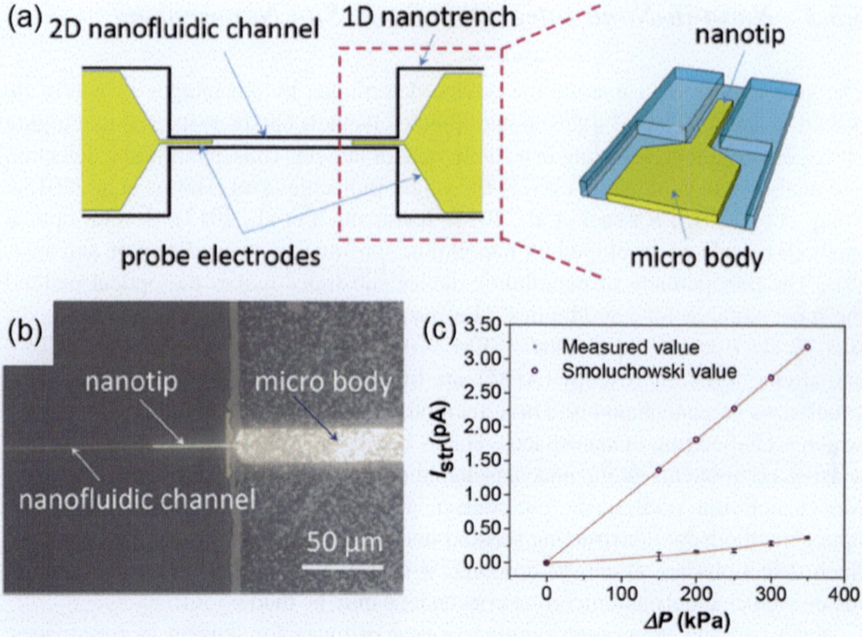

Fig. 5.10 (**a**) Schematic drawing of nanoelectrode in nanochannel. (**b**) Digital image of the nanoelectrode in nanofluidic device. (**c**) A comparison of the experimented streaming current values with the Smoluchowski values. (Adapted with permission from Xu and Xu (2015). Copyright 2015, Wiley-VCH)

solution of sulforhodamine B (50×10^{-6} M) was used to confirm the capability of the nanovalves. As shown in Fig. 5.9b, c, no fluorescence was observed in the nanochannels when the temperature was below 30.8 °C (LCST) at the beginning and 20 min after introduction because the channels were blocked by the swollen polymers. When heating the device at 45 °C (above the LCST), strong fluorescence was observed because the polymers inside the nanochannels were shrunken, allowing the liquids to pass through (Fig. 5.10d, e). The fluorescence images demonstrated the availability of active nanovalves for nanoscale fluid manipulation.

In conclusion, Nano-in-Nano Integration can function as an active nanovalve in nanofluidic devices. The manipulation capability of ultra-small volumes inside the nanochannels was successfully demonstrated in our study. This polymer-based active nanovalve is expected to become a useful tool for the evaluation of nanoscale liquid behaviors in nanospaces, which may be helpful in future single-molecule chemical reaction studies.

5.3.3 Nano-in-Nano Integration for In Situ Nanosensing

The size feature of a nanofluidic device determines its advantages in nanoscale detection and analysis. Targets in nanofluidic channels can be restricted to a highly closed environment, resulting in a stable state of targets. Thus, nanofluidic detection and analysis are promising, even at the single-molecule level (Harms et al., 2015a; Wang et al., 2015; Rassaei et al., 2014; Teerapanich et al., 2017). Several optical methods have been developed for nanofluidic sensing to achieve detection and analysis. The transparency of nanofluidic device substrates makes the optical method the most suitable for nanofluidics. However, other nonoptical detection methods, such as electrochemical methods, SEM, X-ray photoelectron spectroscopy (XPS), and atomic force microscopy (AFM), are limited by the closed features and ultra-small sizes of nanochannels. These methods require direct "contact" with targets, which is challenging in nanospaces. This is because of the difficulty in constructing sensing components (e.g., nanosensors) that can function inside the nanospace. Most nanofluidic studies use "non-contact," indirect detection, such as as-mentioned optical methods, or constructing sensor outside the nanochannels for the detection. Such detections are effective; however, it is difficult to reflect the true situation inside nanochannels and incorrect conclusions may be made.

Nano-in-Nano Integration provides a new pathway for sensing in nanospaces. Owing to the technologies for Nano-in-Nano fabrication, metallic nanocomponents inside nanochannels can be obtained, which can function as sensors or electrodes for in situ sensing. As such, ultra-small targets can be "contacted" directly without interference from the background. This is important for nanoscale sensing because the total volume and content concentration are usually very small, and conventional indirect methods may not have sufficient capability to distinguish the target signals and backgrounds. The Nano-in-Nano components permit direct femtoliter-level liquid detection, which can offer a new angle for the evaluation and understanding of many fluidic phenomena.

We previously reported a proof-of-concept study that utilized a pair of nanoelectrodes in a nanofluidic channel for the in situ electrokinetic probing of water (Xu & Xu, 2015). The streaming currents were directly measured by the Nano-in-Nano electrodes. To achieve the nanoscale sensing, Nano-in-Nano probe electrodes were fabricated in a nanochannel through which water flows (Fig. 5.10a). As shown in Fig. 5.10b, the nanotips of the probe electrodes were precisely located inside the nanochannel to achieve nanoscale sensing. The streaming currents were measured using an electrometer. Consequently, the streaming currents increased with the introduction of pressure (Fig. 5.10c). However, these experimental results are significantly different from the theoretical Smoluchowski values, which indicates that the previously available theories in conventional areas may not be valid for cases at the nanoscale. The discovery of this mismatch can be attributed to in situ sensing using Nano-in-Nano Integration, which is promising in reflecting the situation inside nanospaces.

5.4 Supporting Technologies for Nano-in-Nano Integrated Nanofluidic Devices

There are two important supporting technologies that ensure the functionality of nanofluidic systems with Nano-in-Nano Integration. One is the low-temperature chip bonding technology for nanofluidic devices and the other is the regeneration technology for the used devices. These are significant in practical applications.

5.4.1 Low-Temperature Bonding for Nano-in-Nano Devices

The nanofluidic devices with Nano-in-Nano Integration worked only when chips were formed after bonding processes. The final bonding step of nanofluidic device fabrication involves bonding the cover substrate with the nanomanufacturing substrate to form a closed system. Thus, a liquid can be introduced (Chantiwas et al., 2010; Persson et al., 2012; Abgrall et al., 2007). There are two types of bonding technologies: indirect bonding, which requires an adhesive layer, and direct bonding. Conventional direct bonding technologies, especially glass fusion thermal bonding, require temperatures over 1000 °C under vacuum conditions. The ultra-high-temperature requirement has several limitations. First, the cost of bonding is high. In addition to the requirement of expensive devices such as vacuum ovens, the energy consumption is also considerably high. Second, the entire process is time-consuming because it takes time to finish the temperature rising process and the cooling process. Third, most importantly, the ultra-high-temperature may easily damage the Nano-in-Nano structures. This is similar to the case of gold patterns that are easily damaged when the temperature is too high, resulting in the failure of Nano-in-Nano fabrication. The harsh conditions also limit the practical application in many other studies, such as biological studies, which require mild experimental conditions.

To overcome the problems of conventional bonding technologies, we propose low-temperature bonding technologies. Owing to surface activation using physical/chemical methods, the surface energies of the substrates are increased, significantly reducing the bonding temperatures (Fig. 5.11). In the past, a two-step, low-temperature bonding technology was reported, which could reduce the bonding temperature to approximately 200 °C (Xu et al., 2012). Inspired by the techniques of the semiconductor industry, oxygen reactive ion etching (RIE) plasma treatment and nitrogen microwaves were applied for the surface activation of glass substrates, resulting in a strong bonding strength of the nanofluidic chip. Furthermore, one-step, room temperature bonding technologies for nanofluidic chips have been reported. The O_2/CF_4 gas mixture plasma was used in this method for glass surface treatment (Xu et al., 2013). After bonding at room temperature (~25 °C), a nanofluidic chip was obtained, which can maintain functions under a high introduction pressure of 2.5 MPa.

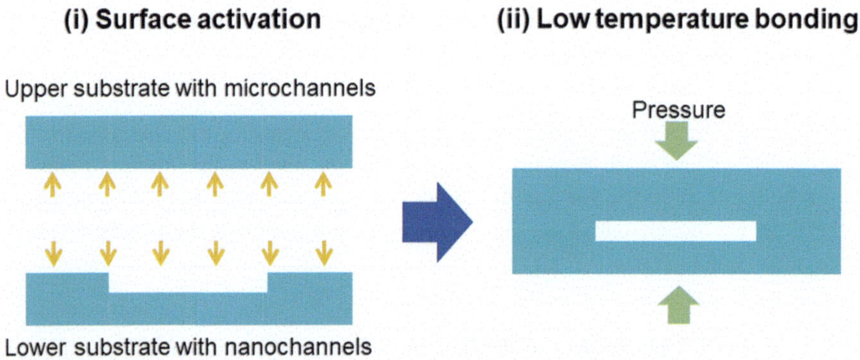

Fig. 5.11 Schematic drawing of low-temperature bonding of nanofluidic device by surface activation of substrates

These early achievements of low-temperature bonding technologies are promising in the "back-end" work of nanofluidic fabrication, which makes nanofluidic devices more suitable for other practical applications. Most importantly, such mild bonding conditions are beneficial for the fabrication of Nano-in-Nano Integration.

5.4.2 *Regeneration of Nano-in-Nano Devices*

The fabrication of Nano-in-Nano Integration is difficult because many processes such as alignment, EB lithography, dry etching, and bonding must be conducted with high precision. Therefore, it would be wasteful and inefficient if the nanofluidic devices only permitted single use (Mukhopadhyay, 2005). In fact, the single use of nanofluidic devices is a real problem, as well as in all the lab-on-a-chip (LOC) field. Obviously, the fouling and clogging of nanochannels after use may lead to the "death" state of the nanofluidic device, which means that the device's functions cannot be maintained. Furthermore, the residual reagents and samples inside the nanochannels may result in an increase in background, which may lead to incorrect conclusions. Unlike conventional tools for bulk-solution level experiments (e.g., test tubes and beakers), nanofluidic channels are difficult to wash and treat owing to their ultra-small sizes and closed features. Currently, "cleaning" inside nanochannels is still challenging. If a single use of nanofluidic devices is still unavoidable, the application of nanofluidics will be further limited.

To achieve the regeneration of nanofluidic devices, especially those with Nano-in-Nano structures, we proposed a thermochemical decomposition-based regeneration method, which consists of four processes: dehydration, high-temperature redox chemical reaction, high-temperature gasification, and cooling (Xu et al., 2015b). Different substances, such as liquid, dust, biomolecules, and organic matter inside the nanochannels, were removed by the programmed processes. As a proof-of-concept demonstration, the repeated formation of fluorescent SAMs was applied to

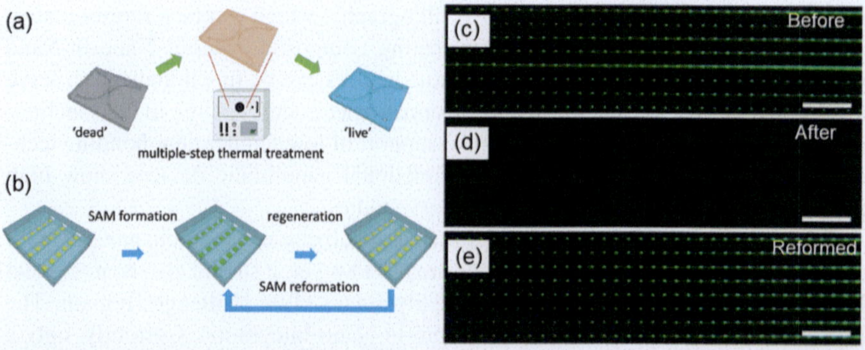

Fig. 5.12 (**a**) Schematic drawing of nanofluidic device regeneration processes. (**b**) Schematic drawing of the evaluation of nanofluidic regeneration by SAM formation. Fluorescence images of (**c**) first SAM formation before regeneration; (**d**) after regeneration; (**e**) second SAM formation using the regenerated chip (scale bar: 10 μm). (Reproduced from Xu et al. (2015b) with permission from the Royal Society of Chemistry)

a nanofluidic device (Fig. 5.10b). First, fluorescent SAMs were formed on the Nano-in-Nano gold array surfaces, which is the state of "before regeneration" (Fig. 5.10c). After the regeneration process, no fluorescence was observed, as shown in Fig. 5.12d, indicating the successful regeneration of the nanofluidic device. Finally, the as-regenerated chip was reused. The fluorescent SAM arrays were formed again after the self-assembly process, confirming the reusability of the regenerated device (Fig. 5.12e).

The regeneration method provides an opportunity for the multiple use of nanofluidic devices. The excellent substance removal capability of this method also ensures a low residual background for the regenerated device, which is important for the repeated use of the device in many fields.

To draw a conclusion, two supporting technologies were introduced. It should be noted that Nano-in-Nano Integration is powerful but fragile. With these supporting technologies, Nano-in-Nano Integration can fully develop its possibilities.

5.5 Conclusion and Outlooks

Owing to the development of nanofluidic studies, manufactured nanofluidic devices are playing an increasingly irreplaceable role in scientific research. However, current nanofluidic devices with only simple functions cannot satisfy the increasing requirements of scientific studies. As a solution, we proposed Nano-in-Nano Integration for the advanced fabrication of functional nanofluidic devices. Nano-in-Nano is used to fabricate nanostructures inside the nanochannels. Generally, there are two types of fabrication strategies. As a representative subtractive strategy, FIB milling technology has the potential to fabricate high-aspect-ratio Nano-in-Nano patterns on glass surfaces. Multiple EB lithography, which can be a representative additive strategy, is promising for fabricating composite-material Nano-in-Nano structures. The as-fabricated internal nanostructures can be functionalized to serve as high-throughput arrays, active valves, or nanoelectrodes, leading to the functionalization of nanofluidic devices. With the support of low-temperature bonding technologies and regeneration technology, functional nanofluidic devices show high potential for application in multidisciplinary studies.

Nano-in-Nano Integration realizes the functionalization of nanochannels. This is a significant development in nanofluidic areas. However, it should also be noted that Nano-in-Nano Integration is still in its early stages. Two challenges remain. The first challenge is the limited types of Nano-in-Nano Integration. Currently, only a few types of structures have been developed. The main reason for this is the difficulty in fabricating precise nanostructures inside nanochannels. The second challenge is the limited application of Nano-in-Nano Integration. As an early exploration, we previously demonstrated several applications. However, these applications are far from satisfying scientific requirements. An increasing number of functional components need to be integrated into nanofluidic devices.

To advance the field of Nano-in-Nano Integration for functional nanofluidic devices, based on current progress, we propose some outlooks as follows:

1. Utilizing new materials in fabrication is promising for the diversification of Nano-in-Nano Integration and simplifying the processes. Glass substrates are relatively difficult to manufacture. Nano-in-Nano fabrication on glass surfaces is more complex because of the requirement of alignment and multiple steps. Alternatively, some polymers, such as PDMS, have the potential for application in the fabrication of Nano-in-Nano Integration, owing to their ease of processing. In the meantime, new additional materials such as carbon nanotubes (CNTs) and silver can be integrated into the interior, which may result in new functions of nanofluidic devices.
2. Designing new types of Nano-in-Nano Integration is helpful in diversifying the functions of nanofluidic devices and extending the application range. Functions such as arrays, valves, and electrodes have been confirmed. Other functions, such as filtration, acoustic separation, and magnetic response, are also achievable when constructing relevant components, which will extend the application range of functional nanofluidic devices.
3. It is necessary to develop the unique functions of Nano-in-Nano Integration. A unique function indicates that the function is unavailable under other conditions. A promising function is the handling of single molecules, which is the true value of Nano-in-Nano Integration. The irreplaceable capability of single-molecule handling has the potential to develop novel functional nanofluidic devices, which are useful in biological or chemical analyses.

We expect that Nano-in-Nano Integration will offer a new angle for the advanced fabrication of nanofluidic devices with multiple functions, leading scientific research to a new generation.

Acknowledgments This work was supported in part by JSPS KAKENHI (Grant Nos. JP19KK0129, JP18H01848, JP20H00497, JP16K13653, JP26706010, JP26630403, JP21H04640, JP20H00497, JP21J14595), MEXT KAKENHI (Grant Nos. JP19H04678, JP17H05468, and JP26107714), JST PRESTO (Grant No. JPMJPR18H5), and the Asahi Glass Foundation.

References

Abgrall, P., & Nguyen, N. T. (2008). Nanofluidic devices and their applications. *Analytical Chemistry, 80*(7), 2326–2341.

Abgrall, P., Low, L. N., & Nguyen, N. T. (2007). Fabrication of planar nanofluidic channels in a thermoplastic by hot-embossing and thermal bonding. *Lab on a Chip, 7*(4), 520–522.

Arayanarakool, R., Shui, L., Van Den Berg, A., & Eijkel, J. C. T. (2011). A new method of UV-patternable hydrophobization of micro- and nanofluidic networks. *Lab on a Chip, 11*(24), 4260–4266.

Au, A. K., Lai, H., Utela, B. R., & Folch, A. (2011). Microvalves and micropumps for BioMEMS. *Micromachines, 2*, 179–220.

Bocquet, L., & Tabeling, P. (2014). Physics and technological aspects of nanofluidics. *Lab on a Chip, 14*(17), 3143–3158.

Branton, D., Deamer, D. W., Marziali, A., Bayley, H., Benner, S. A., Butler, T., Di Ventra, M., Garaj, S., Hibbs, A., Huang, X., Jovanovich, S. B., Krstic, P. S., Lindsay, S., Ling, X. S., Mastrangelo, C. H., Meller, A., Oliver, J. S., Pershin, Y. V., Ramsey, J. M., … Schloss, J. A. (2008). The potential and challenges of nanopore sequencing. *Nature Biotechnology, 26*(10), 1146–1153.

Cao, W., Wang, J., & Ma, M. (2018). Mechano-nanofluidics: Water transport through CNTs by mechanical actuation. *Microfluidics and Nanofluidics, 22*(11), 1–10.

Chantiwas, R., Hupert, M. L., Pullagurla, S. R., Balamurugan, S., Tamarit-López, J., Park, S., Datta, P., Goettert, J., Cho, Y. K., & Soper, S. A. (2010). Simple replication methods for producing nanoslits in thermoplastics and the transport dynamics of double-stranded DNA through these slits. *Lab on a Chip, 10*(23), 3255–3264.

Duan, C., Wang, W., & Xie, Q. (2013). Review article: Fabrication of nanofluidic devices. *Biomicrofluidics, 7*(2), 026507.

Faez, S., Orrit, M., Lahini, Y., Weidlich, S., Weidlich, S., Wondraczek, K., Zeisberger, M., Schmidt, M. A., Garmann, R. F., Manoharan, V. N., & Schmidt, M. A. (2015). Fast, label-free tracking of single viruses and weakly scattering nanoparticles in a nanofluidic optical fiber. *ACS Nano, 9*(12), 12349–12357.

Friedrich, R., Block, S., Alizadehheidari, M., Heider, S., Fritzsche, J., Esbjörner, E. K., Westerlund, F., & Bally, M. (2017). A nano flow cytometer for single lipid vesicle analysis. *Lab on a Chip, 17*(5), 830–841.

Frykholm, K., Alizadehheidari, M., Fritzsche, J., Wigenius, J., Modesti, M., Persson, F., & Westerlund, F. (2014). Probing physical properties of a DNA-protein complex using nanofluidic channels. *Small, 10*(5), 884–887.

Guo, S., Meshot, E. R., Kuykendall, T., Cabrini, S., & Fornasiero, F. (2015). Nanofluidic transport through isolated carbon nanotube channels: Advances, controversies, and challenges. *Advanced Materials, 27*(38), 5726–5737.

Harms, Z. D., Haywood, D. G., Kneller, A. R., & Jacobson, S. C. (2015a). Conductivity-based detection techniques in nanofluidic devices. *Analyst, 140*(14), 4779–4791.

Harms, Z. D., Haywood, D. G., Kneller, A. R., Selzer, L., Zlotnick, A., & Jacobson, S. C. (2015b). Single-particle electrophoresis in nanochannels. *Analytical Chemistry, 87*(1), 699–705.

Harms, Z. D., Selzer, L., Zlotnick, A., & Jacobson, S. C. (2015c). Monitoring assembly of virus capsids with nanofluidic devices. *ACS Nano, 9*(9), 9087–9096.

Hatsuki, R., Yujiro, F., & Yamamoto, T. (2013). Direct measurement of electric double layer in a nanochannel by electrical impedance spectroscopy. *Microfluidics and Nanofluidics, 14*(6), 983–988.

Haywood, D. G., Saha-Shah, A., Baker, L. A., & Jacobson, S. C. (2015). Fundamental studies of nanofluidics: Nanopores, nanochannels, and nanopipets. *Analytical Chemistry, 87*(1), 172–187.

He, Y., Tsutsui, M., Fan, C., Taniguchi, M., & Kawai, T. (2011). Controlling DNA translocation through gate modulation of nanopore wall surface charges. *ACS Nano, 5*(7), 5509–5518.

Holt, J. K. (2009). Carbon nanotubes and nanofluidic transport. *Advanced Materials, 21*(35), 3542–3550.

Kamai, H., & Xu, Y. (2021). Fabrication of ultranarrow nanochannels with ultrasmall nanocomponents in glass substrates. *Micromachines, 12*(7), 775.

Kim, S. J., Song, Y. A., & Han, J. (2010). Nanofluidic concentration devices for biomolecules utilizing ion concentration polarization: Theory, fabrication, and applications. *Chemical Society Reviews, 39*(3), 912–922.

Kim, Y., Kim, K. S., Kounovsky, K. L., Chang, R., Jung, G. Y., Depablo, J. J., Jo, K., & Schwartz, D. C. (2011). Nanochannel confinement: DNA stretch approaching full contour length. *Lab on a Chip, 11*(10), 1721–1729.

Kim, M., Ha, D., & Kim, T. (2015). Cracking-assisted photolithography for mixed-scale patterning and nanofluidic applications. *Nature Communications, 6*, 1–8.

Kim, D. J., Ha, D., Zhou, Q., Thokchom, A. K., Lim, J. W., Lee, J., Park, J. G., & Kim, T. (2017). A cracking-assisted micro-/nanofluidic fabrication platform for silver nanobelt arrays and nanosensors. *Nanoscale, 9*(27), 9622–9630.

Kneller, A. R., Haywood, D. G., & Jacobson, S. C. (2016). AC electroosmotic pumping in nanofluidic funnels. *Analytical Chemistry, 88*(12), 6390–6394.

Kovarik, M. L., & Jacobson, S. C. (2009). Nanofluidics in lab-on-a-chip devices. *Analytical Chemistry, 81*(17), 7133–7140.

Kudr, J., Skalickova, S., Nejdl, L., Moulick, A., Ruttkay-Nedecky, B., Adam, V., & Kizek, R. (2015). Fabrication of solid-state nanopores and its perspectives. *Electrophoresis, 36*(19), 2367–2379.

Lee, Y. S., Bhattacharjee, N., & Folch, A. (2018). 3D-printed quake-style microvalves and micropumps. *Lab on a Chip, 18*(8), 1207–1214.

Lin, L., Mawatari, K., Morikawa, K., Pihosh, Y., Yoshizaki, A., & Kitamori, T. (2017). Micro/extended-nano sampling interface from a living single cell. *Analyst, 142*(10), 1689–1696.

Lin, L., Chen, Q., & Sun, J. (2018). Micro/nanofluidics-enabled single-cell biochemical analysis. *TrAC Trends in Analytical Chemistry, 99*, 66–74.

Liu, S., Pu, Q., Gao, L., Korzeniewski, C., & Matzke, C. (2005). From nanochannel-induced proton conduction enhancement to a nanochannel-based fuel cell. *Nano Letters, 5*(7), 1389–1393.

Louër, A. C., Plecis, A., Pallandre, A., Galas, J. C., Estevez-Torres, A., & Haghiri-Gosnet, A. M. (2013). Pressure-assisted selective preconcentration in a straight nanochannel. *Analytical Chemistry, 85*(16), 7948–7956.

Mao, P., & Han, J. (2005). Fabrication and characterization of 20 Nm planar nanofluidic channels by glass-glass and glass-silicon bonding. *Lab on a Chip, 5*(8), 837–844.

Marie, R., & Kristensen, A. (2012). Nanofluidic devices towards single DNA molecule sequence mapping. *Journal of Biophotonics, 5*(8–9), 673–686.

Mawatari, K., Kubota, S., Xu, Y., Priest, C., Sedev, R., Ralston, J., & Kitamori, T. (2012). Femtoliter droplet handling in nanofluidic channels: A Laplace nanovalve. *Analytical Chemistry, 84*(24), 10812–10816.

Mawatari, K., Kazoe, Y., Shimizu, H., Pihosh, Y., & Kitamori, T. (2014). Extended-nanofluidics: Fundamental technologies, unique liquid properties, and application in chemical and bio analysis methods and devices. *Analytical Chemistry, 86*(9), 4068–4077.

Mawatari, K., Koreeda, H., Ohara, K., Kohara, S., Yoshida, K., Yamaguchi, T., & Kitamori, T. (2018). Nano X-ray diffractometry device for nanofluidics. *Lab on a Chip, 18*(8), 1259–1264.

Menard, L. D., & Ramsey, J. M. (2011). Fabrication of sub-5 nm nanochannels in insulating substrates using focused ion beam milling. *Nano Letters, 11*(2), 512–517.
Mijatovic, D., Eijkel, J. C. T., & Van Den Berg, A. (2005). Technologies for nanofluidic systems: Top-down vs. bottom-up—A review. *Lab on a Chip, 5*(5), 492–500.
Mokkapati, V. R. S. S., Di Virgilio, V., Shen, C., Mollinger, J., Bastemeijer, J., & Bossche, A. (2011). DNA tracking within a nanochannel: Device fabrication and experiments. *Lab on a Chip, 11*(16), 2711–2719.
Mukhopadhyay, R. (2005). When microfluidic devices go bad. *Analytical Chemistry, 77*(21), 429a–432a.
Mukhopadhyay, R. (2006). What does nanofluidics have to offer? *Analytical Chemistry, 78*(21), 7379–7382.
Napoli, M., Eijkel, J. C. T., & Pennathur, S. (2010). Nanofluidic technology for biomolecule applications: A critical review. *Lab on a Chip, 10*(8), 957–985.
Ouyang, H., Xia, Z., & Zhe, J. (2010). Voltage-controlled flow regulating in nanofluidic channels with charged polymer brushes. *Microfluidics and Nanofluidics, 9*(4–5), 915–922.
Öz, R., Kk, S., & Westerlund, F. (2019). A nanofluidic device for real-time visualization of DNA-protein interactions on the single DNA molecule level. *Nanoscale, 11*(4), 2071–2078.
Paik, K., Liu, Y., Tabard-cossa, V., Waugh, M. J., Huber, D. E., Provine, J., Howe, R. T., Dutton, R. W., & Davis, R. W. (2012). Control of DNA capture by nanofluidic transistors. *ACS Nano, 6*(8), 6767–6775.
Park, Y. S., Oh, J. M., & Cho, Y. K. (2018). Non-lithographic nanofluidic channels with precisely controlled circular cross sections. *RSC Advances, 8*(35), 19651–19658.
Peng, R., & Li, D. (2016). Fabrication of polydimethylsiloxane (PDMS) nanofluidic chips with controllable channel size and spacing. *Lab on a Chip, 16*(19), 3767–3776.
Perry, J. L., & Kandlikar, S. G. (2006). Review of fabrication of nanochannels for single phase liquid flow. *Microfluidics and Nanofluidics, 2*(3), 185–193.
Perry, J. M., Zhou, K., Harms, Z. D., & Jacobson, S. C. (2010). Ion transport in nanofluidic funnels. *ACS Nano, 4*(7), 3897–3902.
Persson, F., Fritzsche, J., Mir, K. U., Modesti, M., Westerlund, F., & Tegenfeldt, J. O. (2012). Lipid-based passivation in nanofluidics. *Nano Letters, 12*(5), 2260–2265.
Piruska, A., Branagan, S. P., Minnis, A. B., Wang, Z., Cropek, D. M., Sweedler, J. V., & Bohn, P. W. (2010a). Electrokinetic control of fluid transport in gold-coated nanocapillary array membranes in hybrid nanofluidic-microfluidic devices. *Lab on a Chip, 10*(10), 1237–1244.
Piruska, A., Gong, M., Sweedler, J. V., & Bohn, P. W. (2010b). Nanofluidics in chemical analysis. *Chemical Society Reviews, 39*(3), 1060–1072.
Poddar, A., Maity, D., Bandopadhyay, A., & Chakraborty, S. (2016). Electrokinetics in polyelectrolyte grafted nanofluidic channels modulated by the ion partitioning effect. *Soft Matter, 12*(27), 5968–5978.
Prakash, S., Piruska, A., Gatimu, E. N., Bohn, P. W., Sweedler, J. V., & Shannon, M. A. (2008). Nanofluidics: Systems and applications. *IEEE Sensors Journal, 8*(5), 441–450.
Rassaei, L., Mathwig, K., Kang, S., Heering, H. A., & Lemay, S. G. (2014). Integrated biodetection in a nanofluidic device. *ACS Nano, 8*(8), 8278–8284.
Rauscher, M., & Dietrich, S. (2008). Wetting phenomena in nanofluidics. *Annual Review of Materials Research, 38*, 143–172.
Shirai, K., Mawatari, K., & Kitamori, T. (2014). Extended nanofluidic immunochemical reaction with femtoliter sample volumes. *Small, 10*(8), 1514–1522.
Shirai, K., Mawatari, K., Ohta, R., Shimizu, H., & Kitamori, T. (2018). A single-molecule ELISA device utilizing nanofluidics. *Analyst, 143*(4), 943–948.
Sparreboom, W., Van Den Berg, A., & Eijkel, J. C. T. (2009). Principles and applications of nanofluidic transport. *Nature Nanotechnology, 4*(11), 713–720.
Sui, X., Zhang, Z., Zhang, Z., Wang, Z., Li, C., Yuan, H., Gao, L., Wen, L., Fan, X., Yang, L., Zhang, X., & Jiang, L. (2016). Biomimetic nanofluidic diode composed of dual amphoteric chan-

nels maintains rectification direction over a wide PH range. *Angewandte Chemie International Edition, 55*(42), 13056–13060.

Syed, A., Mangano, L., Mao, P., Han, J., & Song, Y. A. (2014). Creating sub-50 nm nanofluidic junctions in a PDMS microchip via self-assembly process of colloidal silica beads for electrokinetic concentration of biomolecules. *Lab on a Chip, 14*(23), 4455–4460.

Teerapanich, P., Pugnière, M., Henriquet, C., Lin, Y. L., Chou, C. F., & Leïchlé, T. (2017). Nanofluidic fluorescence microscopy (NFM) for real-time monitoring of protein binding kinetics and affinity studies. *Biosensors & Bioelectronics, 88*, 25–33.

Wang, C., Shi, Y., Wang, J., Pang, J., & Xia, X. H. (2015). Ultrasensitive protein concentration detection on a micro/nanofluidic enrichment chip using fluorescence quenching. *ACS Applied Materials & Interfaces, 7*(12), 6835–6841.

Wang, H., Deng, C., Xiao, C., Liu, L., Liu, J., Peng, Y., Yu, B., & Qian, L. (2019a). Fast and maskless nanofabrication for high-quality nanochannels. *Sensors and Actuators, B: Chemical, 288*, 383–391.

Wang, C., Wang, Y., Zhou, Y., Wu, Z. Q., & Xia, X. H. (2019b). High-performance bioanalysis based on ion concentration polarization of micro-/nanofluidic devices. *Analytical and Bioanalytical Chemistry, 411*(18), 4007–4016.

Wu, C., Lin, T. G., Zhan, Z., Li, Y., Tung, S. C. H., Tang, W. C., & Li, W. J. (2017). Fabrication of all-transparent polymer-based and encapsulated nanofluidic devices using nano-indentation lithography. *Microsystems & Nanoengineering, 3*, 16084.

Xu, Y. (2018). Nanofluidics: A new arena for materials science. *Advanced Materials, 30*(3), 1702419.

Xu, Y., & Matsumoto, N. (2015). Flexible and in situ fabrication of nanochannels with high aspect ratios and nanopillar arrays in fused silica substrates utilizing focused ion beam. *RSC Advances, 5*(62), 50638–50643.

Xu, Y., & Xu, B. (2015). An integrated glass nanofluidic device enabling in-situ electrokinetic probing of water confined in a single nanochannel under pressure-driven flow conditions. *Small, 11*(46), 6165–6171.

Xu, Y., Wang, C., Dong, Y., Li, L., Jang, K., Mawatari, K., Suga, T., & Kitamori, T. (2012). Low-temperature direct bonding of glass nanofluidic chips using a two-step plasma surface activation process. *Analytical and Bioanalytical Chemistry, 402*(3), 1011–1018.

Xu, Y., Wang, C., Li, L., Matsumoto, N., Jang, K., Dong, Y., Mawatari, K., Suga, T., & Kitamori, T. (2013). Bonding of glass nanofluidic chips at room temperature by a one-step surface activation using an O_2/CF_4 plasma treatment. *Lab on a Chip, 13*(6), 1048–1052.

Xu, Y., Matsumoto, N., Wu, Q., Shimatani, Y., & Kawata, H. (2015a). Site-specific nanopatterning of functional metallic and molecular arbitrary features in nanofluidic channels. *Lab on a Chip, 15*(9), 1989–1993.

Xu, Y., Wu, Q., Shimatani, Y., & Yamaguchi, K. (2015b). Regeneration of glass nanofluidic chips through a multiple-step sequential thermochemical decomposition process at high temperatures. *Lab on a Chip, 15*(19), 3856–3861.

Xu, Y., Shinomiya, M., & Harada, A. (2016). Soft matter-regulated active nanovalves locally self-assembled in femtoliter nanofluidic channels. *Advanced Materials, 28*(11), 2209–2216.

Zand, K., Pham, T., Davila, A., Wallace, D. C., & Burke, P. J. (2013). Nanofluidic platform for single mitochondria analysis using fluorescence microscopy. *Analytical Chemistry, 85*(12), 6018–6025.

Zhang, B., Wood, M., & Lee, H. (2009). A silica nanochannel and its applications in sensing and molecular transport. *Analytical Chemistry, 81*(13), 5541–5548.

Chapter 6
NEMS Sensors Based on Novel Nanomaterials

Shubham Yadav, Soumya Tripathy, and Deblina Sarkar

6.1 Introduction to NEMS Sensors

The field of nanomechanical systems (NEMS) aims to explore the potential and behavior of mechanical devices which are submicron in size (Loh & Espinosa, 2012; Ekinci & Roukes, 2005). In these devices, the physical motion of the structure is modulated using electrical or optical sources. As the device dimensions start to go down, NEMS show increase in resonant frequency and improvements in sensitivity and quality factor (Chiu et al., 2008, "Atomic-Scale Mass Sensing Using Carbon Nanotube Resonators"; Chaste et al., 2012; Jensen et al., 2008, "An Atomic-Resolution Nanomechanical Mass Sensor") which is not the case with other sensing devices such as transducers, etc. Moreover, these devices offer a high dynamic range (signal-to-noise ratio) which is essential for numerous applications. They also allow integration of novel materials with the traditional materials like Si and SiN for improving the sensitivity of detection to zeptogram mass, creating new kind of sensors which were not possible before like piezoelectric behavior on quantum scale, high-quality factors, changes in conductivity, etc. Most of the work done on NEMS till now has been based on devices made from bulk materials such as silicon and silicon nitrides which can be patterned by lithography and etching processes to create structures like cantilevers and clamped beams (Bagheri et al., 2011). In the recent past, the paradigm has shifted to making use of novel materials like graphene, MoS_2, carbon nanotubes, aluminum nitride, etc. as they hold promise in terms of outstanding mechanical and electrical properties (Geim & Novoselov, 2009; Lee et al., 2012). Further, these materials can be patterned using the CMOS compatible lithographic processes and are intrinsically nanoscale. The use of these novel materials has enabled them to achieve extremely high resonant frequencies (~THz), measuring quantum masses of the order of zg, and having resolution down to atomic mass

S. Yadav · S. Tripathy · D. Sarkar (✉)
Media Arts and Sciences, Massachusetts Institute of Technology, Cambridge, MA, USA
e-mail: deblina@mit.edu

Z. Yang (ed.), *Advanced MEMS/NEMS Fabrication and Sensors*,
https://doi.org/10.1007/978-3-030-79749-2_6

units (Perruisseau-Carrier, 2012; Chen et al., 2009, "Performance of Monolayer Graphene Nanomechanical Resonators with Electrical Readout"). Also, wide frequency tuning can be achieved by controlling the stiffness of the devices (by means of external forces), operating these devices at high temperatures, sensing gas molecules and chemical ions, and tracing biomolecules as well. Even micrometer-size devices can achieve frequencies in the gigahertz regime for strains near 1%. Higher frequencies are possible if we can scale down these devices to submicron levels with higher strains which is possible with these materials and advanced nanofabrication techniques. Further, since these devices can be mass-produced and easily fabricated, their impact would be much more than other devices used for sensing applications. Also, this has made these devices more penetrating in our daily life like the hundreds of sensors currently being placed in our mobile phones, making things smarter through IOT, providing portable medical facilities like testing kits in these trying times, and many more. Thus, in this chapter, we will explore four different material categories, namely, silicon nanowires, carbon nanotubes, 2D materials, and piezoelectric materials, and the recent developments in the NEMS technologies based on them. Further, different sensing applications including mass, temperature, gas, chemical, and biomolecule sensing will be discussed for each of these four different categories of materials.

6.2 Silicon Nanowires in NEMS-Based Sensing

6.2.1 Introduction to Silicon Nanowires

Silicon nanowires (SiNWs) are one-dimensional silicon structures having lateral dimensions less than 100 nm. The nanofabrication methods for synthesis of SiNWs include both top-down and bottom-up approaches Namdari et al., 2016). Some of the significant factors that critically affect the performance of the sensors involving SiNWs can be the carrier densities, the diameters, and surface chemistry of SiNWs. The bottom-up approach involves methods such as vapor–liquid–solid (VLS) technique, oxide-assisted growth (OAG), and photolithography-mediated synthesis. Vapor–liquid–solid (VLS) technique is the most adopted route to synthesize SiNWs. The bottom-up method comprises the techniques to fabricate SiNWs from a bulk silicon such a silicon wafer assisted by a metal catalyst. In contrast, the top-down technique involves trimming down bulk silicon wafer to an applicable and required dimension of SiNWs following photolithography or e-beam lithography.

The synthesis of SiNWs through VLS was first reported in 1964 using a silicon substrate catalyzed by a liquid Au droplet. VLS involves the deposition of metal catalysts such as gold (Au), iron (Fe), platinum (Pt), and aluminum (Al) on a silicon wafer followed by growth of SiNWs through chemical vapor deposition (CVD). Above eutectic temperature, silane (SiH_4) or tetrachlorosilane ($SiCl_4$) is passed over metal catalyst-coated silicon wafer within a tube furnace (Suzuki et al., 2007). In the OAG route of synthesis, SiO is used as starting material for the growth of SiNWs to stimulate the nucleation and the growth of SiNWs without application of metal catalyst, which produces metal-free SiNWs with high purities (Shao et al., 2008). Metal-assisted chemical etching involves the deposition of an electroless metal such as silver, nickel, platinum, and gold following a fluoride-based etching to obtain SiNWs (Shiu et al., 2011).

6.2.2 *Advantages of Silicon Nanowires*

Nanoelectromechanical sensors based on SiNWs involve different electromechanical phenomena, such as shifts in the fundamental resonance frequency of SiNWs-based resonator and piezoresistive change in electrical conductivity of SiNWs due to applied strain. The mathematical representation of the fundamental frequency of vibration of SiNWs can be deduced following the equation of motion of a freely vibrating rod governed by the fourth-order Euler–Bernoulli beam model wave equation (Cleland & Roukes, 2002), i.e.,

$$\rho A \frac{\partial^2 Y}{\partial t^2}(x,t) = -\frac{\partial^2}{\partial x^2} EI \frac{\partial^2 Y}{\partial x^2}(x,t) \tag{6.1}$$

where Y is the transverse displacement of the freely vibrating rod; E is Young's modulus of the rod; I is the bending moment of inertia, or the second moment of the cross-sectional area A; and ρ is the density of the rod material.

The clamped ends, at $x = 0$ and $x = L$, impose the boundary conditions $Y(0) = Y(L) = 0$ and $Y'(0) = Y'(L) = 0$. The solutions have the following form:

$$Y_n(x,t) = \left(C_1\left(\cos k_n - \sinh k_n x\right) + C_{2n}\left(\sin k_n - \sinh k_n x\right)\right)e^{-i\Omega_n t} \tag{6.2}$$

with eigenvectors k_n satisfying cos k_nL.cos*h* $k_nL = 1$. The first four eigenvectors are given by $k_nL = 4.730\,04$, 7.8532, 10.9956, and 14.1372. The angular frequencies Ω_n are given by

$$\Omega_n = \sqrt{\frac{EI}{\rho A}} k_n^{\ 2} \tag{6.3}$$

The fundamental frequency is given by

$$f_1 = \Omega_1 / 2\pi = \frac{1}{2\pi}\sqrt{\frac{EI}{\rho A}} k_1^{\ 2} \tag{6.4}$$

SiNWs can be synthesized with different geometry of cross section such as rectangular, hexagonal, and triangular. To obtain fundamental resonant frequency of SiNWs with different geometry of cross section, we can put values of the second moment of the different cross-sectional area of SiNWs in Eq. (6.4). The second moment of rectangular cross-sectional area and corresponding fundamental frequency are given by

$$I_{\text{rectangular}} = wt^3 / 12 \tag{6.5}$$

$$f_{1,\text{rectangular}} = 1.027 \frac{t}{l^2} \sqrt{\frac{E}{\rho}} \tag{6.6}$$

The second moment of a hexagonal cross section is given by (Feng et al., 2007)

$$I_{\text{hexagonal}} = 5\sqrt{3} d^4 / 144 \tag{6.7}$$

$$f_{1,\text{hexagonal}} = 0.9395 \frac{d}{l^2} \sqrt{\frac{E}{\rho}} \tag{6.8}$$

The piezoresistive property of SiNWs is generally exploited to fabricate different sensors such as strain sensors, pressure sensors, force sensors, and motion sensors, where applied force or pressure creates strain-mediated change in electrical conductance in SiNWs. The piezoresistance coefficient of SiNWs defines the piezoresistive effect of SiNWs. The longitudinal piezoresistance coefficient (π_l) and transverse piezoresistance coefficient (π_t) can be measured and defined as the relative change in conductivity per unit stress (Nakamura et al., 2008):

$$\pi_l = \frac{\Delta\rho_l}{\rho_{0\sigma_l}}, \quad \pi_t = \frac{\Delta\rho_t}{\rho_{0\sigma_t}} \tag{6.9}$$

where σ_l and σ_t are the longitudinal and transverse tensile stresses, respectively. ρ_0 is the resistivity along the wire axis without stress. $\Delta\sigma_l$ and $\Delta\sigma_t$ are variations in ρ_0 due to σ_l and σ_t, respectively.

Furthermore, gauge factor (G_{pr}) is defined as the change in resistance of SiNWs per unit applied strain and per strength of the corresponding unstrained SiNWs, and the gauge factor is an extensively used quantity for the sensor applications:

$$G_{PR} = \frac{1}{\varepsilon_l} \frac{\Delta R}{R_0} \tag{6.10}$$

where ε_l is the longitudinal strain and R_0 is the resistance of SiNWs without strain.

It is apparent from Eq. (6.3) that the resonant frequency of the resonator beam depends on the ratio of Young's modulus (E) to density (ρ) of the material. Achieving higher frequency with a reduced density of the material is critical for the functionality and responsiveness of nanoelectromechanical resonators. In addition, the surface roughness of material used to fabricate resonating beam critically affects its quality factor of sensor involving the resonator. Indeed, silicon nanowire surface is often well-faceted crystalline planes with atomic sharpness which favors the high-quality factor of the silicon nanowire resonators (Seo et al., 2015). Furthermore, high Young's modulus in GPa range with ultrahigh flexibility of SiNWs makes it a suitable candidate to be used in electromechanical system (Hsin et al., 2008). In addition to the mechanical property, electrical and piezoresistive property of SiNWs can be fine-tuned by well-developed technologies to achieve giant piezoresistive properties, which also favor SiNWs to be profoundly used in NEMS-based piezoresistive sensors (Barwicz et al., 2010).

6.2.3 Fabrication and Measurement

6.2.3.1 Fabrication

The electromechanical system generally involves SiNWs used as double- or single-clamped beam resonators, an array of vertically aligned SiNWs. Some of the standard fabrication techniques for SiNWs-based NEMS structure involve the (1) fabrication of suspended SiNWs using focused ion beam implantation following wet etching, (2) fabrication of SiNWs arrays from bulk silicon using photolithography techniques, (3) fabrication of groups of suspended SiNWs using metal catalyst that assisted the growth of nanowire using VLS technique.

The process involving ion implantation using focused ion beam and wet etching is one of the promising methods to synthesize freestanding SiNWs between two posts (Phan et al., 2015). The SiNWs are fabricated from a silicon-on-insulator (SOI) wafer. First, amorphous Si nanowires embedded in the top of the SOI wafer are formed using gallium ions (Ga+) implantation at a dose of approximately 7 × 1015 ions/cm^2. Second, since the etching rate of single-crystalline Si is more significant than that of the amorphous Si, tetramethylammonium hydroxide (TMAH)-mediated etching of the crystalline Si layer underneath amorphous SiNWs creates suspended SiNWs. Third, electrodes are then deposited for the electrical measurement of SiNWs during exposure to external stress.

SiNWs-based NEMS sensors also use the arrays of SiNWs as sensing elements. The SiNWs are synthesized from bulk silicon wafers using bulk micromachining techniques (Wasisto et al., 2013). In this fabrication process, a silicon wafer is covered with the photoresist, and the photoresist is then patterned using UV lithography producing round-shaped structures. The patterned silicon wafer is etched cryogenically in sulfur hexafluoride and oxygen plasma. The photoresist layer is then removed in a solution containing a mixture of hydrogen peroxide and sulfuric acid to expose an array of SiNWs.

Another interesting SiNWs-based NEMS sensor involves arrays of horizontally suspended SiNWs between two posts. This fabrication process involves a metal catalyst that assisted the growth of arrays of horizontal SiNWs starting from a sidewall of one posts and touching the opposite sidewall at the end of the process (Fernández-Regúlez et al., 2010). The process can be divided into four parts: vertically confined growth of SiNWs array (Step 1), laterally confined growth of SiNWs array (Step 2–6), deposition of Au catalyst nanoparticles in enclosed area exposing Si using galvanic displacement method (Step 7), and growth of SiNWs using VLS technique (Step 8). The Au catalyst nanoparticles determine the diameter and growth location of SiNWs. The nanowires grow only at the predefined areas bridging the opposite facing sidewalls of the supporting structure.

Figure 6.1a shows a 3D schematic view of MOSFET comprising a single vertical SiNW holding two stacked gates (Li et al., 2011). Figure 6.1l, m shows the scanning electron micrographic and transmission electron micrographic image of the MOSFET comprising vertically stacked twin gate transistor on a single SiNW. Single SiNW is fabricated using deep ultraviolet (DUV)-mediated dot patterning of hard mask followed by etching of silicon and sacrificial oxidation of etched surface. Arsenic (As) implantation is performed to define bottom electrodes (Fig. 6.1b). The bottom gate is fabricated through oxide deposition through high-density plasma (HDP) process followed by etching (Fig. 6.1c), which comprises growth of gate oxide and deposition of α-Si followed by chemical–mechanical polishing (CMP) and etching using tetramethylammonium hydroxide (TMAH) solution to define the gate length, gate implantation, activation, resist patterning, and gate extension (Fig. 6.1d). Thin (~50 nm) α-Si was then deposited, followed by angle implantation and activation (Fig. 6.1e). Tetraethyl orthosilicate (TEOS) oxide is deposited through low-pressure chemical vapor deposition (LPCVD) (Fig. 6.1f), which is

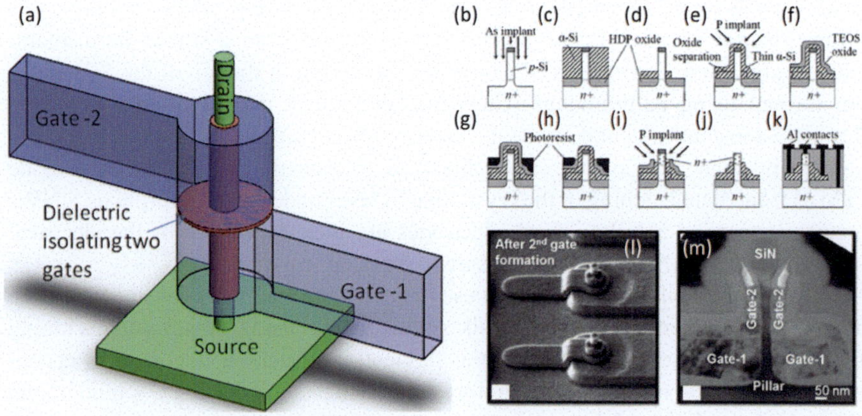

Fig. 6.1 (**a**) 3D schematic view of MOSFET comprising the stacked vertical twin gate SiNW. (**b**) Arsenic (As) implantation after SiNW definition. (**c**) After isolation oxide and stack formation of oxide/α-Si gate, followed by chemical–mechanical polishing of α-Si. (**d**) Wet etching of α-Si, first gate doping, and lithography-mediated etching. (**e**) Deposition of second gate stack and angle implantation and activation. (**f**) Deposition of TEOS through LPCVD. (**g**) Coating of resist and etching. (**h**) Wet etching of oxide. (**i**) Resist striping and isotropic etching of α-Si tip and implantation and activation. (**j**) Patterning of the second gate. (**k**) Metallization with aluminum (Li et al., 2011). (**l**) SEM image of developed MOSFET. (**m**) TEM image of developed MOSFET (Li et al., 2011). [Figure (**a**)–(**m**), adapted with permission from (Li, X. et al. Vertically stacked and independently controlled twin-gate MOSFETs on a single Si nanowire. IEEE Electron Device Lett. 32, 1492–1494 (2011)). Copyright (2011) IEEE]

followed by a coating of resist and blanket etching in an O_2 ambient determining the length of the top gate (Fig. 6.1g). The α-Si tip is then exposed by wet etching exposed oxide (Fig. 6.1h) and isotropically dry-etched. The top electrode is then implanted and activated (Fig. 6.1i). Top gate extension is then fabricated (Fig. 6.1j) following the deposition of the pre-metal dielectric, which is etched through contact lithography, and final metallization is performed for electrodes (Fig. 6.1k).

6.2.3.2 Measurement Principles

The magnetomotive technique is used to detect the amplitude of vibration of an electrically resonating beam such as SiNWs. The magnetomotive detection is based on the induction of electromotive force (emf) across the resonating SiNWs, which is placed in a uniform static magnetic field. The induced emf is picked up by the detection circuit, and the displacement of resonating SiNWs is measured following the equation given by (Ekinci, 2005)

$$\upsilon_0(t) = \xi l B \dot{x}(t) \tag{6.11}$$

where B is the uniform static magnetic field strength, l is the length of the beam, and ξ is a geometric factor ($\xi \approx 0.885$ for a doubly clamped beam). $x(t)$ is the displacement of the midpoint of a beam where the beam is in a harmonic oscillation at a frequency of ω. The bandwidth of the detection in this technique will be mostly limited by the parasitic capacitances present in the detection circuit. When the resonator is actuated through either magnetomotive or electrostatic actuation, the parasitic coupling between the input and the output signals reduces the extent of orthogonality between the actuation and detection.

Optical methods for detecting vibration of a NEMS resonator use the setup of optical interferometry techniques, which generally involves path-stabilized Michelson interferometry and Fabry–Perot interferometry (Ekinci, 2005). Path-stabilized Michelson interferometry involves the interference of a tightly focused laser beam reflecting from the resonator in motion and a stable reference beam. In the case of Fabry–Perot interferometry, the optical cavity formed between the NEMS surface and the substrate modulates the optical signal detected by the photodetector as the resonator moves in the out-of-plane direction.

Capacitive detection of motion in the resonator beam is based on the modulation of the electrical capacitance between the resonating element and a fixed gate due to change in effecting distance between the resonating beam and the gate (Ekinci, 2005). For a capacitor, it is clear that

$$C = Q / V \tag{6.12}$$

where Q is the charge on the capacitor and V is the voltage across it. If we take the derivative of both sides, we can obtain the following equation:

$$dQ = VdC + CdV \tag{6.13}$$

It can be seen from Eq. (6.13) that a change in capacitance at a constant voltage or current will generate current or voltage across the capacitor. The modulation in capacitance can be detected by measuring the change in voltage or current flow in the capacitor being biased by a current or voltage. The limit of capacitive detection of motion is determined by the parasitic element developed in the circuit shown in gray.

Piezoresistive and piezoelectric detections are based on the change in electrical properties of the material such as electrical resistance and electrical polarization developed in a material as a result of the strains generated inside a material due to exerted external stress (Ekinci, 2005). Piezoresistive sensors involve the detection of the resistance changes as it is actuated. For example, in a doubly clamped p-type Si beam, the resistance change (ΔR), a function of the displacement, the beam length l, and the resistance R are given by

$$\frac{\Delta R}{R} \approx \frac{x^2}{l^2} \tag{6.14}$$

In the piezoelectric method of detection of induced strain, the polarization fields produced by the strain fields within the piezoelectric nanomechanical element are measured by detecting the potential drop across the strained element.

6.2.4 *Sensing Applications*

6.2.4.1 Physical Sensing

Fernández-Regúlez et al. have fabricated ultrasensitive strain gauges exploiting the exceptional piezoresistive and electromechanical properties of SiNWs synthesized through vapor–liquid–solid mechanism (Fernández-Regúlez et al., 2010). Transduction of electromechanical response of the sensor is acquired by measuring the variation in resistance of the SiNWs array using a lock-in amplifier connected in a half-Wheatstone bridge setup following the application of a controlled force using an AFM tip perpendicular to the surface of the cantilever. The developed sensor shows a transduction sensitivity of 105 m^{-1} and linear behavior of the resulting curve between deflection of cantilever and change in resistance. Phan et al. have used the piezoresistive effect of p-type SiNWs and fabricated strain sensors through focused ion beam implantation and wet etching (Phan et al., 2015). The developed sensor provides a gauge factor of approximately 47 with good reproducibility to repeated bending cycles.

Soon et al. have developed a pressure sensor with embedded piezoresistive p-type <110> oriented SiNWs in a membrane with a square cross section of 100 nm on a silicon-on-insulator (SOI) wafers (Soon et al., 2010). The sensor provides a piezoresistive response of 13 Pa^{-1}. Shaby et al. have used polysilicon double nanowire piezoresistors, where two nanowires are attached to the diaphragm for sensing pressure exploiting the piezoresistive nature of SiNWs (Shaby & Juliet, 2011). Variable pressure creates strain in silicon nanowires, which reflects in a change in resistance of nanowires. As-fabricated double polysilicon pressure sensor possesses sensitivity of 160 mV/V KPa.

Han et al. have fabricated a ring-shaped force sensor in which SiNWs are embedded at the anchor of each of the four silicon bridges supporting a suspended ring structure located at the center of four suspended beams (Han et al., 2013). Applied external force displaces the suspended ring, which produces bending in silicon bridges. The bending of the bridges induces strain in embedded SiNWs changing their resistance as a result of the piezoresistive nature of SiNWs. The fabricated sensor can respond to displacement in nanometer ranges with a sensitivity of 13.4×10^{-3} μm^{-1} in the z-direction with high linearity (>99.9%) to the applied load with no observed hysteresis. Allain et al. have developed a motion detector based on SiNWs electromechanical system in which SiNWs act as piezoresistive element detecting displacements up to 180 nm with a resolution down to a few Angströms (Allain et al., 2012).

Bartsch et al. have developed a silicon nanowire resonator-based platform that can achieve frequency stability of 10 ppm, which can be translated into a limit of detection of mass lower than 60 kDa (Bartsch et al., 2014). The platform comprises a junction-less field-effect transistor in which a SiNW connects drain and source of the transistor and a gate, which actuates the resonance in SiNWs, providing a very high resonance frequency of 100 MHz. The detection principle is based on the depletion charge modulation in the junction-less field-effect transistor body, which provides a better output signal-to-noise ratio than piezoresistive method detection in the same platform (Bartsch et al., 2014). Mechanical resonance is detected in the drain current of the highly doped silicon wire with a cross section down to ~30 nm. Feng et al. have developed a very-high-frequency single-crystal SiNWs-based nanoelectromechanical resonator (Fig. 6.2a). The resonance in suspended SiNW is both actuated and detected through magnetomotive technique (Feng et al., 2007). Figure 6.2b shows frequency response of magnetomotively transduced resonance in suspended SiNW at different magnetic field. Figure 6.2e shows the magnetomotive response of the sensor at magnetic field of 6T when RF sweeps from −61 to −41 dBm. The signal-to-noise ratio increases with increasing magnetic field as shown in Fig. 6.2c, and dependence of voltage signal on magnetic field is shown in Fig. 6.2d, where the voltage shows a B^2 dependence. As designed, SiNWs resonator can detect mass with a sensitivity approaching 10 zg.

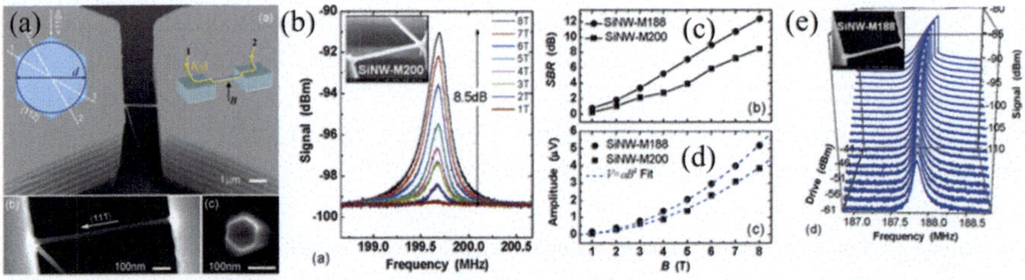

Fig. 6.2 (**a**) SEM images of suspended pristine SiNWs (Feng et al., 2007). (**b**) Magnetomotive transduction of the metalized SiNW containing the device's response at varying B field (Feng et al., 2007). (**c**) Detection efficiency, signal-to-background ratio (SBR) in dB during the measurement (Feng et al., 2007). (**d**) B^2 dependence of voltage signal amplitude of the SiNW observed in magnetomotive transduction (Feng et al., 2007). (**e**) Magnetomotive response of the SiNW under RF drive to the device sweeps from −61 to −41 dBm at B = 6 T (Feng et al., 2007). [For figures (**a**)–(**e**), adapted with permission from (Feng, X. L., He, R., Yang, P. & Roukes, M. L. Very High-Frequency Silicon Nanowire Electromechanical Resonators. NANO Lett. 7, 19 (2007)). Copyright (2007) American Chemical Society]

6.2.4.2 Gas Sensing

Alexandra et al. have fabricated a SiNWs-based electromechanical system, where transduction has been achieved by both the junction-less field-effect transistor and the piezoresistive effect using SiNWs with widths down to ~20 nm (Koumela et al., 2013). Allan deviation (σA) of the SiNW has been measured to be σA ~20 ppm for the junction-less field-effect detection and σA ~3 ppm for the detection through piezoresistive effect at room temperature and low pressure. The Allan deviation of σA ~ 1 ppm can be achieved using this platform, and this σA corresponds to a sub-zeptogram mass resolution at ambient temperature, which can be successfully applied for gas sensing.

6.2.4.3 Chemical Sensing

Wasisto et al. have developed a SiNWs-based resonator to detect aerosol containing nanoparticles by measuring shifts in the resonant frequency of SiNWs resonator induced by the mass of nanoparticles adsorbed on resonator (Wasisto et al., 2013). The sensor comprises an array of gold-coated SiNWs fabricated through the inductively coupled plasma to electrostatically capture aerosol of TiO_2 nanoparticles. The sensor shows a mass sensitivity of 7.1 Hz/fg with a resolution of 31.6 fg. Yong-Jun Seo et al. have developed a SiNWs nano-probe containing a silicon mirror for magnetic resonance force microscopy (MRFM) for radical sensing (Seo et al., 2015). The SiNWs nano-probe is attached to an Nd–Fe–B magnet at the end of the nanowire. The fabricated nanowire probe shows a resonance frequency of 11.256 kHz and a quality factor of 12,800 with a force sensitivity of attonewton, which has been used for 3D imaging of radicals through force mapping based on electron spin resonance. As-developed probe can detect force and magnetic field gradient up to 82 aN and ~ 70.1 G μm^{-1} at room temperature, respectively. The probe has been used to image a radical containing poly-10-(4-vinylbenzyl)-10H-phenothiazine (PVBPT) particles, where the radical density has been calculated as 4.6×10^{18} spins cm^{-3}.

6.2.4.4 Biomolecule Sensing

Guthy et al. have developed a SiNWs-based biosensor that detects analyte through a change in the resonance frequency of SiNWs resonator due to loading of analyte on the SiNWs. The author has followed streptavidin–biotin conjugation chemistry to detect streptavidin using a biotinylated SiNWs electromechanical resonator (Guthy et al., 2013). The functionalization of biotin on the nanowires causes a downward frequency shift of ~25 kHz from its original frequency of naked nanowires. The specific capture of streptavidin causes an additional frequency shift of 120 kHz, corresponding to an attached mass of 300 ag. The molecular weight of a single streptavidin molecule is 99.6 zg, and this added mass corresponds to a streptavidin surface density of 0.3 molecules per 100 nm^2 on the resonator. Fischer et al. have also developed biotinylated SiCN nanowire-based detection of streptavidin with a mass sensitivity of 1.01 ag/Hz to 1.63 ag/Hz (Fischer et al., 2008).

6.3 Carbon Nanotubes in NEMS-Based Sensing

6.3.1 Introduction to Carbon Nanotube (CNT)

CNT is considered as one-dimensional bucky tube shape due to its high aspect ratio of around 1000. Figure 6.3a shows the cross-sectional view of bundle of CNTs having width W and height H. The structure of CNT can be considered as a rolled-up graphene sheet comprising seamlessly joined carbon atoms in sp2 hybridization (Fig. 6.3b) and characterized by a rolled-up vector or chiral vector $C = na_1 + ma_2$, where m and n define the chirality of CNT and a_1 and a_2 are lattice vectors of graphene (Aqel et al., 2012). The length of the rolled-up vector C is the circumference of CNT, which is given by the following relationship: $C=|C| = a\sqrt{n^2 + nm + m^2}$, where the magnitude of a is related to carbon–carbon bond length a_{cc}. The value of a_{cc} is 0.1421 nm in the case of graphite, which is widely used in the structural calculation of CNT. But the better approximation of a_{cc} for CNT is 0.144 4 nm due to the curvature of CNT. The value of a is related to a_{cc} by the following relation: $a = |a_1| = |a_2| = a_{cc}\sqrt{3}$. In addition, the angle between the chiral vector and zigzag nanotube axis is the chiral angle and defined by the following relation: $\theta = \tan^{-1}\left(m\sqrt{3}\right)/\left(m+2n\right)$. The structure of CNT is determined by the value of integers (m, n), which defines the chirality of CNT and affects the electrical, mechanical, and optical properties. If $|m\text{-}n| = 3i$, the CNT is metallic, and if $|m\text{-}n| = 3i \pm 1$,

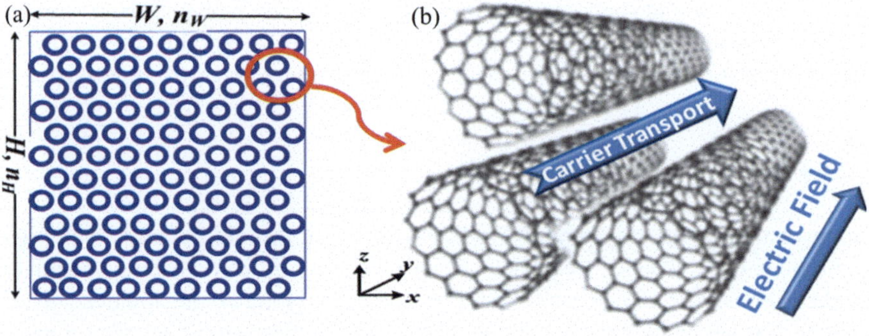

Fig. 6.3 (**a**) Cross-sectional view of a bundle of CNTs having width *W* and height *H* (Sarkar et al., 2011). (**b**) The direction of electric field and carrier transport in single-walled carbon nanotubes (Sarkar et al., 2011). [Figure (**a**)–(**b**) adapted with permission from (Sarkar, D., Xu, C., Li, H. & Banerjee, K. High-frequency behavior of graphene-based interconnects-Part I: Impedance modeling. IEEE Trans. Electron Devices 58, 843–852 (2011)). Copyright (2011) IEEE)

the CNT is semiconducting. Depending on vales of *m*, *n*, and θ, the CNT are of three types.

- $\theta = 30°$ and $m = 0$; the CNT is zigzag type.
- $\theta = 0°$ and $m = n$; the CNT is armchair type.
- $0^{\circ} < \theta < 30^{\circ}$ and $m \neq n$; the CNT is chiral.

However, single-walled carbon nanotube (SWCNT) is considered to a single-layer graphene sheet rolled up, forming a seamless nanotube of diameter close to 1 nm. In contrast, multiwalled carbon nanotube (MWCNT) is a congruence of co-centric tubes of SWCNTs with different diameters. The interlayer spacing in MWCNT is close to graphitic interlayer spacing among graphene, and it is around 3.3 Å (Aqel et al., 2012). Another special case of MWCNT is double-layered carbon nanotubes (DMCNT), which show quite similar properties and morphology like SWCNT.

The methods for synthesis of carbon nanotubes have been dramatically evolved since its discovery. Various methods such as chemical vapor-phase deposition (Kumar & Ando, 2010), laser ablation (Arepalli, 2004), electric arc discharge (Awasthi et al., 2005), flame synthesis (Nakazawa et al., 2005), pyrolysis of plastic (Chung & Jou, 2005), and liquid hydrocarbon synthesis have been developed for the synthesis of CNTs (Zhang et al., 2002).

Rice University pioneered the synthesis of carbon nanotubes using laser ablation of metal–graphite in an argon atmosphere (Guo et al., 1995). The technique uses laser-mediated heating of target carbon precursor with or without a catalyst. High temperature causes a precursor to sublimate and create plasma consisting of ionized carbon vapors. Energetic carbon vapors move to the cooler side of the reactor, and it is cooled down rapidly by the stream of inert gas, which nucleates the growth of nanotubes (Hasnain & Nayak, 2019). In practical setup, a high-power laser beam is focused onto the target, carbon precursor, and kept within a controlled flow of inert gas and the vaporized carbon from laser-induced plasma collected on temperature-controlled soot collector.

Chemical vapor deposition is considered as the most economically viable and promising technology for large-scale synthesis of carbon nanotubes (Tripathi et al., 2015; Awadallah et al., 2012). Different types of CVD have also been investigated for large-scale synthesis of carbon nanotubes such as catalytic chemical vapor deposition (CCVD), plasma-enhanced chemical vapor deposition (PECVD), radio-frequency chemical vapor deposition (RFCVD), microwave plasma chemical vapor deposition (MWPCVD), water-assisted chemical vapor deposition (WCVD), and hot filament chemical vapor deposition (HFCVD) (Shukrullah et al., 2014). In CVD setup, vaporized hydrocarbon and inert gas are purged into a tube with controlled temperature and pressure, where the hydrocarbon is decomposed to form individual energetic carbon atoms. Dissociated carbon atoms diffuse into the catalyst on the substrate and precipitate into carbon nanotube at a certain temperature suitable for the growth of CNT (Hasnain & Nayak, 2019). The metal catalyst generally involves nano-sized nickel or cobalt particles embedded on the substrate. The properties of carbon nanotubes depend on the type of catalyst, the substrate, the precursor of carbon, and the type of CVD used for the synthesis of carbon nanotubes.

Arc discharge-mediated synthesis of carbon nanotube was first employed at NEC's Fundamental Research Laboratory (Iijima, 1991). This method involves generation of arc between two closely placed electrode under a controlled atmosphere under 300 to 700 mbar pressure. A high-temperature plasma is created between two electrodes at a direct current of 50 to 100 A and driving voltage between 15 V and 30 V. Arc discharge chamber consists of two electrodes enclosed in a chamber with controlled atmosphere. Cathode is a pure graphite electrode and anode is consisting of carbon precursor along with catalyst. The arc creates a high temperature of 4000 K to 6000 K, which is enough to ionize gas filled in the chamber into electrons and positive ions forming a glowing plasma. Energetic electrons and ions hit anode at much higher speed and sputter the carbon precursor filled within the anode. In addition, high temperature of the arch sublime carbon precursor and vapors of carbon is further ionized by high temperature, or plasma. Ionized viscous cluster of carbon drifts toward cathode and become liquefied due to lower temperature of cathode. At optimum temperature gradient, added quenching effect on cathode solidifies and crystalizes liquid carbon into carbon nanotubes (de Heer, 2005; Arora & Sharma, 2014).

Nucleation following the growth of CNT can be classified into three steps (Kataura et al., 2000). In the first step, at higher temperature, dissociated carbon atoms form cluster and diffuse into metal catalyst. In the second step, the system is cooled down to a medium temperature that inhibits diffusion of carbon into catalyst, and the surface of metal particles becomes saturated with carbon clusters in the shape of fullerene-like caps. When the temperature of the system is at eutectic point in the third step, the CNT starts to grow from the site of nucleation. However, commonly used catalysts are nickel, cobalt, or palladium, and different hydrocarbons are used as source of carbon atoms such as methane, acetylene, ethanol, carbon monoxide, and ethylene.

6.3.2 Advantage of Carbon Nanotube

For a CNT of length "L" with attached mass, considering boundary conditions for the CNT-based resonator, we can deduce the fundamental resonant frequency of the system from Eq. (6.1) (Chowdhury & Adhikari, 2011):

$$f_n = \frac{1}{2\pi}\sqrt{\frac{k_{eq}}{m_{eq}}} \tag{6.15}$$

where k_{eq} and m_{eq} are equivalent stiffness and mass of CNT with an attached mass in the first mode of vibration, respectively.

For cantilever CNT (Joshi et al., 2011),

$$k_{eq} = 12.36\frac{EI}{L^3} \tag{6.16}$$

$$m_{eq} = \rho AL + 4M \tag{6.17}$$

For bridged CNT (Joshi et al., 2011),

$$k_{eq} = 500.56\frac{EI}{L^3} \tag{6.18}$$

$$m_{eq} = \rho AL + 2.522M \tag{6.19}$$

The resonant frequency of the CNT with the attached mass is given as (Joshi et al., 2011)

$$f_n = \frac{1}{2\pi}\frac{\alpha_k \beta}{\sqrt{1+\alpha_m \Delta M}} \Delta M = \frac{M}{\rho AL} \beta = \sqrt{\frac{EI}{\rho AL^4}} \tag{6.20}$$

where $\alpha_k = 3.51$ and 22.37, $\alpha_m = 4$ and 2.52 for cantilever and bridged configuration, respectively (Joshi et al., 2011). ΔM is the relative mass change. For the free vibration of CNT, substituting $\Delta M = 0$ in Eq. (6.20), the resonant frequency is obtained as

$$f_{on} = \frac{1}{2\pi}\alpha_k \beta \tag{6.21}$$

The resonant frequency shift due to the attached mass can be expressed as

$$\Delta f = f_{on} - \frac{f_{no}}{\sqrt{1+\alpha_m \Delta M}} \tag{6.22}$$

Changes in conductance of CNT as a result of applied deformation are the basis for the piezoresistive property of CNT. Mechanical deformation due to applied stress changes the band structure of carbon nanotubes, which is the relationship between the electronic property and structure of CNT. Due to the fact that CNT is considered to be a rolled-up graphene sheet, also the electronic structure of a CNT is considered as the periodic boundary condition of a rolled-up graphene sheet, which results in the quantization of the wave functions of electrons within the nanotube. The quantization affects the direction of electrons along the axis of the CNT. External stress alters the bond length of the carbon nanotube and shifts the periodicity of the quantization of electrons, altering the bandgap of CNT (Obitayo & Liu, 2012). An unstrained metallic carbon nanotube has no bandgap, while a semiconducting carbon nanotube shows a varying bandgap inverse to the diameter of the nanotube:

$$E_{\mathrm{Gap}}^{0} = \frac{2\gamma a}{\sqrt{3}d} \tag{6.23}$$

E_{Gap}^{0} is ~0.1 to 2 eV for an average semiconducting carbon nanotube, the tight-binding overlap integral γ is 2.6 eV, d is the diameter of the carbon nanotube, and a is the unit vector of graphene (Saito et al., 1998).

A thermally activated transport model is used to measure the change in resistance due to the change in bandgap in the CNT (Minot et al., 2003):

$$R = R^{\mathrm{C}} + \frac{1}{|t|^2}\frac{h}{8e^2}\left(1+\exp\left(\frac{E_{\mathrm{Gap}}^{0}+\left(dE_{\mathrm{gap}}/d\varepsilon\right)\varepsilon}{KT}\right)\right) \tag{6.24}$$

R^{C} is the contact resistance; $|t|$ is the average transmission coefficient; h is Planck's constant; e is the electron's quantity of the electricity; K is the Boltzmann constant; T is the absolute temperature; and $dE_{\mathrm{gap}}/d\varepsilon$ is the strain-dependent bandgap. E is the strain induced in the CNT.

Furthermore, the relationship between shifts in resonant frequency (Δf) of CNT and attached mass (Δm) is described by a responsivity function $R(x)$. The responsivity function is the ratio between the shift in resonant frequency (Δf) and the attached mass (Δm), which is the function of position x of the attached mass. Assuming attached mass is being evenly distributed along the CNT, this relation is simplified to the following equation by averaging over the responsivity function:

$$\Delta f = Avg.R(x).\Delta m = -\frac{f_0}{2m_0}\Delta m \tag{6.25}$$

where f_0 is the resonant frequency of the CNT and m_0 is the initial mass of the CNT. Responsivity function defines the sensitivity of shifts in the fundamental frequency of CNT resonator toward change in mass of the system by adsorption of analyte on CNT. It can be seen from Eq. (6.25) that to maintain a maximized responsivity of a resonator, one should reduce the mass of the resonator by maintaining high resonance frequencies of the resonator. Indeed, CNTs are one-dimensional material with much smaller size and density than other available resonators made of silicon or other dense materials. For instance, CNTs generally have a mass in the order of ~10^{-21} kg, which is more than four orders of magnitude less than micromachined resonators having a mass in the order of ~10^{-17} kg (Yang et al., 2006). In addition, CNTs have the highest elastic modulus, which helps in maintaining high resonance frequencies of a resonator (Treacy et al., 1996). As a result of these properties, carbon nanotubes have been a suitable candidate as an actuator used in NEMS-based sensor applications. In contrast to CNTs, SiNWs have ultrahigh flexibility than that of CNTs. CNTs are more prone to stress-induced structural defects due to nicking or breaking in bonding. At the same time, SiNWs' ultrahigh flexibility makes it more resistant to much higher stress in practical applications such as strain or force sensor (Hsin et al., 2008).

6.3.3 Fabrication and Measurement

Electromechanical CNT-based sensors generally utilize either single-clamped or double-clamped CNT-based cantilever as a resonator, which converts perturbation in mechanical vibration into measurable electrical signals. Electromechanical CNT-based sensors also use CNTs adsorbed unto a flexible substrate or cantilever, where mechanical bending in substrate or cantilever creates strain in CNT, which alters the bandgap of CNTs causing variation in electrical conductance. In fact, the fabrication of electromechanical sensors involves a critical step of integrating CNTs with electrodes and ensuring proper contact between nanotube and electrodes. The integration technique can be broadly divided into three categories. The first method involves dispersing the suspension of CNTs on a chemically functionalized lithographic pattern following locating nanotubes using an AFM, followed by e-beam lithography to crates electrodes to fix located nanotubes. In this case, a hydrophobic self-assembled monolayer (SAM) of trimethylsilyl (TMS) with $-CH_3$ termination is prepared on an oxidized Si sample through chemical vapor deposition of hexamethyldisilazane. The TMS-SAM layers are then patterned to expose underneath SiO_2. A second SAM layer of aminopropyltriethoxysilane is prepared subsequently on the exposed SiO_2 pattern, and individual single-walled nanotubes are randomly deposited on the chemically patterned surfaces by dispersing suspension of CNTs in DMF or sodium dodecyl sulfate (SDS) solution.

The second method involves patterning catalyst through e-beam lithography, or micro-contact printing, followed by in situ growth of CNTs using CVD (Franklin & Dai, 2000; Cassell et al., 1999). In this method, the liquid-phase catalyst is transferred onto a patterned surface containing towers following micro-contact printing through a polydimethylsiloxane (PDMS) stamp inked with a liquid-phase catalyst. The liquid-phase catalyst comprises inorganic chloride precursors of a metal catalyst, a removable triblock copolymer serving as the structure-directing agent for the chlorides, and appropriate alcohol for dissolution of the inorganic chloride and polymer compounds. The transferred catalyst on the patterned substrate is then subjected to heat treatment (at 400 °C for 4 h) followed by calcination at 700 °C to remove polymer component of liquid-phase catalyst, which leaves immobilized catalysts on the towers. The catalyst containing a patterned substrate is subjected to CVD to grow freestanding CNTs on towers. Another technique involves preparing islands of catalyst and growing CNTs from the islands of catalyst through CVD (Kong et al., 1998).

The third technique employs AFM for nanorobotics manipulation to precisely align CNTs. In this method, deposited CNTs on the substrate are precisely aligned by using AFM (Fukuda et al., 2003). Another method for post-growth manipulation of CNTs involves electrophoresis-induced alignment of individual CNT between two electrodes, or on a substrate. In fact, dielectrophoresis is the movement of neutral particles within a liquid medium under the influence of a nonuniform electric field due to the frequency-dependent polarization of particles. The method involves dielectrophoresis of suspension of CNTs in suitable solvents present between two electrodes using a composite DC/AC field. AC electric field attracts CNTs to their desired locations, and DC electric field traps them due to electrostatic attraction (Chung et al., 2004).

As can be observed from the above discussion, CNT-based electromechanical sensors measure shifts in resonance frequency and change electrical conductance from the *I–V* curve in case of the mechanical resonator and piezoresistive-based sensor. One of the ways to measure shifts in resonance frequency is the detection technique based on nanotube radio receiver design (Jensen et al., 2007, 2008, "An Atomic-Resolution Nanomechanical Mass Sensor"). In this method, an electromagnetic wave from an incoming radio transmission impinges upon the nanotube forcing it to physically vibrate when the frequency of the incoming wave coincides with the flexural resonance frequency of nanotube. Here, nanotube serves as both antenna and tuner of a radio, where the amplification and demodulation stages rely on the remarkable field emission properties of carbon nanotubes, which are due to their needlelike geometry that concentrates the electric field. The electrode opposite the nanotube is biased to induce a field emission current from the nanotube, and an amplitude-modulated (AM) frequency-swept using a voltage-controlled oscillator (VCO), radio-frequency (RF) signal is coupled to the nanotube, forcing it into resonance, and consequently modulating the field emission current. The modulated field emission current is recovered by a lock-in amplifier, and the resonance peak is displayed on the oscilloscope or recorded by a computer. Another method is established on frequency-modulated current measurement, where the motion of CNT is measured by frequency-modulated (FM) mixing technique at liquid helium temperature in which motion of CNT is transduced into low-noise electrical signals (Gouttenoire et al., 2010). For piezoresistive *I–V* characteristic of the CNT, actuating element is measured in response to applied strain or pressure (Stampfer et al., 2006a).

6.3.4 Sensing Application

6.3.4.1 Physical Sensing

Jensen et al. have used a single-clamped and double-walled CNT beam as a resonator for detection of mass with a sensitivity of 1.3 × 10–25 kg $Hz^{-1/2}$, which equals to "0.4 gold atom" $Hz^{-1/2}$. Mass is detected from shifts in the resonance frequency of clamped CNT by detecting modulated field emission current by lock-in amplifier (Jensen et al., 2008, "An Atomic-Resolution Nanomechanical Mass Sensor"). Clamped CNT emits electrons, the field emission current, under the influence of field created by biased electrode opposite to CNT, while the field emission current is amplitude modulated by coupling a radio-frequency signal to CNT.

Stampfer et al. have developed a nanoscale force sensor where the mechanical deformation of suspended SWCNT leads to considerable change in its conductance (Stampfer et al., 2006a). A single SWNT is suspended from the substrate by clamping nanotube onto two electrodes, and a cantilever is in contact with suspended nanotube. If now an external out-of-plane force acts on the freestanding cantilever, it will deflect, which finally leads to the mechanical deformation of the clamped nanotube. The electrical measurements of SWCNT show reversibility toward applied mechanical load or vertical deflection to the cantilever using an AFM tip in contact mode. A deflection of 35 nm increases the resistance to 5.5 MΩ, which is approximately 18 times of resistances of SWCNT at zero deflection; the resistance at zero deflection is around 300 kΩ. Thus, the concept of nanoelectromechanical systems has been proven. The observed electromechanical behavior of nanotube can be attributed to strain-dependent bandgap opening, which consequently leads to an exponential increase of the resistance R as a function of applied strain (Obitayo & Liu, 2012).

SWNT-based pressure sensor utilizes the electromechanical piezoresistive behavior of CNT, where a change in strain-induced in nanotubes by applied pressure is translated into variation in electrical resistance. The resistance of the metallic SWNT is determined by measuring the I–V transfer characteristics of the nanotubes for each applied differential pressure. Stampfer et al. have used a circular diaphragm of atomic layer-deposited 100-nm-thick alumina containing adsorbed individual electrically connected SWCNT (Fig. 6.4a), where the deflection of alumina membrane due to pressure creates strains in adsorbed nanotubes. Figure 6.4b shows the resistance of the piezoresistive element (SWCNT) at different pressure applied on the sensor. The pressure-dependent increase in resistance R as a function of the applied differential pressure Δp revealed a gauge factor of 210 (Stampfer et al., 2006b). Thus, the first pressure sensor transfer function of a metallic SWNT pressure sensor has been demonstrated.

Tong et al. have developed a strain sensor using a silicon cantilever with CNTs as strain sensing elements (Tong et al., 2007). MWNTs with an average length of 5 μm and an average diameter of 17 nm are dielectrophoretically assembled between electrodes deposited on a silicon cantilever. The CNT-based cantilevers

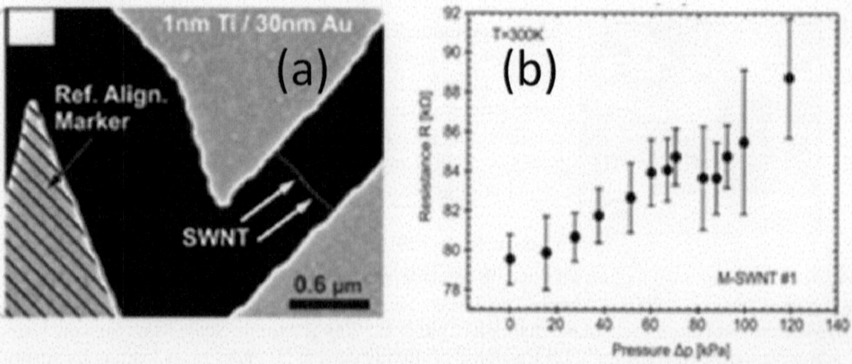

Fig. 6.4 (**a**) Scanning electron micrograph of the device comprising a membrane of alumina with electrically connected SWCNT (Stampfer et al., 2006b). (**b**) The change in resistance of the SWCNT-based evice in response to the applied differential pressure (Stampfer et al., 2006b). [Figure (**a**)–(**b**) adapted with permission from (Stampfer, R. C. et al. Fabrication of single-walled carbon-nanotube-based pressure sensors. Nano Lett. 6, 233–237 (2006)). Copyright (2006) American Chemical Society]

demonstrated a linear relationship between resistance changes and externally applied strain. The gauge factor ranged from 78.84 to 134.40 for different electrode configurations. Su et al. have developed a two-dimensional CNT-based strain sensor comprising two pairs of SWCNTs-cantilever modules which are present on both *X* and *Y* directions with each module of a pair in the opposite direction in *X*- or *Y*-axis. The as-fabricated sensor can detect bending strain in both *X* and *Y*, thus suggesting its two-dimensional nature of sense. The sensing principle is based on the strain-induced change in electrical resistance of CNT, and experimental results show that the sensor achieves a high strain resolution of 0.00099% with a maximum piezoresistivity gauge factor of the sensor of 744 (Su et al., 2012).

6.3.4.2 Gas Sensing

Chiu et al. have used CNT-based single-electron transistor as an electromechanical resonator in which source–drain voltage of $V_{sd} = U_{sd} + V_0 \cos \omega t$ is applied across source and drain, while a detuned AC signal of $V_g = V_{DC} + V_{AC} \cos [(\omega + \delta\omega)t]$ is applied to the gate (Fig. 6.5a) (Chiu et al., 2008, "Atomic-Scale Mass Sensing Using Carbon Nanotube Resonators"). Figure 6.5b shows the schematic diagram of the measurement circuit involving as-developed CNT-based single-electron transistor. By tuning this ω to a nanomechanical resonance of CNT, it produces vibration modulating gate capacitance and changing the measured current in response to amplitude of vibration. As can be seen from Fig. 6.5c, loading of Xe atoms on the CNT resonator significantly shifts the value of mixing current (I) and resonance frequency to lower regime, which shows the atomic-scale resolution of the as-developed sensor.

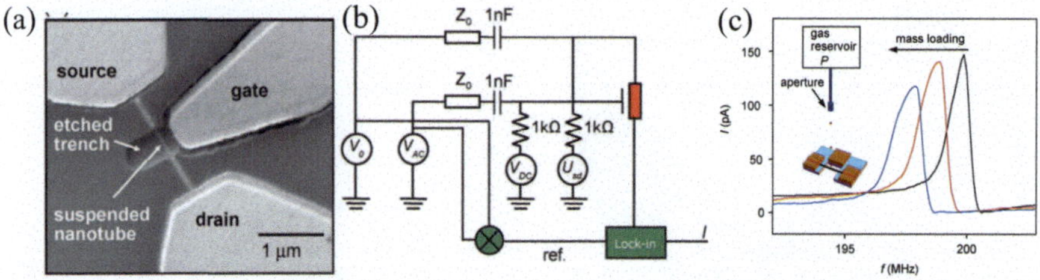

Fig. 6.5 (**a**) Scanning electron micrographs of a suspended CNT with electrodes for source, drain, and gate (Chiu et al., 2008, "Atomic-Scale Mass Sensing Using Carbon Nanotube Resonators"). (**b**) Measurement circuit schematic diagram (Chiu et al., 2008, "Atomic-Scale Mass Sensing Using Carbon Nanotube Resonators"). (**c**) Response of mixing current (*I*) under mass loading by adsorption of Xe atoms. Black curve, the initial resonance; red curve, after adsorption for ≈600 s; blue curve, after adsorption for ≈1600 s (Chiu et al., 2008, "Atomic-Scale Mass Sensing Using Carbon Nanotube Resonators"). [Adapted with permission from (Chiu, H. Y., Hung, P., Postma, H. W. C. & Bockrath, M. Atomic-scale mass sensing using carbon nanotube resonators. Nano Lett. 8, 4342–4346 (2008)). Copyright (2008) American Chemical Society]

6.3.4.3 Chemical Sensing

Chaste et al. have reported a resonator comprising a carbon nanotube of 150 nm in length suspended over a trench with a resonance frequency of ~2 GHz. The detection principle is based on detecting mechanical motion of CNT by frequency-modulated (FM) mixing technique at liquid helium temperature in which motion of CNT is transduced into low-noise electrical signals (Chaste et al., 2012). The resonator provides an ultralow resolution of 1.7 yg, which is used to detect the adsorption of naphthalene and binding energy of xenon on CNT. The sensor can even detect adsorption of a single naphthalene molecule producing a frequency shift of 3×10^5 Hz within 1–2 s. Kuroyanagi et al. have developed a CNT cantilever-based hygrometer and explored the frequency response of open-ended and closed-ended CNT (Kuroyanagi et al., 2014). When open-ended cantilever is used, the adsorption of water molecules produces shifts in resonance frequency. However, when closed-ended CNT is used, the change in viscosity of moist air surrounding CNT produces shifts in resonance frequency. Also open-ended CNT shows a decrease in resonant frequency and quality factor, where the resonant frequency shifts to higher values in case of closed-ended CNT.

6.3.4.4 Biomolecule Sensing

Patel et al. have evaluated the use of DWCNT resonator for the detection of biomolecules, including alanine and deoxyadenosine and infectious agents, including *Coronaviridae* and *Bartonella bacilliformis* (Patel & Joshi, 2015). Aniline with amino terminal residue and deoxyadenosine with free residue have masses of $1.213071E^{-22}$ g and $4.16965E^{-22}$ g, respectively. Different types of DWCNT including zigzag, (8,8)@(6,6), (10,10)@(5,5) and (7,7)@(4,4), and armchair, (14,0)@

(10,0), (18,0)@(9,0), and (12,0)@(7,0), of length 25 nm are observed for their sensitivity by loading eight molecules of aniline with amino terminal residue, seven molecules of deoxyadenosine with free residue, seven *Coronaviridae* particles, and seven cells of *Bartonella bacilliformis.* Figure 6.6a shows the frequency shifts due to attached eight molecules of aniline with amino terminal residue on cantilevers and their sensitivities (Fig. 6.6b).

6.4 2D Materials in NEMS-Based Sensing

6.4.1 Introduction to 2D Materials

2D materials have several features which are either not present in their bulk counterparts (mostly these materials are not found as bulk) or are completely unique to these materials. Categorizing all the 2D materials into one set is challenging as these include materials ranging from graphene to perovskites to TMDs, etc. Figure 6.7 shows different categories of TMDs along with the E–K diagram, symmetry points in the crystal structure. All these materials are rather different but offer tremendous potential in terms of exotic properties like bandgap tunability, high-temperature coefficients, piezoelectricity, etc. which are relevant for different applications (Ling et al., 2016; Akinwande et al., 2017; Cao et al., 2013, 2015a, b; Arjmandi-Tash et al., 2017; Gupta et al., 2015; Sarkar et al., 2014, 2015a, b). In particular, graphene has low bandgap and extremely high mobilities—founding its use as an electrode (Han et al., 2007)—while MoS_2 shows piezoelectricity (Zhou et al., 2016, "Theoretical Study on Two-Dimensional MoS2 Piezoelectric Nanogenerators") which opens up an entirely different field of piezoelectricity-based NEMS (discussed later in Sect. 6.5 of this chapter) and so on. Material-specific properties for some of them can be found in the following literatures (Han et al., 2007; Gupta et al., 2015; Akinwande et al., 2017; Novoselov et al., 2016). Briefly, we can say that

(a)

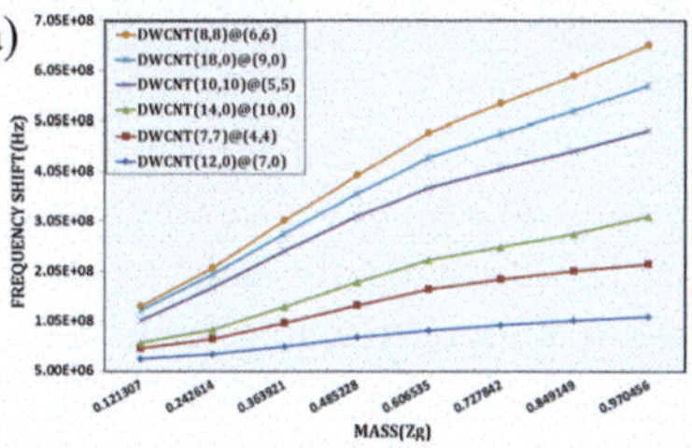

(b)

DWCNT type	Outer diameter (nm)	Inner diameter (nm)	Sensitivity (zg/GHz)
(8,8)@(6,6)	10.78	8.092	0.1424
(10,10)@(5,5)	13.42	6.74	0.1221
(7,7)@(4,4)	9.43	5.39	0.1011
(14,0)@(10,0)	10.81	7.78	0.1359
(18,0)@(9,0	14.012	7.0063	0.1251
(12,0)@(7,0)	9.341	5.44	0.0989

Fig. 6.6 (**a**) Frequency shift of DWCNT-based cantilever due to eight molecules of alanine with amino terminal vs. mass (Zg). (**b**) Mass sensitivity of cantilevered DWCNT (Patel & Joshi, 2015). [Adapted with permission from (Patel, A. M. & Joshi, A. Y. Detection of biological objects using dynamic characteristics of double-walled carbon nanotubes. Appl. Nanosci. 5, 681–695 (2015)), Copyright (2015) Springer Nature]

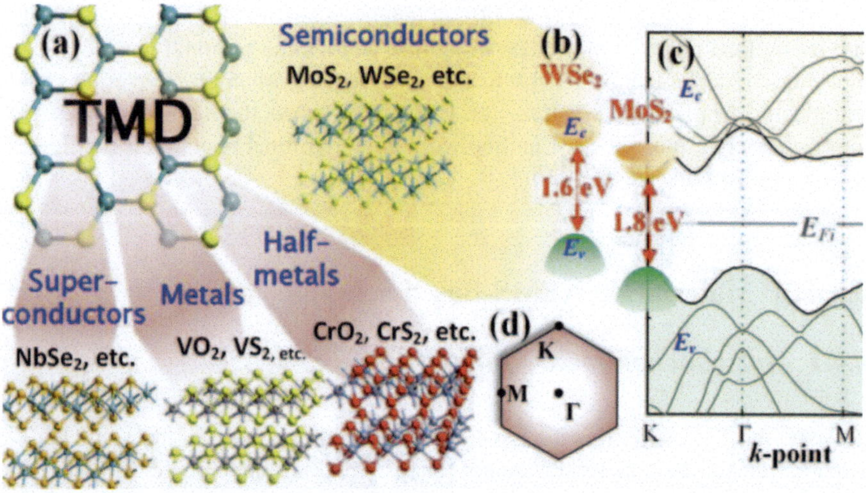

Fig. 6.7 (**a**) Different categories of TMDs, (**b**) energy diagram for WSe2 and MoS_2, (**c**) E–K diagram for single-layer MoS_2. E_C, E_V, and E_{FI} represent the conduction band, valence band, and Fermi level, respectively, and (**d**) first Brillouin zone for TMDs with all the symmetry points in the crystal. [Adapted with permission from Ref [80], © IEEE, 2013]

2D materials offer outstanding physical properties like high-temperature coefficients, extremely high Young's modulus, and other properties including tunable bandgap, high conductivity and mobilities, and high absorption coefficients and unique magnetic properties as well.

For NEMS based on suspended structures, the in-plane stiffness and breaking strength of 2D materials are typically characterized via nano-indentation. These measurements have yielded a very high Young's modulus like 1TPa for graphene, high strain limits (30% for phosphorene), and low mass density (7.4e–19 kg/μm^2 for graphene) which lead to high and tunable resonant frequencies and smaller mass detection limits.

6.4.2 Advantages of 2D Materials

As already mentioned, 2D materials provide features like high resonant frequency, higher strains, high-temperature coefficients, etc. This has led to the development of NEMS devices which can offer options like resonant frequency tuning, high-quality factors for better sensitivity, and a few other advantages to say the least. Since the 2D materials can withstand extremely high strains, it can lead to extremely wide range of frequency tuning. Further, the DC gate bias can modify the tension in the beam by pulling the beam toward the gate. Electrostatic frequency tuning of up to 400% has been observed using similar configurations (Chen et al., 2009,

"Performance of Monolayer Graphene Nanomechanical Resonators with Electrical Readout"). The gate bias-based tuning can be described by a simplified model that includes three contributions: built-in strain, the additional strain upon deformation, and the electrostatic attraction to the gate (for rigorous calculations, see Sazonova et al., 2004). Based on the origins of strain, we can design NEMS corresponding to the particular applications. For example, applications requiring large frequency tuning need devices with low built-in tension, applications requiring higher frequencies and high-quality factor can be designed with materials having large built-in strains, etc. Moreover, the built-in strain in 2D materials can be modified during the fabrication process, using methods like chemical modification, as well as strain mismatch upon heating or cooling to provide application-based tuning.

Moving on, damping is an important parameter while considering sensitive mass detection in the NEMS devices. The damping of the mechanical resonators is quantified by quality factor, Q, which is defined as $Q = 2\pi E/\delta E$, where E is the average energy stored in the resonator over the time period and δE is the energy dissipated to the environment during one cycle. Usually, the resonant frequencies increase with shrinking NEMS size, but the quality factors Q usually suffer, which is not favorable in many applications—Si–NEMS have $Q = 10^3$ at room temperature. Similarly, doubly clamped MoS_2 resonators have displayed $Q = 10^3$ at room temperature (Zhang et al., 2017; Chen et al., 2009, "Performance of Monolayer Graphene Nanomechanical Resonators with Electrical Readout").

Usually, frequency-Q product is often cited as the figure of merit (FOM) when comparing different resonators. For smaller Si-based flexural mode resonators, the FOM is of the order of 10^{11}. For carbon-based resonators, the FOM can exceed 3×10^{13}. Thus, 2D material-based NEMS may be able to exceed Si and other materials in combining high frequency, quality factor, as well as force sensitivity (Kaur et al., 2016; Li & Östling, 2015).

Further, mechanical nonlinearity becomes an increasingly important consideration as NEMS size starts scaling down. Therefore, it is important to characterize and understand nonlinearity in NEMS. At large vibration amplitudes, two additional higher-order terms become significant in the equation of motion for NEMS. The first one is the nonlinear restoring force:

$$F = -\mathrm{k}\,\delta z - \alpha\left(\delta z\right)^3 \tag{6.26}$$

where α is the nonlinear spring constant. The other nonlinear term appears as a combination of displacement δz and velocity $\dot{z} : \eta\left(\delta z\right)^2 \dot{z}$. Since this term is proportional to the velocity, it can be related to "frictional damping," and η is the damping coefficient. However, these two effects can be reduced as illustrated by Eichler et al. (2011). They were able to achieve the highest quality factor to date for a graphene resonator by careful adjustments in driving power, in order to avoid nonlinear damping.

6.4.3 Fabrication/Measurement

2D materials can be fabricated either via mechanical exfoliation from bulk materials or by using large-scale growth techniques like CVD/PVD or epitaxial growth. The details on fabrication of individual materials can be found elsewhere (Eletskii et al., 2011; Zhang et al., 2017; Kaur et al., 2016; Li & Östling, 2015; Gomez et al., 2018, "Nitrogen-Doped Carbon Nanodots for Bioimaging and Delivery of Paclitaxel"; Zhang et al., 2013). 2D material-based NEMS usually have the following device structure – a sheet released from the substrate (Wang et al., 2013; Bunch et al., 2008) and connected via two electrodes on the sides to measure the changes in terms of resonance, strain, conductivity, etc. This common configuration is called trench-based NEMS and can be fabricated in two ways—either trenches are created first and then the 2D material is transferred onto them or the material is transferred/grown first and then these trenches are created. The first approach includes prefabricating a large number of trenches with corresponding patterned electrodes and then transferring the exfoliated material on top. The suitably suspended devices are searched optically thereafter. This approach suffers from low yield, but as this approach does not require processing of the 2D materials, this is considerably faster. Sometimes, a gate electrode is added in the etched trench to reduce the parasitic capacitance and enable electrical measurements later on (Xu et al., 2010).

The other approach to fabricating these structures is to exfoliate/grow the 2D material onto a blank Si/SiO_2 wafer, followed by patterning of the electrodes (typically Cr/Au or Cr/Pd) by lithography techniques, and then releasing the structure by wet etching of oxide in the presence of hydrofluoric acid (HF). Because it requires custom-designing and alignment of electrodes for each flake, this approach is slow. However, it has good yield and allows patterning the devices into specific dimensions using additional lithography steps. Also, this method allows fabrication of very high aspect ratio structures (micrometers in width with only 100 nm spacing from the gate).

For large area applications (like making multiple arrays), these materials can be grown using CVD or epitaxial growth techniques. The specific methods corresponding to various materials can be found elsewhere (Li & Östling, 2015; Kaur et al., 2016; Zhang et al., 2013). CVD process yields high-quality films of 2D materials. The material is either grown directly on the substrate or is grown separately before being transferred onto the substrates for device fabrication.

Once the devices are fabricated, the next obvious step is to characterize them. For this, we need some actuation and detection techniques which can detect the motion of the beams with high resolution. So far, detection has been one of the key challenges in the NEMS technology due to the tremendous scaling down of the dimensions of the devices. Typically, optical and electrical techniques are used for detection of the 2D material motion. We will briefly describe optical and electrical detection schemes here.

6.4.3.1 Optical Detection

In the most common implementation scheme, a high-energy laser (usually 405 nm) is focused onto the material and modulated at the driving frequency. This results in changes in the temperature of the cantilever and causes periodic contraction/expansion of the layer that leads to motion. Simultaneously, another low-energy laser beam (typically 633 nm or depending on the material) is used for detection: the light gets reflected from the suspended beam, and also passes through the suspended beam and then gets reflected from the substrate beneath, thus forming an interference pattern; the intensity of the reflected signal is sensitive to the position of the material. The motion of the cantilever is detected by tracking the intensity modulation of the reflected signal using a photodiode (Robinson et al., 2008; Shivaraman et al., 2009; Lee & Feng, 2014; Lee et al., 2018).

Another way of optically detecting the motion is to use an atomic force microscope (AFM) tip to measure the cantilever deflection. In this method, the drive signal applied to the gate is amplitude modulated (AM), with the modulation frequency f_{mod} set to the resonant frequency of the AFM cantilever. This in effect provides a drive to the AFM cantilever at f_{mod} due to the interaction between the cantilever and the 2D material (not shown here). This technique can be used to image the resonance spatially, to map out the oscillation mode shape. However, resonant quality factors are quite low, since the AFM imaging is typically carried out in air (Zhu et al., 2015; Garcia-Sanchez et al., 2008).

6.4.3.2 Electrical Detection

The 2D material-based sheet is supported by source and drain electrodes, and suspended above a gate electrode. A DC voltage is applied to the gate which causes a static deflection of the sheet toward the gate due to charge induction phenomena and leads to changes in capacitance between the gate electrode and the sheet. Resonant motion is actuated by applying an additional RF signal δV_{g} to the gate, which leads to an RF force $\delta F = C_{\mathrm{g}}^{'} V_{\mathrm{g}} \delta V_{\mathrm{g}}$. This drive scheme can be used in conjunction with either electrical or optical readout.

One of the significant advantages of using semiconducting materials (like MoS_2 and graphene) for NEMS applications is that its charge-dependent conductance G can be used to transduce mechanical motion to a time-varying current. Similarly, the sensitivity of the source–drain current on the deflection of the sheet is usually high enough to employ electrical readout mechanism. The *S*/*D* current as a function of deflection of the sheet can be expressed as

$$\frac{dI_{\mathrm{d}}}{dz} = V_{\mathrm{d}} V_{\mathrm{g}} \frac{dG}{dV_{\mathrm{g}}} \frac{C_{\mathrm{g}}^{'}}{C_{\mathrm{g}}} \tag{6.27}$$

In high-mobility graphene, $\frac{dG}{dV_g}$ can be very large, leading to a current that is many orders of magnitude higher than expected for a device of similar dimensions made from a material lacking similar transconductance.

However, in this configuration, the parasitic capacitance between the gate and the bonding pads could be very large and may lead to a large RF signal noise that makes the mechanical signal (translated into *S*/*D* current) hard to detect. To improve the signal-to-noise ratio, a two-source frequency mixing technique is used to down-mix the RF mechanical signal to a much lower frequency (~kHz). Briefly, an RF voltage at frequency f is applied to the gate for drive, resulting in a time-varying conductance of the graphene at f. A second RF voltage at $f + \Delta f$ is applied to the drain, with Δf typically in the kilohertz range. The drain current I_D then has components at both $2f + \Delta f$ and Δf in the frequency domain. The component at Δf is then tracked by lock-in techniques, avoiding direct pickup of the RF capacitive background. The magnitude of this current is given as

$$I_d^{2-\text{source}} = \delta V_d \frac{dG}{dV_g}\left(\delta V_g + V_g \frac{C_g'}{C_g}\delta z\right) \tag{6.28}$$

There are a few more variations in mixing techniques like single source amplitude modulation (Van Der Zande et al., 2010) and frequency modulation (Gouttenoire et al., 2010). Although these mixing techniques are quite effective for device characterization, they do suffer from a few limitations. While mixing the mechanical signal down from the megahertz to the kilohertz regime, the measurement bandwidth needs to be greatly reduced, which in turn seriously impedes real-time applications such as high-speed detection and mechanical signal processing. To overcome these limitations, techniques like direct RF readout (Xu et al., 2010) have also been developed with much lower RF signal noise and higher speed. In an attempt to increase the detection speed further, Song et al. (2012) have demonstrated measurements with local gates coupled to an LC circuit, but this requires optimization in terms of impedance matching which requires further improvements.

6.4.4 Sensing Applications

2D material based NEMS have found applications in all sorts of sensing areas. The basic principles behind the working of the design includes detecting changes due to temperature, strain, and effective mass changes due to adsorption of gases, biomolecules, chemical, ions, etc. on the device surface. Here we will discuss each of these applications in slightly more detail with examples taken from latest research articles.

6.4.4.1 Physical Sensing

Physical sensing includes detecting changes like strain, temperature, pressure, etc. The principle behind detecting temperature changes lies in the fact that when the temperature of the substrate changes, due to the difference in temperature coefficient (α) between the substrate and the 2D material, the strain in the suspended beam changes, which in turn leads to the shifts in resonant frequency. The temperature can be increased (decreased) using a laser source incident on the beam or by heating up the substrate. Depending on the sensitivity to thermal changes, extremely low-temperature changes can be measured. Similarly, strain can also lead to changes in material properties. As an example, the bandgap in MoS_2 monolayers and multilayers can be modulated by applying strain. A 1.9% compressive strain increases the bandgap from 1.73 eV to 1.86 eV and changes it from direct to indirect. Moreover, a tensile strain of around 11% causes the bandgap energy to be zero, essentially making the material to transition from semiconductor to metallic. Further, the strain can be created by temperature differences, gate-biased conduction changes, and some other means as well. The devices made out of 2D materials are typically ultrasensitive to changes in strain and can be easily employed as various kinds of sensors. Rui et al. (Yang et al., 2015) designed a new method to characterize the thermal expansion coefficient based on MoS_2-based resonators. In their study, they calibrated the temperature coefficient (α) of frequency (*TCF*) for their devices using laser heating and found the *TCF* to be −0.396%/K and found out α to be 3.7–4 ppm/K for MoS_2 which supports other experimental findings.

On the other hand, Fan Ye et al. (2018) made graphene-based NEMS and demonstrated ultrawide frequency tuning up to $\Delta f/f_0 = 1300\%$. This was the first experimental demonstration of electrothermally tuned few-layer graphene NEMS operating at very high frequencies. Further, they found that the device temperature can be tuned between 300 K and 2650 K, which is among the highest operating temperature known to date for electromechanical resonators. Also, when the device temperature increases beyond 1800 K, thermal emission can be detected which can find applications in timing functions as well. A lot of research is also ongoing to find ways to tune the bandgap using mechanical means. In this regard, Harihar et al. (Behera & Mukhopadhyay, 2012) demonstrated that strain can lead to changes in bandgap of 2D h-BN and h-AlN single layers. They demonstrated that by applying strains as high as 10%, bandgap could be tuned by more than 2 eV in AlN and around 1 eV in h-BN (shown in Fig. 6.8a, b) which leads to a significant change in electrical properties of these materials including conductivity, mobility, absorption spectra, etc. This is opening up the pathways for materials in NEMS applications as well as transistors, LEDs, etc. Moreover, Abhilash et al. (Abhilash et al., 2012) fabricated a NEMS device on sapphire substrate with InAs as the 2D material in which they used a local gate configuration to reduce the parasitic capacitance. This led them to achieve wide bandwidth actuation and detection. Using this local gate technique, they observed variation of nonlinearity with temperature and also studied the variations of thermal strains due to heating in the nanostructures which are shown in Fig. 6.8c, d.

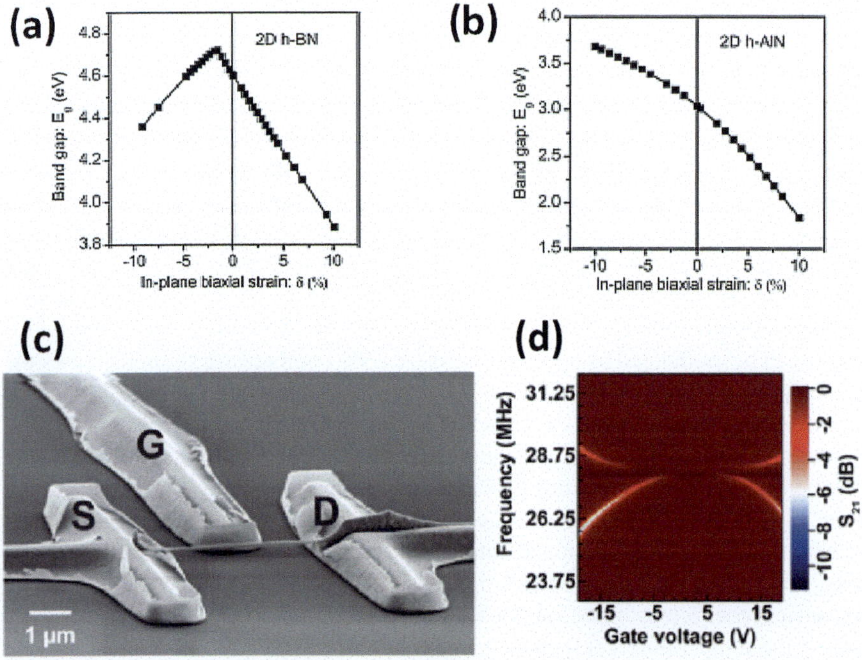

Fig. 6.8 Variations in bandgap for in-plane biaxial strains in (**a**) 2D h-BN and (**b**) 2D h-AlN, (**c**) SEM image of a side-gated InAs resonator, and (**d**) resonance frequency shifts for the device as a fucntion of side-gate bias levels. [(**a**), (**b**) adapted with permission from Ref. [106], ©AIP Publishing, 2012, (**c**), (**d**) adapted with permission from Ref. [107], American Chemical Society, 2012]

6.4.4.2 Gas Sensing

For gas sensing applications, the usual technique lies in detecting the changes in mass due to adsorption of gas particles on the device which leads to a shift in the resonant frequency. The challenge lies in detecting these small changes in mass which may not be detected if the changes are small enough (which is usually the case with bulk material-based NEMS). 2D materials have really low mass density and high-quality factor which leads to large changes in effective mass and high sensitivity. Thus, they can be easily detected using either optical or electrical detection techniques discussed in the beginning of the chapter. In one of these efforts, Chen et al. (2009) ("Performance of Monolayer Graphene Nanomechanical Resonators with Electrical Readout") studied the effects of mass loading/unloading on graphene resonators experimentally, by measuring the resonant frequency as a function of applied gate bias before and after deposition of a known amount of mass. They used electrical readout mechanism to characterize their device. Also, they characterized the negative thermal expansion coefficient of graphene by letting the device cool down and observing upshifts in the resonance. Further, the quality factor also improved with decreasing temperature, to ~10^4 at 5 K. The results show the promise of graphene for mass sensing as the device had a mass sensitivity of 1 zg/$Hz^{0.5}$. Similarly, Manoharan et al. (Muruganathan et al., 2018) designed graphene resonators with a top-gated configuration for mass sensing of hydrogen and argon mixture gases. When they allowed the flow of the gas mixture described above, the changes in resonant frequency and quality factor were noticed which indicated adsorption of gas molecules onto the resonator surface. Using their device, they were able to measure at least ~15 attogram changes for the optimized pressure level of the system. As an example of another mass sensor, Sakhaee-Pour et al. (2008) designed graphene-based device which could detect atomistic dust particles with resolutions up to 10^{-6} fg. They used single-layer graphene sheets as resonators and operated them in the ultrahigh-frequency range, 1.5 GHz–2.9 THz, and found that the frequency shifts nonlinearly with increase in length of the resonators.

6.4.4.3 Chemical and Biomolecule Sensing

Several studies have been done revolving around adsorption principle to detect chemical molecules as well. Adsorption leads to induced surface stress which can be measured by cantilever's flexural deflection. This method is also known as "Stoney's formula" which provides the relationship between the surface stress and flexural deflection (Ibach, 1997). The detection principle based on Stoney's formula has also led to the label-free detection of specific molecules. For label-free detection, the surface of the NEMS has to be chemically modified by receptor molecules which will eventually bind to the specific chemical species. Detection of biomolecules is also based on label-free detection using NEMS resonators. The 2D materials are functionalized using specific chemicals which will allow the binding of the biomolecules onto the surface of the resonators, and depending on the sensitivity of the device, very small concentration of these molecules can be detected. As an example, Hanay et al. (2012) reported the first realization of single-molecule NEMS-based mass spectrometry in real time. In their device, as each molecule in the sample was getting adsorbed onto the resonator surface, its mass and position of adsorption were continuously tracked using two driven vibrational modes in the device. They have demonstrated multimode NEMS-based mass spectrometry by analyzing IgM antibody complexes in real time.

Mass spectrometry using NEMS has been shown to resolve neutral species as well, providing a resolving power that increases with increase in mass, and thus allows the acquisition of spectra, molecule-by-molecule, in real time. For detecting biomolecules, Haixin et al. (Chang et al., 2010) developed a highly sensitive and specific fluorescence resonance energy transfer (FRET) aptasensor for thrombin detection which is based on the dye-labeled aptamer-assembled graphene as shown in Fig. 6.9a, b. Due to the noncovalent assembly between aptamer and graphene,

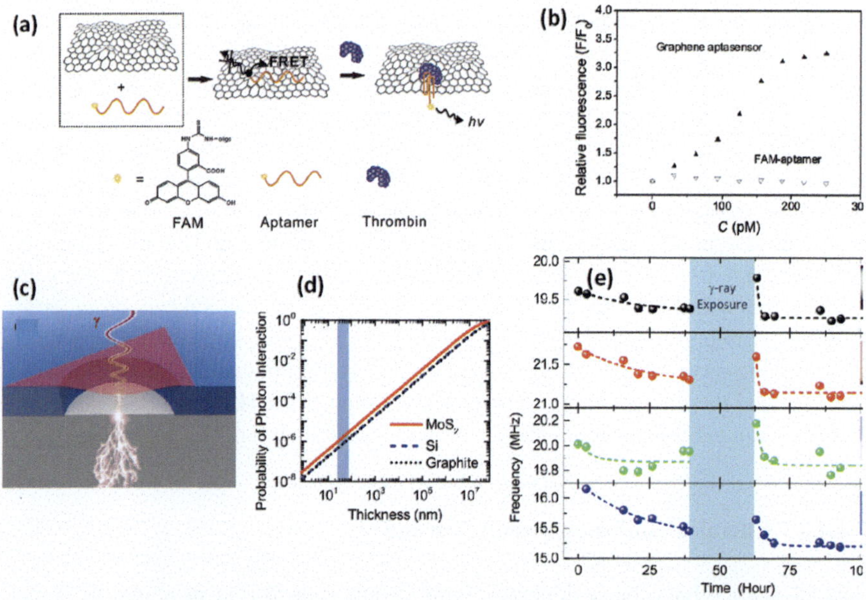

Fig. 6.9 (**a**) Schematic showing fluorescence changes in FAM–aptamer via thrombin concentrations happening via FRET mechanism. Fluoroscence of dye quenches due to binding with graphene, and it recovers due to thrombin binding to the aptamers causing FAM away from the graphene surface, (**b**) fluorescence changes of FAM–aptamer and the graphene aptasensor for different thrombin concentrations, (**c**) schematic showing γ-radiation on a MoS_2 resonator, (**d**) photon interaction probability of γ-radiation on MoS_2, Si, and graphite, and (**e**) resonant frequencies plotted for various devices before and after exposure to γ-rays for some duration (inset shows optical images of the differnet MoS_2 resonators). Scale bar = 3 μm. [(**a**), (**b**) adapted with permission from Ref. [112], ©American Chemical Society, 2010, (**c**)–(**e**) adapted with permission from Ref [113], © AIP Publishing, 2016]

fluorescence gets quenched in the dye due to FRET. However, the addition of thrombin leads to the formation of quadruplex–thrombin complexes which have weak affinity to graphene and thus leads to the fluorescence recovery. This device had a detection limit of concentration as low as 31.3 pM. In a different application, Jaesung et al. (Lee et al., 2016) investigated MoS_2-based drumhead resonators for detecting gamma radiations (Fig. 6.9c–e). They exposed the MoS_2 layer with radiation of ~5000 photons at an energy of 662 keV for over 24 h or 12 and measured the upshifts in the resonance frequency. The upshifts were found to be between 0.5 and 2.1% of the resonant frequency. The MoS_2-based resonators showed photon responsivity of 30–82 Hz/photon, with an intrinsic gamma-ray sensitivity of around 0.02–0.05 photon. These kinds of devices open up a potential usage for early detection of very-low-dosage radiations.

6.5 Piezoelectrics in NEMS-Based Sensing

6.5.1 Introduction to Piezoelectric Materials

Piezoelectricity was discovered in 1880s by Jacques and Pierre Curie. It is defined as the generation of dipole charges in response to applied mechanical stress or vice versa. This effect results from the electromechanical interaction between the mechanical and electrical states in a crystalline material with no inversion symmetry. The physics behind this mechanism can be found elsewhere (Rosen et al., 1992; Mason, 1981; Iwasaki et al., 1976; Martin, 1972). These materials have found their applications as power sources (Jung et al., 2015; Wu et al., 2014), sensors (Tressler et al., 1998; Gupta et al., 2010; Chee et al., 1998), actuators (Khan et al., 2016; Panda & Sahoo, 2015; Hall, 2001), resonators (Pohanka, 2017; Muralt et al., 2009; Lakin, 2003, "A Review of Thin-Film Resonator Technology"), filters (Tadigadapa & Mateti, 2009; Lakin, 2002, "Coupled Resonator Filters"; Lakin, 2003, "A Review of Thin-Film Resonator Technology"), etc. The earliest materials used for these applications were mostly quartz and Rochelle salts. Since these were bulk materials, they had limited applications (Eom & Trolier-McKinstry, 2012, "Thin-Fi Lm Piezoelectric MEMS"; Trolier-Mckinstry & Muralt, 2004). Once modern materials like AlN, ZnO, GaN, etc. were discovered (shown in Fig. 6.10), and since they could be integrated with semiconductor industry, these materials found themselves being used in unprecedented applications (Mohn et al., 2016; Huong Giang et al., 2013;

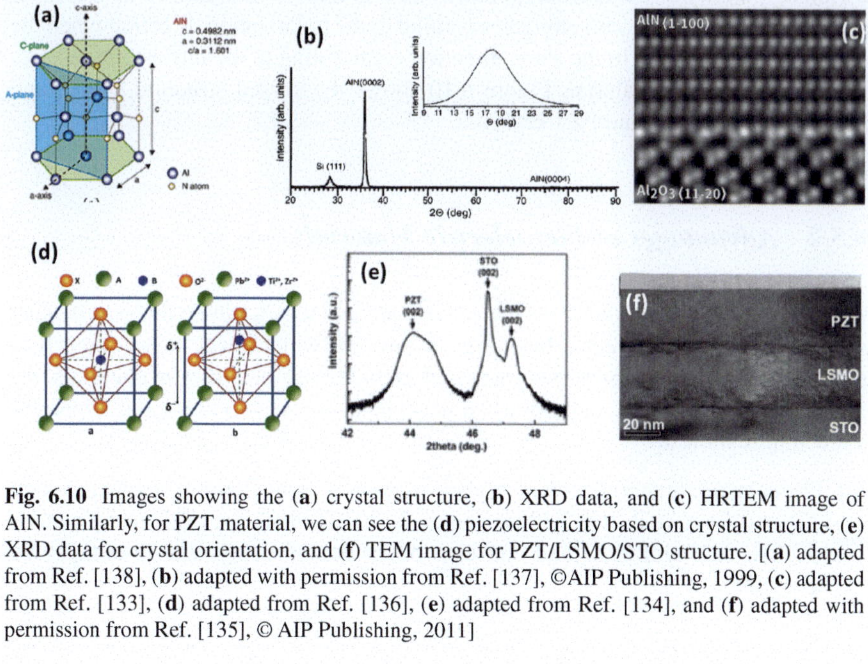

Fig. 6.10 Images showing the (**a**) crystal structure, (**b**) XRD data, and (**c**) HRTEM image of AlN. Similarly, for PZT material, we can see the (**d**) piezoelectricity based on crystal structure, (**e**) XRD data for crystal orientation, and (**f**) TEM image for PZT/LSMO/STO structure. [(**a**) adapted from Ref. [138], (**b**) adapted with permission from Ref. [137], ©AIP Publishing, 1999, (**c**) adapted from Ref. [133], (**d**) adapted from Ref. [136], (**e**) adapted from Ref. [134], and (**f**) adapted with permission from Ref. [135], © AIP Publishing, 2011]

Arredondo et al., 2011; Pramanik et al., 2013; Auner et al., 1999; Taniyasu & Kasu, 2010). Recently, these materials have found applications as NEMS and being used as resonators, sensors, oscillators, and filters in semiconductor industry and have completely dominated the RF market including the 4G/5G market (Wierer et al., 2010; Guldiken et al., 2006; Briqech et al., 2017). Piezoelectrics are also being used to generate and sense ultrasonic waves as required in ultrasonic imaging, and in nondestructive testing owing to their high transduction efficiency. Similarly, as filters, piezoelectric resonators play a dominant role in the high-frequency range due to their high-quality factors, thus outperforming LC resonators (Gupta et al., 2010; Lakin, 2003, "A Review of Thin-Film Resonator Technology"; Hall, 2001). With the advent of piezoelectricity in 2D materials, new opportunities will open up for highly sensitive sensors, ultralow-power energy harvesters, miniaturization of antennas, filters for electronic circuits, NEMS, etc.

All this has led researchers to explore thin-film piezoelectric materials based on two important crystal structures—wurtzites (Guo et al., 2002; Kisi & Elcombe, 1989) and perovskites (Bokov & Ye, 2006). The two most important materials within wurtzite structure are ZnO and AlN. All of the atoms for wurtzites are tetrahedrally coordinated and are arranged in puckered hexagonal rings perpendicular to the crystallographic c-axis. On application of a stress parallel to the c-axis, the tetrahedra deform primarily by changing the N–Al–N (O–Zn–O) bond angle, instead of changing the Al–N (Zn–O) bond length. This leads to a relative displacement of the center of the positive and negative charges within the crystal, which leads to the origin of piezoelectric d_{33} coefficient in AlN (ZnO). On the other hand, in perovskites, there are multiple mechanisms available to contribute to the piezoelectric constants, including polarization extension, polarization rotation, and domain wall motion. As an example, in PZT, the cell elongates parallel to the polarization direction when the electric field is applied in the same direction contributing to the intrinsic piezoelectric response in the material. Figure 6.10 shows the crystal structure, XRD, and TEM images for AlN and PZT crystals.

6.5.2 *Advantages of Piezoelectric Materials*

Given that there are not many piezoelectric materials which can be integrated using semiconductor's planar technology, the first question which needs answering is how these materials are advantageous compared to traditional electrostatic materials like Si, SiC, graphene, MoS_2, etc. So, when we compare these two classes of materials, a few things show up:

(a) Piezoelectric materials offer higher frequency, temperature stable resonant devices (Levy et al., 1998; Tabrizian et al., 2010) which are made using materials like AlN and ZnO and can also be fabricated using CMOS-compatible technology.

(b) Charges are generated whenever the device undergoes mechanical changes which would lead to overall low-power requirements (Adams et al., 2003).
(c) These materials can be used as energy harvesters as well (though small power is generated). Thus, these can provide multimodal applications at the same time.
(d) Piezoelectric devices offer higher energy densities. For actuators, the force per unit area increases linearly with the electric field for the piezoelectric drive and quadratically for the electrostatic materials. In other words, the required voltages to achieve a given displacement in structures of the same stiffness need at least an order of magnitude lower voltages for piezoelectric actuators.
(e) Piezoelectricity is a stress-induced effect, as opposed to capacitive displacement transduction. Thus, its transduction gain is unchanged when scaling down. As an example, materials like AlN, ZnO, etc. show bulk piezoelectric response even when scaled down to nm scales.

Thus, based on the above discussions, we can conclude that it is advantageous to explore NEMS based on piezoelectric materials.

6.5.3 Fabrication/Measurement

Various piezoelectric materials have been explored so far for NEMS applications including ZnO, AlN, PZT, PMN-PT, $LbNiO_3$, etc. In the early 1990s, PZT was explored a lot owing to its high piezoelectricity for actuators (Furukawa et al., 1979; Shrout & Zhang, 2007). However, for higher frequency applications such as filters and sensors, AlN and ZnO have various advantages including high-quality factor and compatibility with the thin-film fabrication process. Moreover, integration and process compatibility with the rest of the on-chip devices has been demonstrated with these wurtzite materials. The strong polarity of their crystalline structure allows for a polar growth and a stable piezoelectric response with time as well. In recent times, AlN has stepped up as the material of choice for practically all RF devices based on piezoelectric thin films (Elfrink et al., 2009; Iriarte et al., 2010). This is mainly because the deposition process of AlN thin films is reproducible, and it exhibits a high thermal conductivity for an electrically insulating material. Further, it has a high-quality factor, and it is CMOS-compatible. Thus, we will briefly explain the fabrication process for AlN (Yarar et al., 2016; Iqbal & Mohd-Yasin, 2018) here, and for other materials, excellent references can be looked upon (Ramadan et al., 2014; Wang et al., 2004; Zhou et al., 2016, "Theoretical Study on Two-Dimensional MoS2 Piezoelectric Nanogenerators"; Kim et al., 2016; Pan et al., 2010; Emanetoglu et al., 1999). In principle, AlN can be grown using any major thin-film technique which keeps the process free of oxygen. However, it is only magnetron sputtering that enables deposition at rather low temperatures as the c-axis crystal growth in AlN requires certain growth conditions which are not thermodynamically favorable to achieve at low temperatures otherwise. Thus, for a (002) crystal growth, typically a seed crystal (usually 111 Pt) is needed. Once we have a highly oriented Pt (111) (confirmed via XRD and FWHM of the rocking curves), one can start with the process of sputter deposition of AlN. The fabrication conditions include pre-deposition Ar^+ plasma etching for 5 min, ~10^{-7} Torr base pressure, less than 3 mT deposition pressure, and a 99.9995% purity aluminum target, with reactive sputtering in nitrogen environment with a high ratio of N_2/Ar. Also, the sputtering needs to be done in two steps—initial growth of 50 nm at ~400 °C followed by the deposition at 300 °C—in order to provide higher energy for the molecules to align in the preferred crystal orientation.

A lot of progress has been made in piezoelectric NEMS structures (Fang et al., 2013; Karabalin et al., 2009; Khan et al., 2016; Eom & Trolier-McKinstry, 2012, "Thin-Film Piezoelectric MEMS") like BAW, SMR, FBARs, etc. However, while using piezoelectric devices for sensor applications, detection of really small concentrations of molecules (either gases or chemicals) is employed. This happens by adsorption of the gas molecules/ions and thus the effective mass is proportional to the device surface. This leads to a trade-off between mass sensitivity and capture area. Due to this, micromechanical resonators including FBARs, SAWs, BAWs, CMUTs, and contour mode resonators are not employed as often since they suffer from lower sensitivities. However, flexural resonators (Verbridge et al., 2006; Witkamp et al., 2006) offer high mass sensitivity, and cantilevers in particular offer a larger dynamic range which becomes crucial for detecting extremely low concentrations of materials.

Flexural resonators are typically multilayer laminates with at least one layer exhibiting piezoelectricity (Fig. 6.11c). These devices can be broadly categorized into one-dimensional bars, or tuning forks, or two-dimensional square or round diaphragms. The exact mathematics governing the mechanics can be found somewhere else (Chang & Kirby, 2017). Here we will just give the equations pertaining to resonance frequency, piezoelectricity, and the lumped-element equivalent circuit as a reference for predicting the resonator behavior. The natural frequency for a 2D plate resonator is given as

$$\omega_{i,j} = \frac{\lambda_{i,j}^2}{l^2}\sqrt{\frac{D}{\mu}} \tag{6.29}$$

where *I*, *j* are the mode indices, $\lambda_{i,j}$ is a constant that depends on the vibration mode, and l is the length or the diameter of the square or the circular resonator. In a thin-film device, the piezoelectric film is free to deform in the vertical direction but is clamped in the plane of the substrate. Therefore, the out-of-plane stress σ_3 is equal to zero, and the in-plane stress can be described as

$$\sigma_1 = -\frac{d_{31}E_3}{s_{11}^E + s_{12}^E} = e_{31} - v_{12}e_{33}E_3 = -e_{31,f}E_3 \tag{6.30}$$

where $s_{i,j}^E$ are the coefficients of the piezoelectric layer's compliance matrix S which is related to the stiffness matrix C via $S = C^{-1}$ and d_{31}, e_{31}, e_{33} are the piezoelectric coefficients.

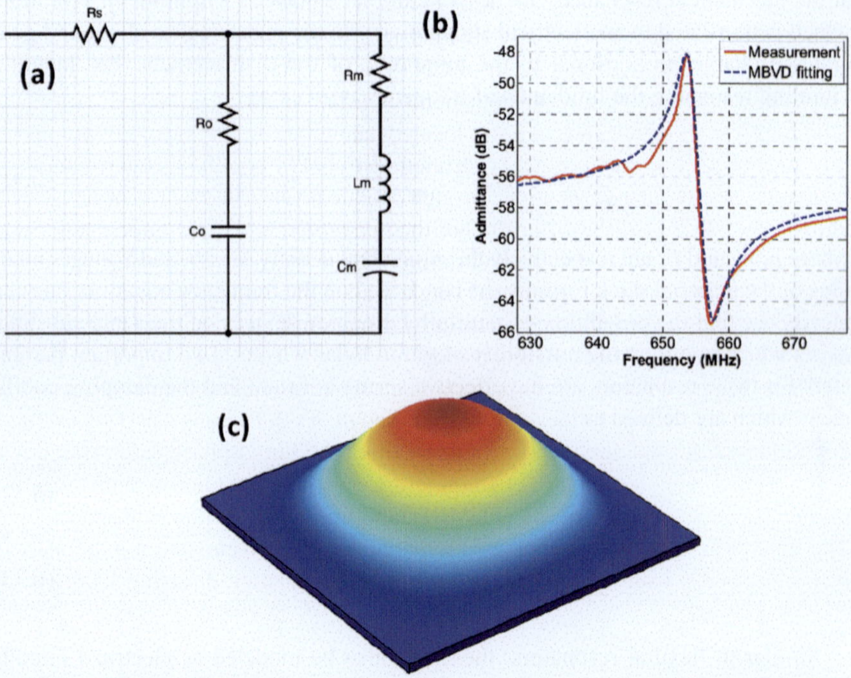

Fig. 6.11 (**a**) MBVD model for the resonator, (**b**) experimental and fitted (BVD model) admittance curves for an AlN resonator, and (**c**) COMSOL simulation for flexural mode square-membrane resonators. [(**b**) adapted with permission from Ref. [166], ©Optical Society of America, 2011, (**c**) adapted from (Abdolvand, Reza, et al. "Micromachined resonators: A review." Micromachines 7.9 (2016): 160)]

For these resonators, usually a modified Butterworth-Van Dyke (MBVD) (Larson et al., 2000; Ghosh & Piazza, 2015) equivalent circuit model (Fig. 6.11a, b) is derived to get various parameters of the resonator circuit. The equivalent circuit parameters are developed in the mechanical domain and represent the mass (L_m), stiffness (k_m), and damping (R_m) of a particular vibration mode. These mechanical parameters can be converted to electrical parameters using the electromechanical coupling parameter to transform mechanical impedances to their respective electrical impedances. The MBVD model is described for a simpler case of FBAR resonators ahead in the book. However, for a complete derivation, one can follow any of these excellent references (Larson et al., 2000; Trolier-Mckinstry & Muralt, 2004; Bhugra & Piazza, 2017).

Another resonator device which can be of interest for sensing applications is the laterally vibrating resonator (Yang, 2006; Gong & Piazza, 2013). This device is formed by sandwiching a thin-film piezoelectric material between metallic electrodes. When a bias is applied across the thickness of the piezoelectric film, a lateral strain is induced in the plane of the device. When the excitation signal is modulated at the mechanical resonance, the device exhibits enhanced amplitude of vibration, which is translated to an electrical signal owing to the piezoelectric effect and generates a charge proportional to the amplitude of the displacement. For laterally vibrating resonator, the fundamental frequency is given as

$$f_0 = \frac{1}{2W}\sqrt{\frac{E_\mathrm{p}}{\rho}} \tag{6.31}$$

where ρ, W, and E_p are respectively the equivalent density, width, and Young's modulus of the material stack forming the bar. Based on the frequency equation, one can clearly see that the possibility of multiple frequency devices on the same substrate exists which opens up the possibility of multimodal sensors. Two important parameters for these resonators are the effective spring constant and the damping coefficient which are defined as

$$k_\mathrm{eff} = \frac{\pi^2}{2}\frac{E_\mathrm{p}TL}{W} \tag{6.32}$$

$$b_\mathrm{eff} = \frac{m_\mathrm{eff}\omega_0}{Q} \tag{6.33}$$

Similar to flexural resonators, these can also be modeled as electrical circuits with a translation between mechanical and electrical parameters which depends on the transduction factor (η). One substantial advantage of laterally vibrating piezoelectric resonators is that the sensitivity to external perturbations can be set independently of its vibration frequency. As such, these resonators have been employed for high-resolution and low-power chemical, physical, and biological sensors.

6.5.3.1 Detection Techniques

The piezoelectric resonator can be modeled as a lumped electric circuit model with parameters being correlated to their mechanical counterparts. The force can be translated as applied voltage, displacement of the resonator can be modeled as the charge stored on the capacitor, and mass of the system can be associated with the inductance and damping to resistance. For piezoelectric resonators, modified Butterworth-Van Dyke model (Lakin, 1992, "Modeling of Thin Film Resonators and Filters"; Larson et al., 2000) is widely used for characterization. The schematic diagram of the model along with the admittance plot for a resonator is shown in Fig. 6.11a, b. The BVD model contains a static capacitance (C_0) which is parallelly connected to a series connection of L_M, C_M, and R_M representing the motion of the resonator. The equivalent circuit parameters relating mechanical parameters to electrical parameters for thin-film bulk acoustic resonators (FBARs) are shown here. These relationships vary for different resonators and can be found elsewhere (Chang & Kirby, 2017). However, equations and the methods needed to extract the electrical parameters remains the same and can be applied for any piezoelectric NEMS resonator and are summarized below:

$$C_M = \frac{8C_0 k_{\text{eff}}^2}{N^2 \pi^2} \tag{6.34}$$

$$L_M = \frac{v}{64 \cdot f_s^3 \cdot \varepsilon \cdot A \cdot k_{\text{eff}}^2} \tag{6.35}$$

$$R_M = \frac{\eta \cdot \varepsilon}{16 \cdot f_s \cdot \rho \cdot A \cdot v \cdot k_{\text{eff}}^2} \tag{6.36}$$

$$C_0 = \frac{\varepsilon \cdot A}{t} \tag{6.37}$$

where k_{eff}^2 is the electromechanical coupling coefficient, A and t are the electrodes area and dielectric thickness, ε is the permittivity, η is the acoustic viscosity of the wave vector k, ν is the wave velocity, and N is the acoustic wave number ($N = 1,3,5\ldots$). In addition to these terms, MBVD also has additional parameters such as R_s and R_0 for better fitting and accounts for acoustic losses. The resonant frequencies are given as

$$w_s = \frac{1}{\sqrt{L_M C_M}} \tag{6.38}$$

$$w_p = \sqrt{\frac{C_m + C_0}{L_M C_M C_0}} = w_p \sqrt{1 + \frac{C_M}{C_0}} \tag{6.39}$$

and quality factor

$$Q_s = \frac{w_s \cdot L_M}{R_M} \tag{6.40}$$

$$Q_p = Q_s \frac{w_p / w_s}{1 + R_0 / R_M} \tag{6.41}$$

At resonance, the real part of the impedance $|Z| = R_1$. The complex impedance (reactance) of the resonator, $Z \sim 1/wC_0$. Therefore, the peak of the resonance in the admittance plot is approximately the inverse of the motional resistance, ($1/R_1$), of the resonator, and the static capacitance C_o is derived from the value of the admittance off resonance (wC_o).

Now that we have established that the resonator can be modeled as a lumped electrical circuit (BVD model), it can be characterized by finding out the equivalent admittance plot using a vector network analyzer. The scattering parameter, S_{11}, of this one-port electrical network is first measured and then converted into the admittance, Y, according to

$$Y = \frac{1}{50}\frac{(1+S_{11})}{(1-S_{11})} \tag{6.42}$$

All the values of the equivalent circuit elements including the quality factor and the electromechanical coupling coefficient can then be extracted from the measurement by MBVD model fitting. This makes the characterization of piezoelectric NEMS simple when compared to the techniques used for 2D material-based NEMS.

6.5.4 *Sensing Applications*

Piezoelectric material-based NEMS have found applications in all sorts of sensing areas. The basic principle behind the working of the design includes detecting mass and strain changes due to temperature, energy transfer, and adsorption of gases, biomolecules, chemical, ions, etc. Here we will discuss each of these applications in slightly more detail with examples taken from ongoing research.

6.5.4.1 Physical Sensing

As already discussed in Sect. 6.4.4.1, the principle behind physical sensing (temperature, light detection) lies in the fact that when the physical parameters are changed, it induces variations in the piezoelectric layer (in terms of mass or strain change) which results in shifts in the resonance frequency. The heat transfer can be assisted using additional layers, temperature can induce additional strains in the piezoelectric, and based on the sensitivity to these changes, extremely small changes can be measured. Further, the strain can be created by temperature coefficient differences, gate-biased conduction changes, etc. The resonators made using piezoelectric materials are highly sensitive to these physical changes and can be employed as various kinds of sensors. Yu et al. (Hui & Rinaldi, 2013a) designed a miniaturized device in which a heat absorbing element and a temperature-sensitive microelectromechanical system (MEMS) resonator were overlapped but separated by a micrometer distance air gap (Fig. 6.12a–c). This design enabled a fast and an efficient heat transfer between the two layers and led to high-resolution thermal power detection owing to the low noise and high-quality factor of the resonator. Noise-equivalent power of 6.5 nW $Hz^{-0.5}$ was achieved for this device. The thermal time constant of 350 μs was measured for the device (which is the smallest ever reported for MEMS resonant thermal detectors), indicating the potential of this device for the implementation of ultrafast and high-resolution resonant thermal detectors. On the other hand, Priyanka et al. (Joshi et al., 2019) designed a microcantilever structure with layer stack consisting of Si/SiO_2/Pt/ZnO/Pt materials for mass sensing applications. ZnO was used as the piezoelectric material and the mass sensing was done using laser Doppler vibrometer. Mass sensitivities as low as 0.3138 Hz/pg were obtained for

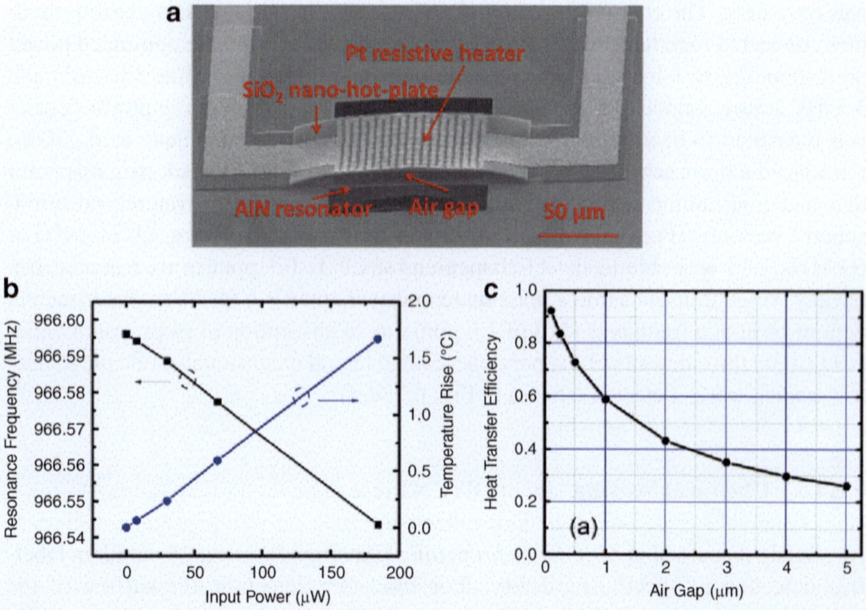

Fig. 6.12 (**a**) SEM image of the resonant thermal detector, (**b**) frequency shifts and temperature changes for different levels of input power for the thermal detector, and (**c**) heat transfer efficiency of the thermal detector for different air gap thicknesses. [Adapted with permission from Ref. [170], ©AIP Publishing, 2013]

this device, and a quality factor (Q) of ~234 was measured at room temperature. For infrared detection, Hui et al. (Hui & Rinaldi, 2013b) integrated an AlN nano-plate resonator with a thin-film Si_3N_4-based absorber. The device had high electromechanical performance parameters (quality factor $Q = 1062$ and electromechanical coupling coefficient $k_t^2 = 1.62\%$). Further, they demonstrated the detector's application for ultrasensitive thermal imaging by characterizing its performance which indicated a noise-equivalent power of ~2.4 nW $Hz^{-0.5}$.

6.5.4.2 Gas Sensing

For gas sensing, the working principle is detecting changes in mass due to gas particles which leads to a shift in the resonant frequency. The challenge lies in detecting these small changes in mass which may not be detected if the changes are small enough (which is usually the case with bulk material-based NEMS). Piezoelectric materials have higher mass sensitivities and a high-quality factor which leads to changes in resonant frequencies which are easily detectable. In an attempt to design an airflow sensor, Huicong et al. (Liu et al., 2012) developed a microcantilever-based sensor which also worked as a wind-driven energy harvester for a self-sustained flow-sensing microsystem. A flow-sensing sensitivity of 0.9 mV/(m/s) was measured. The energy harvester was essentially an array of flow-sensing modules connected together (Fig. 6.13a–c). The output voltage and the optimized power corresponding to a load resistance of 100 k ohm were measured as 18.1 mV and 3.3 nW at flow velocity of 15.6 m/s, respectively. The corresponding power density was measured to be around 0.36 mW/cm^3. Similarly, Wei et al. (Pang et al., 2006) introduced a mass sensor based on lateral extensional-mode piezoelectric resonator that had a minimum detectable mass of 10^{-15} g at room temperature and atmospheric pressure (Fig. 6.13d). The resonator had a quality factor, Q, of 1400 at 60 MHz, and was able to detect changes as small as 0.1 ppm in its resonant frequency which translates into a mass uncertainty of about 4.6 fg. As an experimental demonstration, a frequency shift of 1.6 ppm due to absorption of isopropanol vapor (~73 fg) on the sidewalls of the perylene-coated lateral extensional-mode piezoelectric resonator was detected (shown in Fig. 6.13e, f).

6.5.4.3 Chemical Sensing

As already discussed in Sect. 6.4.4.3, chemical sensing is essentially based on label-free detection of specific molecules. For label-free detection, the surface of the NEMS has to be chemically modified by receptor molecules which will eventually bind to the specific chemical species. The piezoelectric materials are functionalized using specific chemicals which will allow the binding of the chemicals onto the

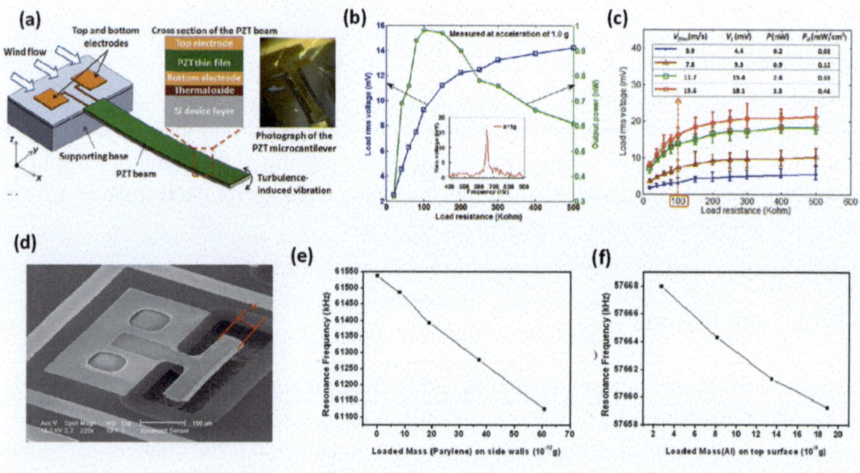

Fig. 6.13 (**a**) Schematic of a PZT cantilever used for flow sensing and energy harvesting, (**b**) output voltage and power plots as a function of load resistance for an excitation frequency of 650 Hz (shown in inset) and at acceleration of 1.0 g, (**c**) load voltages measured against load resistance for different flow velocities, (**d**) SEM image of the lateral extensional-mode resonator. Resonant frequency shifts in the device for different loaded mass of the perylene on (**e**) the side walls and (**f**) on the top surface of the device. [(**a**)–(**c**) adapted with permission from Ref. [173], ©AIP Publishing, 2012, (**d**)–(**f**) adapted with permission from Ref. [174], ©AIP Publishing, 2006]

surface of the resonators, and depending on the sensitivity of the device, very small concentration of these molecules can be detected. In some of the developments, Rinaldi et al. (2009) designed a contour-mode resonant sensor (CMR-S) using AlN as the piezoelectric material. They functionalized the top gold electrode with a single-stranded DNA sequence which increased the adsorption affinity for 2,4-dinitrotoluene (DNT) to 1.5 parts per billion. They characterized their devices and found that it had an improved mass sensitivity of 41.6 KHz um^2/fg, a quality factor of 1400 (in air), and a high electromechanical coupling coefficient of 1.6%. As a mass-sensing application, Amir et al. (Heidari et al., 2013) designed an aluminum nitride (AlN)-based resonators with checker-patterned electrode architecture (Fig. 6.14a, b). The experiments indicated that the resonator had a mass sensitivity of 175.42 m^2/ng and a detection limit of 18.7 ng/cm^2 which translated into a small sample volume needed for sensing applications. Furthermore, this AlN MEMS resonator had a very-low-power consumption which is useful for applications in vivo and network sensors. For a similar application, Hao et al. (Zhang et al., 2005) designed a thin-film bulk acoustic resonator (FBAR) mass sensor for detecting Hg^{2+} in liquid. A thin Au film was deposited on the surface of the resonator for enhanced adsorption of Hg^{2+} amalgamates on the surface. The FBAR sensor was demonstrated to be capable of detecting concentration as low as 10^{-9} M Hg^{2+} (i.e., 0.2 ppb Hg^{2+}). Further, when they used thiol compound-modified Au film instead of a pure Au film, significant improvements in the sensitivity of the FBAR were observed. For selectivity, other cations such as K^+, Ca^{2+}, Mg^{2+}, Zn^{2+}, and Ni^{2+} were also tested, and no effect on the resonant frequency of the FBAR was detected for these cations.

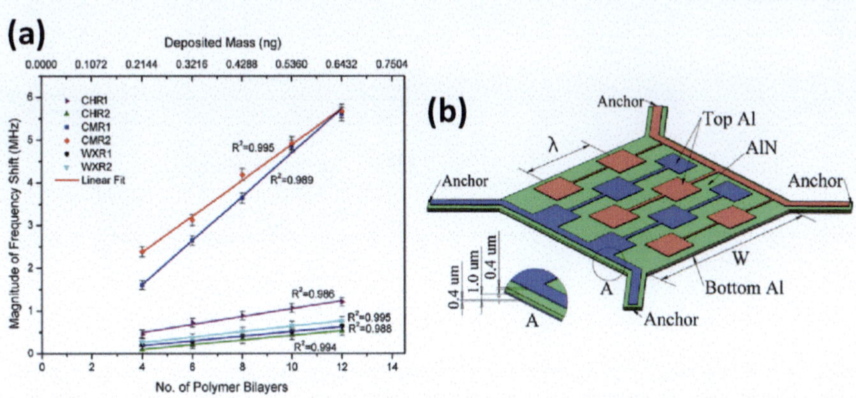

Fig. 6.14 (**a**) Frequency shift for different number of deposited polymer bilayers and (**b**) schematic view of the checker-mode resonator. [Adapted with permission from Ref. [175], ©Elsevier B.B., 2013]

6.5.4.4 Biomolecule Sensing

The detection principle of biomolecules is similar to chemical sensing. Here the surface of the resonator device is chemically functionalized to attach itself to the targeted biomolecules only. This creates a change in effective mass or creates a strain in the device which can be detected as shift in resonant frequency. Some of the papers demonstrating biomolecule sensing are discussed here. Akram et al. (Sadek et al., 2010) designed a piezoelectric-based resonator beam using GaAs. They used multiplexing technique using laser interferometry to monitor variations in transverse oscillation amplitude from the two closely spaced NEMS devices. The devices were operated in vacuum to maintain high-quality factor and resonance frequency stability. Using their device, the direct transduction of neuronal action potentials from extracellular electrodes was demonstrated as shown in Fig. 6.15a. On the other hand, Mario et al. (DeMiguel-Ramos et al., 2017) designed an Ir/AlN/It piezoelectric material-based shear-mode SMR as another biosensor. They tested their resonators on thrombin-binding aptamers. This sensor was based on the principle of detecting differential mass which occurred when the different aptamers (having different molecular weights) attached themselves to the resonator. Surface functionalization was done by binding silane groups on plasma-oxidized SiO_2 surfaces (results shown in Fig. 6.15b). The resonator was characterized to have a mass sensitivity of 1800 KHz/pg·cm^2. As another biosensing device, Martin et al. (Nirschl

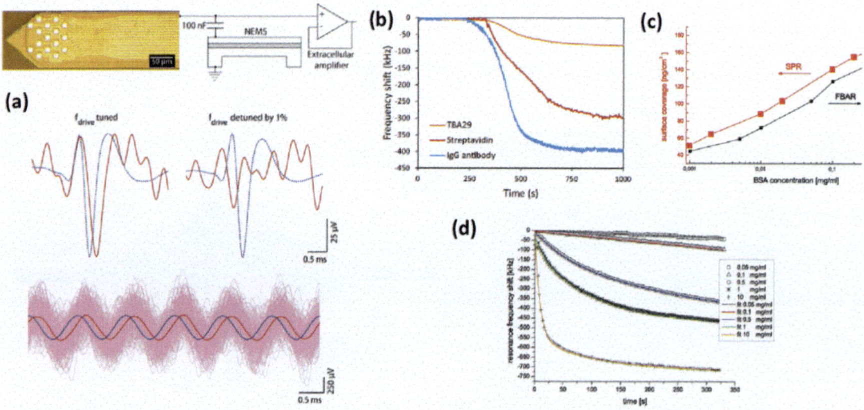

Fig. 6.15 (**a**) Detection of extracellular action potentials using the shown setup. Also, demodulated signals from the NEMS device are shown in red, and the corresponding signal recorded by the amplifier is indicated by blue line, (**b**) frequency shifts for different species on the functionalized sensors, (**c**) surface coverage and frequency shifts plotted for different BSA concentrations for the FBAR sensor device, and (**d**) frequency shifts for adsorption and recrystallization of S-layer monomers of *Bacillus sphaericus* NCTC 9602 into 2D protein layers at the gold surface for various concentrations. [(**a**) adapted with permission from Ref. [178], ©American Chemical Society, 2010, (**b**) adapted with permission from Ref. [179], © Elsevier B.V., 2016, and (**c**), (**d**) adapted with permission from Ref. [180], ©Elsevier B.V., 2009]

et al., 2009) reported the design of an FBAR for the label-free, multiplexed biosensing of DNA and proteins. The resonators were operated in shear mode at around 800 MHz frequency. A mass sensitivity of ~2 KHz cm^2/ng and a minimum detectable mass of~1 ng/cm^2 were achieved for the device. The results for surface coverage and frequency shifts are plotted for different protein concentrations in Fig. 6.15c, d. Lastly, Peng et al. (Gomez et al., "Nitrogen-Doped Carbon Nanodots for Bioimaging and Delivery of Paclitaxel") designed an AlN-based FBAR for detection of epithelial tumor marker mucin-1 (MUC1) based on aptamers–AuNP conjugates. The detection was enabled by the specific bonding between biotin and streptavidin. Streptavidin was assembled on the sensitive area of FBAR. After the recognition between aptamers–AuNP conjugates and MUC1, biotin, along with the conjugates, was captured by streptavidin onto the surface of FBARs. The biosensor exhibited a linear relationship between the frequency shifts and the concentrations of MUC1 ranging from 30 to 500 nM, with a sensitivity of about 818.6 Hz nM^{-1} (Fig. 6.15e). The selectivity experiment demonstrated that they can precisely detect MUC1 with good specificity as well.

6.6 Concluding Remarks

In this chapter, we considered various materials (CNT, SNW, 2D and Piezoelectric) for NEMS based sensing applications. The primary focus of this chapter was on explaining the principles governing the actuation and detection techniques employed for measuring the shifts in fundamental frequency in response to changes in physical properties of the system. Each of these materials have their own unique properties which makes them exciting for various applications. Carbon Nano Tubes are highly conductive, elastic as well as have a very low temperature coefficient which makes them ideal for applications where stress is used as the detection principle. Further, current modulation techniques can be used for tracking these changes. In addition, very stable and precise resonators can be made using CNTs owing to their low temperature coefficients. Similarly, fabrication of Silicon nanowires can be controlled precisely which can result in very high aspect ratio. SNWs also show high electrical conductivity and piezoresistivity properties which can be utilized for making highly sensitive NEMS-based sensors. Recently, 2D materials have emerged as the ideal candidate for NEMS-based applications owing to their unique features including high conductivity, high flexibility, high aspect ratio, excellent thermal conductivity, strain-induced bandgap tunability, etc. This has led researchers to fabricate sensors which can detect single atoms (~1 zg of mass). These devices have also found applications in high-temperature sensing, and high frequency measurements as well. Finally, piezoelectric materials come with their own unique property of charge induction due to applied stress which can be exploited for designing ultralow-power NEMS sensors. These materials have a highly stable temperature coefficient which is suitable for thermal (including infrared sensing), temperature-based sensing. Also, these materials can be operated in different modes which can result in device operation at extremely high-frequency ranges. With the advent of materials like AlN and ZnO which are compatible with CMOS fabrication techniques, these materials seem more promising.

Based on the performance of all these materials, 2D materials and SiNWs are found to be the most sensitive. But they require complicated circuitry for sensing applications as they cannot transduce mechanical motions into electrical properties readily. Piezoelectric materials have this inherent advantage and thus they need much simpler electronics and low power for sensing applications. Further, new 2D materials which demonstrate piezoelectricity (MoS_2, WSe_2) could be the ideal candidates for ultrasensitive, highly miniaturized, and extremely low-power NEMS-based applications.

References

Abhilash, T. S., et al. (2012). Wide bandwidth nanowire electromechanics on insulating substrates at room temperature. *Nano Letters, 12*(12), 6432–6435. https://doi.org/10.1021/nl303804e

Adams, J. D., et al. (2003). Nanowatt chemical vapor detection with a self-sensing, piezoelectric microcantilever array. *Applied Physics Letters, 83*(16), 3428–3430. https://doi.org/10.1063/1.1621733

Akinwande, D., et al. (2017). A review on mechanics and mechanical properties of 2D materials—Graphene and beyond. *Extreme Mechanics Letters, 13*, 42–77. https://doi.org/10.1016/j.eml.2017.01.008

Allain, P. E., et al. (2012). Large range MEMS motion detection using integrated piezo-resistive silicon nanowire. In *2012 IEEE 25th International Conference on micro electro mechanical systems (MEMS)* (pp. 1320–1323). IEEE. http://ieeexplore.ieee.org/document/6170401/

Aqel, A., et al. (2012). Carbon nanotubes, science and technology part (I) structure, synthesis and characterisation. *Arabian Journal of Chemistry, 5*(1), 1–23. https://doi.org/10.1016/j.arabjc.2010.08.022

Arepalli, S. (2004). Laser ablation process for single-walled carbon nanotube production. *Journal of Nanoscience and Nanotechnology, 4*(4), 317–325. https://doi.org/10.1166/jnn.2004.072

Arjmandi-Tash, H., et al. (2017). Large scale integration of CVD-graphene based NEMS with narrow distribution of resonance parameters. *2D Materials, 4*(2), 025023. https://doi.org/10.1088/2053-1583/aa57c6

Arora, N., & Sharma, N. N. (2014). Arc discharge synthesis of carbon nanotubes: Comprehensive review. *Diamond and Related Materials, 50*, 135–150. https://doi.org/10.1016/j.diamond.2014.10.001

Arredondo, M., et al. (2011). Chemistry of Ruddlesden-Popper planar faults at a ferroelectric-ferromagnet perovskite interface. *Journal of Applied Physics, 109*(8), 084101. https://doi.org/10.1063/1.3569858

Auner, G. W., et al. (1999). Microstructure of low temperature grown AlN thin films on Si(111). *Journal of Applied Physics, 85*(11), 7879–7883. https://doi.org/10.1063/1.370600

Awadallah, A. E., et al. (2012). Synthesis of carbon nanotubes by CCVD of natural gas using hydrotreating catalysts. *Egyptian Journal of Petroleum, 21*(2), 101–107. https://doi.org/10.1016/j.ejpe.2012.11.005

Awasthi, K., et al. (2005). Synthesis of carbon nanotubes. *Journal of Nanoscience and Nanotechnology, 5*(10), 1616–1636. https://doi.org/10.1166/jnn.2005.407

Bagheri, M., et al. (2011). Dynamic manipulation of nanomechanical resonators in the high-amplitude regime and non-volatile mechanical memory operation. *Nature Nanotechnology, 6*(11), 726–732. https://doi.org/10.1038/nnano.2011.180

Bartsch, S. T., et al. (2014). Junctionless silicon nanowire resonator. *IEEE Journal of the Electron Devices Society, 2*(2), 8–15. https://doi.org/10.1109/JEDS.2013.2295246

Barwicz, T., et al. (2010). Silicon nanowire piezoresistance: Impact of surface crystallographic orientation. *Applied Physics Letters, 97*(2), 023110. https://doi.org/10.1063/1.3463456

Behera, H., & Mukhopadhyay, G. (2012). Strain tunable band-gaps of two-dimensional hexagonal BN and AlN: An FP-(L)APW+lo study. *AIP Conference Proceedings, 1447*(1), 273–274. https://doi.org/10.1063/1.4709985

Bhugra, H., & Piazza, G. (2017). *Piezoelectric MEMS resonators* (pp. 3–37). https://doi.org/10.1007/978-3-319-28688-4.pdf

Bokov, A. A., & Ye, Z. G. (2006). Recent progress in Relaxor ferroelectrics with perovskite structure. *Journal of Materials Science, 41*(1), 31–52. https://doi.org/10.1007/s10853-005-5915-7

Briqech, Z., et al. (2017). Low-cost wideband Mm-wave phased array using the piezoelectric transducer for 5G applications. *IEEE Transactions on Antennas and Propagation, 65*(12), 6403–6412. https://doi.org/10.1109/TAP.2017.2730239

Bunch, J. S., et al. (2008). Impermeable atomic membranes from graphene sheets. *Nano Letters, 8*(8), 2458–2462. https://doi.org/10.1021/nl801457b

Cao, W., Kang, J., Liu, W., et al. (2013). 2D Electronics: Graphene and beyond. In *2013 Proceedings of the European Solid-State Device Research Conference (ESSDERC)* (pp. 37–44). IEEE. https://ieeexplore.ieee.org/abstract/document/6818814/

Cao, W., Jiang, J., et al. (2015a). Designing band-to-band tunneling field-effect transistors with 2D semiconductors for next-generation low-power VLSI. In *Technical Digest—International Electron Devices Meeting, IEDM* (Vol. 2016-Febru, pp. 12.3.1–12.3.4). IEEE. https://doi.org/10.1109/IEDM.2015.7409682

Cao, W., Kang, J., Sarkar, D., et al. (2015b). Performance evaluation and design considerations of 2D semiconductor based FETs for Sub-10 Nm VLSI. In *Technical Digest - International Electron Devices Meeting, IEDM* (Vol. 2015-Febru, pp. 30.5.1–30.5.4). IEEE. https://doi.org/10.1109/IEDM.2014.7047143

Cassell, A. M., et al. (1999). Directed Growth of Free-Standing single-Walled Carbon Nanotubes [23]. *Journal of the American Chemical Society, 121*(34), 7975–7976. https://doi.org/10.1021/ja992083t

Chang, D. T., & Kirby, D. J. (2017). Shear piezoelectric MEMS resonators. In *Piezoelectric MEMS resonators* (pp. 221–242). https://doi.org/10.1007/978-3-319-28688-4_9

Chang, H., et al. (2010). Graphene fluorescence resonance energy transfer Aptasensor for the thrombin detection. *Analytical Chemistry, 82*(6), 2341–2346. https://doi.org/10.1021/ac9025384

Chaste, J., et al. (2012). A nanomechanical mass sensor with Yoctogram resolution. *Nature Nanotechnology, 7*(5), 301–304. https://doi.org/10.1038/nnano.2012.42

Chee, C. Y. K., et al. (1998). A review on the modelling of piezoelectric sensors and actuators incorporated in intelligent structures. *Journal of Intelligent Material Systems and Structures, 9*(1), 3–19. https://doi.org/10.1177/1045389X9800900101

Chen, C., et al. (2009). Performance of monolayer graphene nanomechanical resonators with electrical readout. *Nature Nanotechnology, 4*(12), 861–867. https://doi.org/10.1038/nnano.2009.267

Chiu, H. Y., et al. (2008). Atomic-scale mass sensing using carbon nanotube resonators. *Nano Letters, 8*(12), 4342–4346. https://doi.org/10.1021/nl802181c

Chowdhury, R., & Adhikari, S. (2011). Boron-nitride nanotubes as Zeptogram-scale bionanosensors: Theoretical investigations. *IEEE Transactions on Nanotechnology, 10*(4), 659–667. https://doi.org/10.1109/TNANO.2010.2060492

Chung, Y. H., & Jou, S. (2005). Carbon nanotubes from catalytic pyrolysis of polypropylene. *Materials Chemistry and Physics, 92*(1), 256–259. https://doi.org/10.1016/j.matchemphys.2005.01.023

Chung, J., et al. (2004). Toward large-scale integration of carbon nanotubes. *Langmuir, 20*(8), 3011–3017. https://doi.org/10.1021/la035726y

Cleland, A. N., & Roukes, M. L. (2002). Noise processes in nanomechanical resonators. *Journal of Applied Physics, 92*(5), 2758–2769. https://doi.org/10.1063/1.1499745

de Heer, W. A. (2005). Liquid carbon, carbon-glass beads, and the crystallization of carbon nanotubes. *Science, 307*(5711), 907–910. https://doi.org/10.1126/science.1107035

DeMiguel-Ramos, M., et al. (2017). Gravimetric biosensor based on a 1.3 GHz AlN shear-mode solidly mounted resonator. *Sensors and Actuators, B: Chemical, 239*, 1282–1288. https://doi.org/10.1016/j.snb.2016.09.079

Eichler, A., et al. (2011). Nonlinear damping in mechanical resonators made from carbon nanotubes and graphene. *Nature Nanotechnology, 6*(6), 339–342. https://doi.org/10.1038/nnano.2011.71

Ekinci, K. L. (2005). Electromechanical transducers at the nanoscale: Actuation and sensing of motion in nanoelectromechanical systems (NEMS). *Small, 1*(8–9), 786–797. https://doi.org/10.1002/smll.200500077

Ekinci, K. L., & Roukes, M. L. (2005). Nanoelectromechanical systems. *Review of Scientific Instruments, 76*(6), 061101. https://doi.org/10.1063/1.1927327

Eletskii, A. V., et al. (2011). Graphene: Fabrication methods and thermophysical properties. *Uspekhi Fizicheskih Nauk, 181*(3), 233. https://doi.org/10.3367/ufnr.0181.201103a.0233

Elfrink, R., et al. (2009). Vibration energy harvesting with aluminum nitride-based piezoelectric devices. *Journal of Micromechanics and Microengineering, 19*(9), 094005. https://doi.org/10.1088/0960-1317/19/9/094005

Emanetoglu, N. W., et al. (1999). Epitaxial ZnO piezoelectric thin films for saw filters. *Materials Science in Semiconductor Processing, 2*(3), 247–252. https://doi.org/10.1016/S1369-8001(99)00022-0

Eom, C. B., & Trolier-McKinstry, S. (2012). Thin-Fi Lm piezoelectric MEMS. *MRS Bulletin, 37*(11), 1007–1017. https://doi.org/10.1557/mrs.2012.273

Fang, X. Q., et al. (2013). Fundamental formulations and recent achievements in piezoelectric nano-structures: a review. *Nanoscale, 5*(5), 1716–1726. https://doi.org/10.1039/c2nr33531j

Feng, X. L., et al. (2007). Very high frequency silicon nanowire electromechanical resonators. *Nano Letters, 7*(7), 1953–1959. https://doi.org/10.1021/nl0706695

Fernández-Regúlez, M., et al. (2010). Lithography Guided Horizontal Growth of Silicon Nanowires for the Fabrication of Ultrasensitive Piezoresistive Strain Gauges. *Microelectronic Engineering, 87*(5–8), 1270–1273. https://doi.org/10.1016/j.mee.2009.10.050

Fischer, L. M., et al. (2008). Specific detection of proteins using nanomechanical resonators. *Sensors and Actuators, B: Chemical, 134*(2), 613–617. https://doi.org/10.1016/j.snb.2008.06.003

Franklin, N. R., & Dai, H. (2000). Enhanced CVD approach to extensive nanotube networks with directionality. *Advanced Materials, 12*(12), 890–894. https://doi.org/10.1002/1521-4095(200006)12:12<890::AID-ADMA890>3.0.CO;2-K

Fukuda, T., et al. (2003). Assembly of nanodevices with carbon nanotubes through nanorobotic manipulations. *Proceedings of the IEEE, 91*(11), 1803–1818. https://doi.org/10.1109/JPROC.2003.818334

Furukawa, T., et al. (1979). Piezoelectric properties in the composite systems of polymers and PZT ceramics. *Journal of Applied Physics, 50*(7), 4904–4912. https://doi.org/10.1063/1.325592

Garcia-Sanchez, D., et al. (2008). Imaging mechanical vibrations in suspended graphene sheets. *Nano Letters, 8*(5), 1399–1403. https://doi.org/10.1021/nl080201h

Geim, A. K., & Novoselov, K. S. (2009). The rise of graphene. *Nanoscience and Technology: A Collection of Reviews from Nature Journals*, 11–19.

Ghosh, S., & Piazza, G. (2015). Piezoelectric actuation of aluminum nitride contour mode optomechanical resonators. *Optics Express, 23*(12), 15477. https://doi.org/10.1364/oe.23.015477

Gomez, I. J., et al. (2018). Nitrogen-doped carbon nanodots for bioimaging and delivery of paclitaxel. *Journal of Materials Chemistry B, 6*(35), 1–3. https://doi.org/10.1039/x0xx00000x

Gong, S., & Piazza, G. (2013). Figure-of-merit enhancement for laterally vibrating Lithium Niobate Mems resonators. *IEEE Transactions on Electron Devices, 60*(11), 3888–3894. https://doi.org/10.1109/TED.2013.2281734

Gouttenoire, V., et al. (2010). Digital and FM demodulation of a doubly clamped single-walled carbon-nanotube oscillator: towards a nanotube cell phone. *Small, 6*(9), 1060–1065. https://doi.org/10.1002/smll.200901984

Guldiken, R., et al. (2006). Dual-Annular-Ring CMUT array for forward-looking IVUS imaging. In *Proceedings - IEEE Ultrasonics Symposium* (Vol. 1, pp. 698–701). https://doi.org/10.1109/ULTSYM.2006.187

Guo, T., et al. (1995). Catalytic growth of single-walled nanotubes by laser vaporization. *Chemical Physics Letters, 243*(1–2), 49–54. https://doi.org/10.1016/0009-2614(95)00825-O

Guo, L., et al. (2002). Regularly shaped, single-crystalline ZnO nanorods with Wurtzite Structure. *Journal of the American Chemical Society, 124*(50), 14864–14865. https://doi.org/10.1021/ja027947g

Gupta, V., et al. (2010). Optimization criteria for optimal placement of piezoelectric sensors and actuators on a smart structure: a technical review. *Journal of Intelligent Material Systems and Structures, 21*(12), 1227–1243. https://doi.org/10.1177/1045389X10381659

Gupta, A., et al. (2015). Recent development in 2D materials beyond graphene. *Progress in Materials Science, 73*, 44–126. https://doi.org/10.1016/j.pmatsci.2015.02.002

Guthy, C., et al. (2013). Large-scale arrays of nanomechanical sensors for biomolecular fingerprinting. *Sensors and Actuators, B: Chemical, 187*, 111–117. https://doi.org/10.1016/j.snb.2012.09.070

Hall, D. A. (2001). Nonlinearity in piezoelectric ceramics. *Journal of Materials Science, 36*(19), 4575–4601. https://doi.org/10.1023/A:1017959111402

Han, M. Y., et al. (2007). Energy band-gap engineering of graphene nanoribbons. *Physical Review Letters, 98*(20), 206805. https://doi.org/10.1103/PhysRevLett.98.206805

Han, B., et al. (2013). Silicon nanowire based ring shape force sensor for sensorized guidewires. In *2013 Transducers & Eurosensors XXVII: The 17th international conference on solid-state sensors, Actuators and Microsystems International Conference on Solid-State Sensors, Actuators and Microsystems* (pp. 718–721). IEEE. http://ieeexplore.ieee.org/document/6626867/

Hanay, M. S., et al. (2012). Single-protein nanomechanical mass spectrometry in real time. *Nature Nanotechnology, 7*(9), 602–608. https://doi.org/10.1038/nnano.2012.119

Hasnain, M. S., & Nayak, A. K. (2019). Synthesis of carbon nanotubes. In *SpringerBriefs in applied sciences and technology* (pp. 17–20). Springer Verlag. https://doi.org/10.1007/978-981-15-0910-0_3

Heidari, A., et al. (2013). A novel checker-patterned AlN MEMS Resonator as Gravimetric Sensor. *Sensors and Actuators, A: Physical, 189*, 298–306. https://doi.org/10.1016/j.sna.2012.09.027

Hsin, C. L., et al. (2008). Elastic properties and buckling of silicon nanowires. *Advanced Materials, 20*(20), 3919–3923. https://doi.org/10.1002/adma.200800485

Hui, Y., & Rinaldi, M. (2013a). High performance NEMS resonant infrared detector based on an aluminum nitride nano-plate resonator. In *2013 Transducers and Eurosensors XXVII: The 17th international conference on solid-state sensors, actuators and microsystems, Transducers and Eurosensors 2013* (pp. 968–971).

Hui, Y., & Rinaldi, M. (2013b). Fast and high resolution thermal detector based on an aluminum nitride piezoelectric microelectromechanical resonator with an integrated suspended heat absorbing element. *Applied Physics Letters, 102*(9), 093501. https://doi.org/10.1063/1.4794074

Huong Giang, D. T., et al. (2013). Electric field-controlled magnetization in exchange biased IrMn/Co/PZT multilayers. *Advances in Natural Sciences: Nanoscience and Nanotechnology, 4*(2), 025017. https://doi.org/10.1088/2043-6262/4/2/025017

Ibach, H. (1997). The role of surface stress in reconstruction, epitaxial growth and stabilization of mesoscopic structures. *Surface Science Reports, 29*(5–6), 195–263. https://doi.org/10.1016/s0167-5729(97)00010-1

Iijima, S. (1991). Helical microtubules of graphitic carbon. *Nature, 354*(6348), 56–58. https://doi.org/10.1038/354056a0

Iqbal, A., & Mohd-Yasin, F. (2018). Reactive sputtering of aluminum nitride (002) thin films for piezoelectric applications: a review. *Sensors (Switzerland), 18*(6), 1797. https://doi.org/10.3390/s18061797

Iriarte, G. F., et al. (2010). Synthesis of C-axis oriented AlN thin films on different substrates: a review. *Materials Research Bulletin, 45*(9), 1039–1045. https://doi.org/10.1016/j.materresbull.2010.05.035

Iwasaki, H., et al. (1976). Piezoelectricity and pyroelectricity of polymers. *Japanese Journal of Applied Physics, 15*, 2041.

Jensen, K., Weldon, J., et al. (2007). Nanotube Radio. *Nano Letters, 7*(11), 3508–3511. https://doi.org/10.1021/nl0721113

Jensen, K., Kim, K., et al. (2008). An atomic-resolution nanomechanical mass sensor. *Nature Nanotechnology, 3*(9), 533–537. https://doi.org/10.1038/nnano.2008.200

Joshi, A. Y., Sharma, S. C., & Harsha, S. P. (2011). Zeptogram scale mass sensing using single walled carbon nanotube based biosensors. *Sensors and Actuators, A: Physical, 168*(2), 275–280. https://doi.org/10.1016/j.sna.2011.04.031

Joshi, P., et al. (2019). Distributed MEMS mass-sensor based on piezoelectric resonant micro-cantilevers. *Journal of Microelectromechanical Systems, 28*(3), 382–389. https://doi.org/10.1109/JMEMS.2019.2908879

Jung, W. S., et al. (2015). High output piezo/triboelectric hybrid generator. *Scientific Reports, 5*(1), 1–6. https://doi.org/10.1038/srep09309

Karabalin, R. B., et al. (2009). Piezoelectric nanoelectromechanical resonators based on aluminum nitride thin films. *Applied Physics Letters, 95*(10), 103111. https://doi.org/10.1063/1.3216586

Kataura, H., et al. (2000). Diameter control of single-walled carbon nanotubes. *Carbon, 38*(11), 1691–1697. https://doi.org/10.1016/S0008-6223(00)00090-7

Kaur, H., et al. (2016). Large area fabrication of semiconducting phosphorene by Langmuir-Blodgett Assembly. *Scientific Reports, 6*(1), 1–8. https://doi.org/10.1038/srep34095

Khan, A., et al. (2016). Piezoelectric thin films: An integrated review of transducers and energy harvesting. *Smart Materials and Structures, 25*(5), 053002. https://doi.org/10.1088/0964-1726/25/5/053002

Kim, S. K., et al. (2016). Directional dependent piezoelectric effect in CVD grown monolayer MoS2 for flexible piezoelectric nanogenerators. *Nano Energy, 22*, 483–489. https://doi.org/10.1016/j.nanoen.2016.02.046

Kisi, E. H., & Elcombe, M. M. (1989). U parameters for the Wurtzite structure of ZnS and ZnO using powder neutron diffraction. *Acta Crystallographica, Section C: Crystal Structure Communications, 45*(12), 1867–1870. https://doi.org/10.1107/s0108270189004269

Kong, J., et al. (1998). Synthesis of individual single-walled carbon nanotubes on patterned silicon wafers. *Nature, 395*(6705), 878–881. https://doi.org/10.1038/27632

Koumela, A., et al. (2013). High frequency top-down junction-less silicon nanowire resonators. *Nanotechnology, 24*(43), 435203. https://doi.org/10.1088/0957-4484/24/43/435203

Kumar, M., & Ando, Y. (2010). Chemical vapor deposition of carbon nanotubes: A review on growth mechanism and mass production. *Journal of Nanoscience and Nanotechnology, 10*(6), 3739–3758. https://doi.org/10.1166/jnn.2010.2939

Kuroyanagi, T., et al. (2014). Cantilevered carbon nanotube hygrometer. *Applied Physics Letters, 104*(19), 193104. https://doi.org/10.1063/1.4876959

Lakin, K. M. (1992). Modeling of Thin Film Resonators and Filters. In *IEEE MTT-S International Microwave Symposium Digest* (Vol. 1, pp. 149–152). IEEE. https://doi.org/10.1109/mwsym.1992.187931

Lakin, K. M. (2002). Coupled resonator filters. In *Proceedings of the IEEE Ultrasonics Symposium* (Vol. 1, pp. 901–908). https://doi.org/10.1109/ultsym.2002.1193543

Lakin, K. M. (2003). A review of thin-film resonator technology. *IEEE Microwave Magazine, 4*(4 SPEC), 61–67. https://doi.org/10.1109/MMW.2003.1266067

Larson, J. D., et al. (2000). Modified Butterworth-Van Dyke circuit for FBAR resonators and automated measurement system. In *2000 IEEE Ultrasonics Symposium. Proceedings. An International Symposium (Cat. No.00CH37121)* (Vol. 1, pp. 863–868). IEEE. https://ieeexplore.ieee.org/abstract/document/922679/

Lee, J., & Feng, P. X. L. (2014). Atomically-thin MoS2 resonators for pressure sensing. In *IFCS 2014–2014 IEEE International Frequency Control Symposium, Proceedings* (pp. 1–4). IEEE. https://ieeexplore.ieee.org/stamp/stamp.jsp?arnumber=6859918

Lee, Y. H., et al. (2012). Synthesis of large-area MoS2 atomic layers with chemical vapor deposition. *Advanced Materials, 24*(17), 2320–2325. https://doi.org/10.1002/adma.201104798

Lee, J., Krupcale, M. J., et al. (2016). Effects of γ-ray radiation on two-dimensional molybdenum disulfide (MoS2) nanomechanical resonators. *Applied Physics Letters, 108*(2), 023106. https://doi.org/10.1063/1.4939685

Lee, J., Wang, Z., et al. (2018). Electrically tunable single- and few-layer MoS2 nanoelectromechanical systems with broad dynamic range. *Science Advances, 4*(3), eaao6653. https://doi.org/10.1126/sciadv.aao6653

Levy, M., et al. (1998). Fabrication of single-crystal Lithium Niobate films by crystal ion slicing. *Applied Physics Letters, 73*(16), 2293–2295. https://doi.org/10.1063/1.121801

Li, J., & Östling, M. (2015). Scalable fabrication of 2D semiconducting crystals for future electronics. *Electronics, 4*(4), 1033–1061. https://doi.org/10.3390/electronics4041033

Li, X., et al. (2011). Vertically stacked and independently controlled Twin-Gate MOSFETs on a single Si nanowire. *IEEE Electron Device Letters, 32*(11), 1492–1494. https://doi.org/10.1109/LED.2011.2165693

Ling, X., et al. (2016). Parallel stitching of 2D materials. *Advanced Materials, 28*(12), 2322–2329. https://doi.org/10.1002/adma.201505070

Liu, H., et al. (2012). Development of piezoelectric microcantilever flow sensor with wind-driven energy harvesting capability. *Applied Physics Letters, 100*(22), 223905. https://doi.org/10.1063/1.4723846

Loh, O. Y., & Espinosa, H. D. (2012). Nanoelectromechanical contact switches. *Nature Nanotechnology, 7*(5), 283–295. https://doi.org/10.1038/nnano.2012.40

Martin, R. M. (1972). Piezoelectricity. *Physical Review B, 5*, 1607–1613.

Mason, W. P. (1981). Piezoelectricity, its history and applications. *Journal of the Acoustical Society of America, 70*(6), 1561–1566. https://doi.org/10.1121/1.387221

Minot, E. D., et al. (2003). Tuning carbon nanotube band gaps with strain. *Physical Review Letters, 90*(15), 4. https://doi.org/10.1103/PhysRevLett.90.156401

Mohn, S., et al. (2016). Polarity control in Group-III nitrides beyond Pragmatism. *Physical Review Applied, 5*(5), 054004. https://doi.org/10.1103/PhysRevApplied.5.054004

Muralt, P., et al. (2009). Piezoelectric thin films for sensors, actuators, and energy harvesting. *MRS Bulletin, 34*(9), 658–664. https://doi.org/10.1557/mrs2009.177

Muruganathan, M., et al. (2018). Zeptogram Level mass sensing of light weight gas molecules using graphene nanomechanical (GNEM) resonator. In *2018 IEEE SENSORS, vol. 2018-Octob* (pp. 1–4). IEEE. https://doi.org/10.1109/ICSENS.2018.8589705

Nakamura, K., et al. (2008). First-principles study on piezoresistance effect in silicon nanowires. *Japanese Journal of Applied Physics, 47*(6), 5132–5138. https://doi.org/10.1143/JJAP.47.5132

Nakazawa, S., et al. (2005). Flame synthesis of carbon nanotubes in a Wall Stagnation flow. *Chemical Physics Letters, 403*(1–3), 158–162. https://doi.org/10.1016/j.cplett.2004.12.091

Namdari, P., et al. (2016). Recent advances in silicon nanowire biosensors: synthesis methods, properties, and applications. *Nanoscale Research Letters, 11*(1), 406. https://doi.org/10.1186/s11671-016-1618-z

Nirschl, M., et al. (2009). Film bulk acoustic resonators for DNA and protein detection and investigation of in vitro bacterial S-layer formation. *Sensors and Actuators, A: Physical, 156*(1), 180–184. https://doi.org/10.1016/j.sna.2009.02.021

Novoselov, K. S., et al. (2016). 2D materials and van Der Waals heterostructures. *Science, 353*(6298), aac9439. https://doi.org/10.1126/science.aac9439

Obitayo, W., & Liu, T. (2012). A review: Carbon nanotube-based piezoresistive strain sensors. *Journal of Sensors, 2012*, 15. https://doi.org/10.1155/2012/652438

Pan, C. T., et al. (2010). Design and fabrication of flexible piezo-microgenerator by depositing ZnO thin films on PET substrates. *Sensors and Actuators, A: Physical, 159*(1), 96–104. https://doi.org/10.1016/j.sna.2010.02.023

Panda, P. K., & Sahoo, B. (2015). PZT to lead free piezo ceramics: A review. *Ferroelectrics, 474*(1), 128–143. https://doi.org/10.1080/00150193.2015.997146

Pang, W., et al. (2006). Femtogram mass sensing platform based on lateral extensional mode piezoelectric resonator. *Applied Physics Letters, 88*(24), 243503–243503-3. https://doi.org/10.1063/1.2213975

Patel, A. M., & Joshi, A. Y. (2015). Detection of biological objects using dynamic characteristics of double-walled carbon nanotubes. *Applied Nanoscience (Switzerland), 5*(6), 681–695. https://doi.org/10.1007/s13204-014-0364-8

Perruisseau-Carrier, J. (2012). Graphene for antenna applications: Opportunities and challenges from microwaves to THz. In *LAPC 2012–2012 Loughborough antennas and propagation conference* (pp. 1–4). IEEE. https://ieeexplore.ieee.org/abstract/document/6402934/

Phan, H. P., et al. (2015). Piezoresistive effect of P-Type silicon nanowires fabricated by a top-down process using FIB implantation and wet etching. *RSC Advances, 5*(100), 82121–82126. https://doi.org/10.1039/c5ra13425k

Pohanka, M. (2017). The piezoelectric biosensors: Principles and applications, a review. *International Journal of Electrochemical Science, 12*(1), 496–506. https://doi.org/10.20964/2017.01.44

Pramanik, S., et al. (2013). Developments of immobilized surface modified piezoelectric crystal biosensors for advanced applications. *International Journal of Electrochemical Science, 8*(6), 8863–8892. http://www.academia.edu/download/32055681/Developments_of_immobilized_surface_modified_piezoelectric_crystal_biosensors_for_advanced_applications.pdf

Ramadan, K. S., et al. (2014). A review of piezoelectric polymers as functional materials for electromechanical transducers. *Smart Materials and Structures, 23*(3), 033001. https://doi.org/10.1088/0964-1726/23/3/033001

Rinaldi, M., et al. (2009). 5–10 GHZ ALN Contour-Mode nanoelectromechanical resonators. In *Proceedings of the IEEE International Conference on Micro Electro Mechanical Systems (MEMS)* (pp. 916–919) https://ieeexplore.ieee.org/abstract/document/4805533/

Robinson, J. T., et al. (2008). Wafer-scale reduced graphene oxide films for nanomechanical devices. *Nano Letters, 8*(10), 3441–3445. https://doi.org/10.1021/nl8023092

Rosen, C., et al. Piezoelectricity. 1992., https://books.google.com/books?hl=en&lr=&id=HmNZuy053foC&oi=fnd&pg=PA1&dq=piezoelectricity+&ots=vp77Wu8Y_O&sig=RwsAUR1Lvq1V5KBdx-6c6C0GPRU.

Sadek, A. S., et al. (2010). Wiring nanoscale biosensors with piezoelectric nanomechanical resonators. *Nano Letters, 10*(5), 1769–1773. https://doi.org/10.1021/nl100245z

Saito, R., et al. (1998). Physical properties of carbon nanotubes. In *Physical properties of carbon nanotubes*. https://doi.org/10.1142/p080

Sakhaee-Pour, A., et al. (2008). Applications of single-layered graphene sheets as mass sensors and atomistic dust detectors. *Solid State Communications, 145*(4), 168–172. https://doi.org/10.1016/j.ssc.2007.10.032

Sarkar, D., Xu, C., et al. (2011). High-frequency behavior of graphene-based interconnects—Part I: Impedance modeling. *IEEE Transactions on Electron Devices, 58*(3), 843–852. https://doi.org/10.1109/TED.2010.2102031

Sarkar, D., Liu, W., et al. (2014). MoS2 field-effect transistor for next-generation label-free biosensors. *ACS Nano, 8*(4), 3992–4003. https://doi.org/10.1021/nn5009148

Sarkar, D., Xie, X., Liu, W., et al. (2015a). A subthermionic tunnel field-effect transistor with an atomically thin channel. *Nature, 526*(7571), 91–95. https://doi.org/10.1038/nature15387

Sarkar, D., Xie, X., Kang, J., et al. (2015b). Functionalization of transition metal dichalcogenides with metallic nanoparticles: Implications for doping and gas-sensing. *Nano Letters, 15*(5), 2852–2862. https://doi.org/10.1021/nl504454u

Sazonova, V., et al. (2004). A tunable carbon nanotube electromechanical oscillator. *Nature, 431*(7006), 284–287. https://doi.org/10.1038/nature02905

Seo, Y. J., et al. (2015). Si nanowire probe with Nd-Fe-B magnet for Attonewton-Scale Force Detection. *Journal of Micromechanics and Microengineering, 25*(4). https://doi.org/10.1088/0960-1317/25/4/045015

Shaby, S. M., & Juliet, A. V. (2011). Improving the sensitivity of MEMS piezoresistive pressure sensor using polysilicon double nanowire. *International Journal of Computer Applications, 34*(9). https://doi.org/10.5120/4124-5933

Shao, M. W., et al. (2008). Ag-Modified Silicon Nanowires Substrate for Ultrasensitive Surface-Enhanced Raman Spectroscopy. *Applied Physics Letters, 93*(23), 233118. https://doi.org/10.1063/1.2969292

Shiu, S. C., et al. (2011). Influence of pre-surface treatment on the morphology of silicon nanowires fabricated by metal-assisted etching. *Applied Surface Science, 257*(6), 1829–1834. https://doi.org/10.1016/j.apsusc.2010.08.086

Shivaraman, S., et al. (2009). Free-standing epitaxial graphene. *Nano Letters, 9*(9), 3100–3105. https://doi.org/10.1021/nl900479g

Shrout, T. R., & Zhang, S. J. (2007). Lead-free piezoelectric ceramics: Alternatives for PZT? *Journal of Electroceramics, 19*(1), 111–124. https://doi.org/10.1007/s10832-007-9047-0

Shukrullah, S., et al. (2014). Mass production of carbon nanotubes using fluidized bed reactor: A short review. *Trends in Applied Sciences Research, 9*(3), 121–131. https://doi.org/10.3923/tasr.2014.121.131

Song, X., et al. (2012). Stamp transferred suspended graphene mechanical resonators for radio frequency electrical readout. *Nano Letters, 12*(1), 198–202. https://doi.org/10.1021/nl203305q

Soon, B. W., et al. (2010). Ultrasensitive nanowire pressure sensor makes its debut. *Procedia Engineering, 5*, 1127–1130. https://doi.org/10.1016/j.proeng.2010.09.309

Stampfer, C., et al. (2006a). Fabrication of discrete nanoscaled force sensors based on single-walled carbon nanotubes. *IEEE Sensors Journal, 6*(3), 613–617. https://doi.org/10.1109/JSEN.2006.874490

Stampfer, R. C., et al. (2006b). Fabrication of single-walled carbon-nanotube-based pressure sensors. *Nano Letters, 6*(2), 233–237. https://doi.org/10.1021/nl052171d

Su, C. C., et al. (2012). Two dimensional carbon nanotube based strain sensor. *Sensors and Actuators, A: Physical, 176*, 124–129. https://doi.org/10.1016/j.sna.2011.10.001

Suzuki, H., et al. (2007). Formation of silicon nanowires by CVD using gold catalysts at low temperatures. *Materials Transactions*, (8), 48, 2202–2206. https://doi.org/10.2320/matertrans.MRA2007059

Tabrizian, R., et al. (2010). Temperature-stable high-Q AlN-on-silicon resonators with embedded array of oxide pillars. In *Technical Digest—Solid-state sensors, actuators, and microsystems workshop*.

Tadigadapa, S., & Mateti, K. (2009). Piezoelectric MEMS sensors: State-of-the-Art and perspectives. *Measurement Science and Technology, 20*(9). https://doi.org/10.1088/0957-0233/20/9/092001

Taniyasu, Y., & Kasu, M. (2010). Improved emission efficiency of 210-Nm deep-ultraviolet aluminum nitride light-emitting diode. *NTT Technical Review, 8*(8).

Tong, J., et al. (2007). Carbon nanotube-based strain sensing cantilevers. In *2007 IEEE 20th International conference on micro electro mechanical systems (MEMS)* (pp. 843–846). IEEE. http://ieeexplore.ieee.org/document/4433135/

Treacy, M. M. J., et al. (1996). Exceptionally high Young's modulus observed for individual carbon nanotubes. *Nature, 381*(6584), 678–680. https://doi.org/10.1038/381678a0

Tressler, J. F., et al. (1998). Piezoelectric sensors and sensor materials. *Journal of Electroceramics, 2*(4). https://doi.org/10.1023/A:1009926623551

Tripathi, N., et al. (2015). Fine-tuning control on CNT diameter distribution, length and density using thermal CVD growth at atmospheric pressure: an in-depth analysis on the role of flow rate and flow duration of acetylene (C2H2) gas. *Applied Nanoscience (Switzerland), 5*(1), 19–28. https://doi.org/10.1007/s13204-013-0288-8

Trolier-Mckinstry, S., & Muralt, P. (2004). Thin film piezoelectrics for MEMS. *Journal of Electroceramics, 12*, 7–17.

Van Der Zande, A. M., et al. (2010). Large-scale arrays of single-layer graphene resonators. *Nano Letters, 10*(12), 4869–4873. https://doi.org/10.1021/nl102713c

Verbridge, S. S., et al. (2006). Optically driven resonance of nanoscale flexural oscillators in liquid. *Nano Letters, 6*(9), 2109–2114. https://doi.org/10.1021/nl061397t

Wang, R., et al. (2004). Fabrication and characterization of potassium-sodium niobate piezoelectric ceramics by Spark-Plasma-Sintering Method. *Materials Research Bulletin, 39*(11), 1709–1715. https://doi.org/10.1016/j.materresbull.2004.05.007

Wang, Y., et al. (2013). Raman spectroscopy study of lattice vibration and crystallographic orientation of monolayer Mos2 under uniaxial strain. *Small, 9*(17), 2857–2861. https://doi.org/10.1002/smll.201202876

Wasisto, H. S., et al. (2013). Silicon nanowire resonators for aerosol nanoparticle mass sensing. In *8th Annual IEEE International Conference on nano/micro engineered and molecular systems, IEEE NEMS 2013* (pp. 506–509). IEEE. http://ieeexplore.ieee.org/document/6559781/

Wierer, J. J., et al. (2010). The impact of piezoelectric polarization and nonradiative recombination on the performance of (0001) Face GaN/InGaN photovoltaic devices. *Applied Physics Letters, 96*(5), 051107. https://doi.org/10.1063/1.3301262

Witkamp, B., et al. (2006). Bending-mode vibration of a suspended nanotube resonator. *Nano Letters, 6*(12), 2904–2908. https://doi.org/10.1021/nl062206p

Wu, W., et al. (2014). Piezoelectricity of single-atomic-layer MoS2 for energy conversion and piezotronics. *Nature, 514*(7253), 470–474. https://doi.org/10.1038/nature13792

Xu, Y., et al. (2010). Radio frequency electrical transduction of graphene mechanical resonators. *Applied Physics Letters, 97*(24), 243111. https://doi.org/10.1063/1.3528341

Yang, J. (2006). A review of a few topics in piezoelectricity. *Applied Mechanics Reviews, 59*(1–6), 335–345. https://doi.org/10.1115/1.2345378

Yang, Y. T., et al. (2006). Zeptogram-scale nanomechanical mass sensing. *Nano Letters, 6*(4), 583–586. https://doi.org/10.1021/nl052134m

Yang, R., et al. (2015). Calibrating temperature coefficient of frequency (TCf) and thermal expansion coefficient (α) of MoS2 nanomechanical resonators. In *2015 Joint Conference of the IEEE International Frequency Control Symposium and the European Frequency and Time Forum, FCS 2015—Proceedings* (pp. 198–201). IEEE. https://ieeexplore.ieee.org/stamp/stamp.jsp?arnumber=7138822

Yarar, E., et al. (2016). Low temperature aluminum nitride thin films for sensory applications. *AIP Advances, 6*(7), 075115. https://doi.org/10.1063/1.4959895

Ye, F., et al. (2018). Glowing graphene nanoelectromechanical resonators at ultra-high temperature up to 2650K. In *2018 IEEE international Electron devices meeting (IEDM)* (Vol. 2018-Decem, pp. 4.4.1–4.4.4). IEEE. https://doi.org/10.1109/IEDM.2018.8614604

Zhang, Y. F., et al. (2002). Liquid phase synthesis of carbon nanotubes. *Physica B: Condensed Matter, 323*(1–4), 293–295. https://doi.org/10.1016/S0921-4526(02)01026-8

Zhang, H., et al. (2005). Mercury ion sensing by film-bulk-acoustic-resonator mass sensor. *Proceedings of IEEE Sensors, 2005*, 203–206. https://doi.org/10.1109/ICSENS.2005.1597671
Zhang, W., et al. (2013). High-gain phototransistors based on a CVD MoS2 monolayer. *Advanced Materials, 25*(25), 3456–3461. https://doi.org/10.1002/adma.201301244
Zhang, Y., et al. (2017). Exfoliation of stable 2D Black Phosphorus for device fabrication. *Chemistry of Materials, 29*(15), 6445–6456. https://doi.org/10.1021/acs.chemmater.7b01991
Zhou, Y., et al. (2016). Theoretical study on two-dimensional MoS2 piezoelectric nanogenerators. *Nano Research, 9*(3), 800–807. https://doi.org/10.1007/s12274-015-0959-8
Zhu, H., et al. (2015). Observation of piezoelectricity in free-standing monolayer MoS2. *Nature Nanotechnology, 10*(2), 151–155. https://doi.org/10.1038/nnano.2014.309

Chapter 7
Evolution of Wafer Bonding Technology and Applications from Wafer-Level Packaging to Micro/Nanofluidics-Enhanced Sensing

Jikai Xu, Zhihao Ren, Bowei Dong, Chenxi Wang, Yanhong Tian, and Chengkuo Lee

7.1 Introduction

Microelectromechanical systems (MEMS) are one of the most successful commercial technologies in the world. They have caught widespread attention since the first functioning silicon motor was entirely fabricated by the standard integrated circuit (IC) manufacturing techniques in the 1980s (Suga et al., 1992). MEMS are composed of components ranging from 1 to 100 μm in size, and the corresponding device is usually distributed between 20 and 1000 μm. Leveraging the design that combines with microelectronics and micro micromachines, the integrated microsystem can achieve the same functions as macroscopic devices. The common products include MEMS-based accelerometers, microphones, micromotors, micropumps, gyroscopes, pressure and humidity sensors, micromachined ultrasound transducers, digital micromirror, etc. However, due to the large surface area to volume ratio,

J. Xu
Department of Electrical and Computer Engineering, National University of Singapore, Singapore, Singapore

State Key Laboratory of Advanced Welding and Joining, Harbin Institute of Technology, Harbin, China

Z. Ren · B. Dong · C. Lee
Department of Electrical and Computer Engineering, National University of Singapore, Singapore, Singapore

Center for Intelligent Sensors and MEMS (CISM), National University of Singapore, Singapore, Singapore
e-mail: elec@nus.edu.sg

C. Wang (✉) · Y. Tian
State Key Laboratory of Advanced Welding and Joining, Harbin Institute of Technology, Harbin, China
e-mail: wangchenxi@hit.edu.cn

Z. Yang (ed.), *Advanced MEMS/NEMS Fabrication and Sensors*,
https://doi.org/10.1007/978-3-030-79749-2_7

environmental electromagnetic effects (e.g., electrostatic charges and magnetic moments) and fluid dynamics (e.g., surface tension and viscosity) need to be considered when designing MEMS devices (Lee & Xie, 2009). Although MEMS technology is originated from IC techniques, lots of new techniques (e.g., bulk micromachining, surface micromachining, thermal oxidation, and high aspect ratio silicon micromachining) have been derived and developed to satisfy its device fabrication. With the continuous development and improvement of technology, MEMS products possess the advantages of small size, lightweight, low power consumption, low cost, high reliability, excellent durability, on-chip integration, and batch fabrication. Moreover, since MEMS technology is compatible with the IC process, the integration of MEMS with consumer electronics and biomedical devices to achieve miniaturization and multifunction has become an irreversible trend.

Among such many characteristics, batch fabrication is an essential feature of MEMS products to control cost and market orientation. One of the most important reasons is attributed to the high-yield wafer-level packaging. However, MEMS devices have stricter packaging requirements (e.g., ultralow stress, high vacuum, strong sealing, high isolation, and precise alignment) than IC electronics due to the complex microsystem. Hundreds of micro/nanometer-scale components have high accuracy but are very fragile in MEMS devices. MEMS packaging should minimize or avoid the stress to enable devices to work normally. The high vacuum can reduce friction and make devices work stably for a long time. Strong sealing is the prerequisite for the operation of some devices (e.g., micro gyroscopes), the requirement for the packaging sealing is as high as 1×10^{12} Pa m^3 s^{-1}. High isolation and precise alignment are to prevent the signal cross talk (e.g., RF switch) and reduce the insertion loss. These requirements increase the cost of MEMS packaging, making it account for 50% to 90% in MEMS device fabrication. Besides, nitrides of silicon, aluminum, and titanium, as well as silicon carbide and other ceramics, are increasingly applied in MEMS fabrication because of the advantageous combinations of materials properties. This is an opportunity to expend the product category, but it is also full of challenges. Therefore, the progress of high-yield wafer-level packaging techniques for different materials is also crucial to MEMS development.

This chapter firstly describes the traditional wafer-level MEMS packaging method, i.e., eutectic bonding. Interfacial studies for several commonly used systems (i.e., Au-Sn, Au-In, In-Sn, and In-Ag) have been reviewed. Secondly, we show the progress of advanced wafer-level packaging, i.e., wafer bonding including room-temperature and low-temperature bonding technologies. We focus on the material applicability of various bonding methods. Finally, we introduce the latest developments of wafer bonding in the emerging field of micro/nanofluidics-enhanced sensors combined with plasmonics.

7.2 Eutectic Hermetic Bonding for the Wafer-Level MEMS Packaging

Eutectic bonding usually uses metal layers that can form a eutectic system to achieve the packaging. Therefore, as least two metals are needed. One is the low-melting metal (solder), such as indium (In) or tin (Sn). The other one is high-melting metal, such as noble metals like gold (Au), silver (Ag), or copper (Cu). Eutectic alloys can be deposited by sputtering, dual-source evaporation, or electroplating. The bonding is usually performed at a temperature of more than 20 °C above the melting temperature of the solder. The solder transforms to be liquid, and the solid-liquid interdiffusion begins to conduct during the heating treatment. According to the principle of phase equilibrium, metallurgical joints with higher temperatures are formed in the cooling process. Due to eutectic bonding that can produce hermetically sealed packaging and electrical interconnection within a single process at low temperatures, it is popular in the MEMS device fabrication (Brand & Baltes, 2002).

Up to now, two approaches have been developed for the MEMS vacuum packaging. One is component-level vacuum packaging, and the other one is wafer-level vacuum packaging (Fig. 7.1a) (Gooch & Schimert, 2003). The main difference between them is the packaging that is performed before or after the wafer dicing process. MEMS devices on the same wafers are sealed in one step in the wafer-level packaging and subsequently tested at a wafer level, thereby yielding substantial savings in time, materials, and labor compared with the component-level packaging method. Various wafer bonding technologies have been investigated for the MEMS packaging. They can be categorized into two approaches, i.e., direct bonding and intermediate-layer bonding (Fig. 7.1b). The commonly used intermediate layers are polymer, glass frit, and eutectic metals. However, the polymer has a gas-permeable interface and cannot meet the hermetic sealing requirement, while glass frit has a relatively high bonding temperature up to 400 °C and problems of outgassing and contaminations. Thus, metal-intermediated eutectic hermetic bonding is preferred. For the wafer-level eutectic bonding, one of the biggest challenges is the total thickness variations caused by the wafer warpage and material deposition. This problem becomes more severe with an increase in wafer size. As a remedy, sufficient solder (e.g., several micrometers in thickness) is required to reduce the nonplanarity issue. The interfacial evolution becomes a critical factor for device reliability. Therefore, we put a particular emphasis on the introduction of interfacial structures in several commonly used eutectic systems.

7.2.1 Au-Sn Eutectic Bonding

The Au-Sn eutectic system has been used for flip-chip bonding in many applications (e.g., optoelectronic and RF device packaging) due to its excellent wetting behavior, good thermal conductivity, and high bonding strength (Lee et al., 1993; Wang et al.,

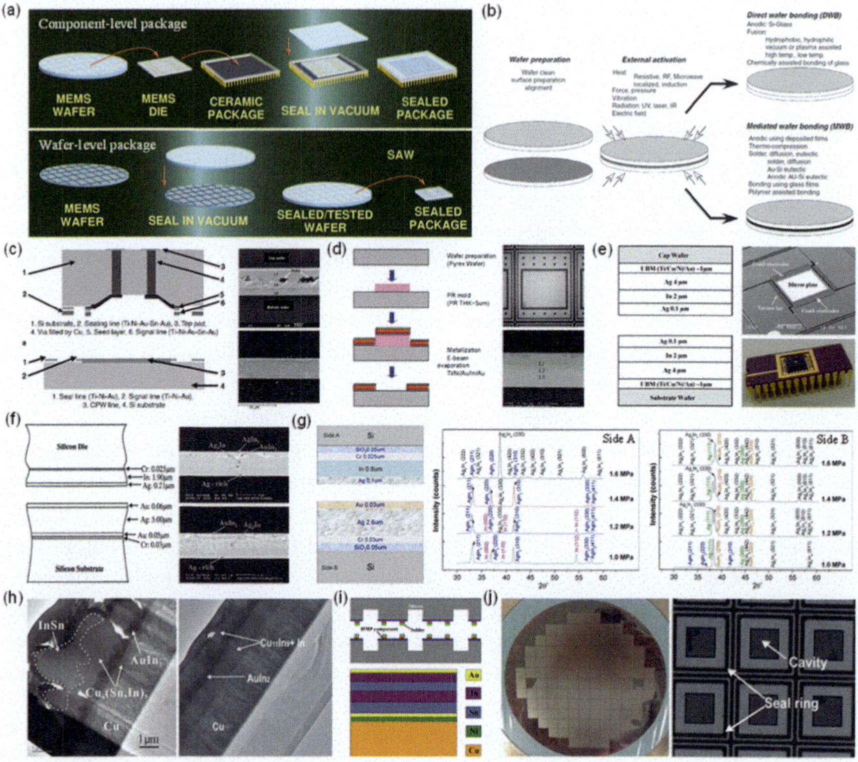

Fig. 7.1 Eutectic hermetic bonding for the wafer-level MEMS/NEMS packaging. (**a**) Schematic diagram of the component-level and wafer-level packaging (Gooch & Schimert, 2003). (**b**) Bonding categorizations for the wafer-level packaging (Dragoi et al. 2008). (**c**) and (**d**) Schematic diagram and interfacial SEM images for the Au-Sn eutectic bonding (Wang et al., 2006; Qian et al., 2007). (**e**) Ag-In eutectic bonding for the hermetic MEMS packaging (Lee et al., 2009). (**f**) and (**g**) Interfacial studies for the Ag-In eutectic bonding (Chuang & Lee, 2002; Made et al., 2009). (**h**) TEM studies of the In-Sn eutectic bonding interfaces (Yu et al., 2008). (**i**) and (**j**) Schematic diagram and wafer-level bonding for the In-Sn eutectic bonding (Yu et al., 2008, 2009a)

2006). There are four different stable intermetallic compounds (i.e., Au_5Sn, AuSn, $AuSn_2$, and $AuSn_4$) in the Au-Sn system at room temperature. Among them, Au_5Sn also refers to Au-20 wt% Sn has a melting point of 280 °C which is considered to be suitable for MEMS packaging. Because of the low eutectic melting point, the Au-Sn system can be used for special devices that are sensitive to high process temperatures (e.g., GaAs dice). A fast diffusion between low-temperature solders and Au/Cu metallization will occur after solder thin film deposition (Qian et al., 2007). The generated high-temperature phases usually cause low yield of good dies after bonding (Welch et al., 2005). Therefore, a thin buffer layer, such as titanium (Ti) or nickel (Ni), needs to be introduced on both cap and bottom wafers (Fig. 7.1c). During the heating process, the Sn layer melts first because of the low melting point. The molten Sn has intimate contact with the Au layer and dissolves Au layers on both sides, causing a series of intermetallic compounds (IMCs) according to the phase diagram. With sufficient heating time, Au and Sn continue to diffuse into each other to form a stable phase because of the concentration gradient. The entire eutectic bonding is considered to be a complex process combined with a chemical reaction and physical interdiffusion.

Because the thickness of deposited metal layers is several micrometers, the quality and reliability of the hermetic eutectic bonding strongly depend on the microstructures formed in the bonding joint. Except for the layer design of the final metallurgical interface, surface cleaning is also important for the formation of the high-quality joint. The oxide layer and contaminants on surfaces to be bonded will generate defects at the interface and result in a degradation of the bonding quality. Without additional surface cleaning, large voids at the joint can be found (Fig. 7.1c). Moreover, the contamination and oxides will also block the interdiffusion process between Au and Sn elements, causing a nonuniform metallurgical joint. However, with O_2 plasma treatment or other surface cleaning method, a uniform and void-free bonding interface is obtained.

7.2.2 Au-In Eutectic Bonding

Au-In alloys have been recognized as hard solders with a melting temperature of more than 450 °C. Au-In alloys have rarely been reported because the high melting temperature is harmful to most semiconductor devices. However, In is a kind of widely used soft solder for the bonding of delicate devices with a low processing temperature (157 °C). It also has excellent adhesive properties and the ability of plastic deformation. Using the transient liquid phase (TLP) method, eutectic bonding between Au and In can be achieved at low temperatures.

In the Au-In equilibrium phase diagram, there are two stable intermetallic compounds in all the 14 phases. One is $AuIn_2$ which has 53.8 wt% of In, and the other one is AuIn which has 36.79 wt% of In. To perform the Au-In TLP bonding, multilayers of Au and In thin films have been deposited and etched on the cap wafer (Fig. 7.1d) in previous studies. The whole metallization process is operated in a high vacuum environment to prevent In oxidation. For the bottom wafer, Cr-Au is deposited on the Si wafer. A void-free interface is formed in the bonding process of 180 °C with the static force for 30 min. When the temperature is raised above the melting point of In during the bonding process, Au_2In is generated first within seconds. It is because the liquid diffusion speed is several orders of magnitude higher than solid-state diffusion. If Au and In films have the stoichiometry to form $AuIn_2$, the compound will form completely and remain stable. For films with higher Au concentrations, $AuIn_2$ will be converted into AuIn. A higher thickness of Au film will eventually result in γ and β phases. The X-ray image and interfacial SEM image after Au-In eutectic bonding indicate that the bonding quality is high (Fig. 7.1d). Since Au-In alloys have higher melting points than Au-Sn systems, they can be used to join materials at low temperatures while working at high temperatures.

7.2.3 In-Ag Eutectic Bonding

In the In-Ag system, the melting point of different IMCs increases with the Ag content according to the phase diagram, i.e., eutectic composition (144 °C) < $AgIn_2$ (166 °C) < γ (300 °C) < ζ (670 °C) < β (695 °C) (Lee et al., 2008). In practical mass production, the industry requires that the bonding interfaces can be survived at a temperature of 285 °C (Lee et al., 2009), because this temperature is the peak temperature in the surface mounting technology (SMT) manufacturing line. Therefore, by the component design, In-Ag eutectic bonding can satisfy this requirement (Chuang & Lee, 2002). To explore the feasibility of MEMS packaging based on this eutectic system, the intermediate bonding layer (IBL) including 2 μm In and 4 μm Ag has been proposed (Fig. 7.1e). Moreover, a thin Ag layer is coated on the IBL surface to prevent the oxidation of In layer and avoid the flux treatment before the wafer bonding process. Mass micromirror chips that can meet the production requirements have been successfully achieved (Fig. 7.1e).

To further improve its bonding reliability, effects of bonding parameters on the interfacial phases have been investigated. Usually, a low-temperature annealing process is beneficial for the atom interdiffusion to make phases uniform (Fig. 7.1f). To study the influence of bonding pressure on the interface, 800 nm In and 2.6 μm Ag multilayers are designed. In-rich layers and Ag-rich layers are deposited on sides A and B, respectively (Fig. 7.1g) (Made et al., 2009). XRD characterization indicates that high bonding pressure tends to make the Ag-rich phase dominate across the bonding interface (Fig. 7.1g). The increase in force can flatten more asperities on the rough surface and increase the contact between two bonding surfaces. Studies also show that the phase-transition temperature for $AgIn_2$ to Ag_9In_4 will reduce with the increase of bonding pressure. It is mainly due to $AgIn_2$ that has a lower mass density (8.23 g/cm^3) compared with Ag_9In_4 (9.90 g/cm^3). The melting temperature is proportional to changes in system volume. Therefore, an increase in the bonding pressure can reduce the melting temperature.

7.2.4 In-Sn Eutectic Bonding

In the In-Sn system, the most important application feature is the 118 °C eutectic point at a composition of 51.7 at.% In and 48.3 at.% Sn (Lee & Choe, 2002). Because of the high diffusion rates between low-temperature solder and Cu as well as Au, parts of solder will be consumed during the layer deposition process (Yu et al., 2009c). The interfacial microstructure of Sn/In/Au/Cu metallization has been studied using the transmission electron microscope (Fig. 7.1h). Au-rich phase leaves the Cu surface and moves into the In layer (Yan et al., 2009). At the same time, In diffuses into the Cu surface and reacts with Au and Cu to generate CuInSn phase (Chuang et al., 2006). Consequently, only a small part of the low-temperature InSn phase is left for the bonding. Therefore, it is important to introduce a buffer layer on the substrate to improve the bonding quality.

To realize the commercial MEMS packaging using the In-Sn eutectic system, a buffer layer used to prevent the interdiffusion between the solder and metal layer is indispensable (Yu et al., 2009a). Both Ti and Ni are optional. However, it has been reported that the chip-to-chip bonding parameters cannot directly be used for the wafer-level hermetic bonding (Lee et al., 2000). One of the important differences is that the bonding stress in the wafer level is larger than that of chip size. Here, we demonstrate a wafer-level packaging process using Ni as the buffer layer (Yu et al., 2009b). The patterned wafers are first metalized by orderly sputtering Ti-Cu-Ni-Au layers (Fig. 7.1i). Among them, Ti acts as the adhesive layer, while Ni is used for the buffer layer to perform the bonding process. The thin Au layer is necessary for the wetting and to prevent the solder from oxidation when exposed to the air. High-quality wafer-level packaging has been achieved with 5.5 MPa bonding pressure and a peak temperature of 180 °C heating for 20 min (Fig. 7.1j).

In addition to the aforementioned eutectic bonding systems, there are other systems (e.g., Au-Si, Al-Si, and Al-Ge) that are also used in practical production and research. According to the materials to be bonded, different eutectic bonding systems need to be designed to realize high-quality wafer-level packaging.

7.3 Low-Temperature Wafer Direct Bonding

For the traditional wafer-level eutectic hermetic bonding, the thickness of the bonding interface is usually several micrometers because thick intermediate layers need to be used to overcome the wafer warpage. With the gradual improvement of wafer quality, commercial wafers with a surface roughness of several angstroms (Å) have been put into production. Therefore, advanced wafer-level packaging technologies, including wafer direct bonding and nanometer-scale intermediate-layer bonding, have been developed. Wafer-level eutectic bonding has been studied earlier in MEMS packaging, and various processes for different devices have been standardized and modeled. However, with the rapid development of structural fabrication technology, eutectic bonding cannot satisfy the requirement of advanced device design. The accuracy of the 3D structure in the longitudinal direction of the device is limited because of the micrometer-scale intermediate layer. Therefore, intermediate-free low-temperature wafer direct bonding is an attractive technology for device packaging (Masteika et al., 2014). Eutectic bonding is utilizing the reaction among the deposited layers to generate interfacial compounds to realize the material joining. Wafer direct bonding is using chemical solutions, high-energy atom or ionic beam, or plasma to clean and activate surfaces to be bonded; the bonding can be achieved by the formation of chemical bonds across the interface. Cleaning is to remove the particles and organic contaminants adsorbed on the surfaces. Activation is to increase the density of functional groups that can spontaneously form chemical bonds. The cleaning and activation processes can usually be completed in one step. Due to the bonding that is intermediate-free, the surface roughness plays an important role in the number of chemical bonds generated at the bonding interface. Smoother surfaces can lead to more direct contact areas, allowing more functional groups on mating surfaces to react to generate the chemical bonds that bridge the interface.

Low-temperature wafer direct bonding can be divided into two categories according to the surface conditions: hydrophobic bonding and hydrophilic bonding. Surfaces treated by diluted hydrofluoric acid or ammonium fluoride are temporarily covered with covalently bonded hydrogen. This kind of surface cannot be wetted with water and is thus hydrophobic. The hydrogen is bonded to the Si surface in the form of mono-hydrides (e.g., Si_3-Si-H, Si_2-Si-H_2, and Si-Si-H_3). The different hydrides form depending on HF concentration and silicon wafer orientation due to the fact that they attach to surface steps differently. Hydrophobic surfaces tend to be contaminated by hydrocarbons more easily than hydrophilic surfaces. Thus, they should be bonded as quickly as possible after the HF treatment. After adhesion, the wafers bond due to van der Waals forces between the weak polar Si-H bonds. These bonds are much weaker than hydrogen-bridged bonds between H and O so that the bond energy is only 20–30 mJ/m^2. Much higher bonding energy can be obtained by a low-temperature annealing process (Fig. 7.2a). Thereby, the Si-H covalent bonds are replaced by the covalent Si-Si bonds.

Under usual atmospheric conditions, silicon is covered with a native oxide of 1–2 nm. Upon cleaning the silicon wafer surface with strongly oxidizing solutions, the native oxide is removed, and chemical oxide forms. This oxide contains covalently bound hydrogen and is thus not stoichiometric so that it reacts instantaneously with water and forms so-called silanol groups (Si-OH) at the wafer surface, which renders the surface hydrophilic. The silanol groups are covered with a few monolayers of water. Two surfaces with a water film adhere to hydrogen bridge bonds. The bond energy at this point is weak and in the order of 100 mJ/m^2. A strong bonding interface can be obtained by a subsequent annealing step. As a result, covalent bonds, i.e., Si-O-Si, are generated across the bonding interface.

7.3.1 *Wet Chemical Activated Bonding*

Wet chemistry treatments are cost-effective methods to clean wafer surfaces without requiring a vacuum system (Wang et al., 2018c; Xu et al., 2019d). To eliminate organic and particulate contaminants on surfaces, the mixture of sulfuric acid and hydrogen peroxide (e.g., H_2SO_4/H_2O_2 mixture, known as SPM or piranha solution) is used for the glass surface cleaning. The mixture of $NH_4OH/H_2O_2/H_2O$ (i.e., RCA solution) is used as a standard cleaning step for Si wafers. Although the surfaces are cleaned by the SPM or RCA solutions, the bonding strength (<1.0 J/m^2) is insufficient for the subsequent fabrication even after a low-temperature annealing process.

To obtain strong bonding strength at low temperatures, modified wet chemical treatments have been performed (e.g., SPM and RCA treatment under heating and

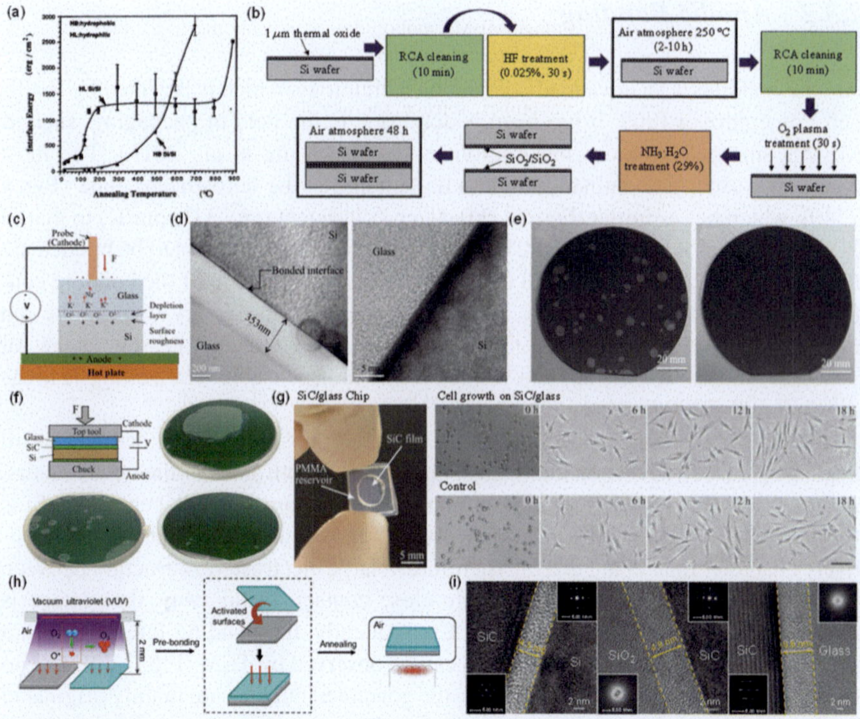

Fig. 7.2 Low-temperature bonding methods: wet chemical activated bonding, anodic bonding, and UV activated bonding. (**a**) Comparison of surface energies for the hydrophobic and hydrophilic bonding at different annealing temperatures. (**b**) Procedures of ZiBond™ methods. (**c**) Schematic diagram of the anodic bonding (Hsu et al., 2017). (**d**) TEM observation of the Si/glass bonding interface (Howlader et al., 2010b). (**e**) Effects of surface pre-treatments on the anodic bonding (Howlader et al., 2010b). (**f**) Anodic bonding for the SiC thin film transfer (Phan et al., 2017). (**g**) Cell culture in the SiC sealed device. (**h**) Ultraviolet (UV) and vacuum ultraviolet (VUV) activation and bonding processes (Xu et al., 2019b). (**i**) SiC heterogeneous integration with Si-based substrates via VUV activated direct bonding method (Wang et al., 2019)

sequential wet chemical treatments) (Masteika et al., 2014). The direct bonding for Si/Si, glass/glass, and Si/glass pairs has been reported (Mai et al., 2015, 2016). Besides, Ziptronix Company from the United States develops a low-temperature bonding technology, called ZiBond™ technology, for oxide wafers based on the multistep surface treatments. The process flow is shown in Fig. 7.2b. The key process includes two steps: HF treatment and RCA cleaning. After 60 h of storage in an air atmosphere at room temperature, the bonding strength for the Si/Si bonded pairs can reach the fracture energy of Si (~2.5 J/m^2). Because the silicon oxide layer can be prepared on a variety of semiconductor materials by the thermal oxidation or e-beam evaporation, ZiBond™ technology is suitable for bonding multiple wafer materials.

7.3.2 Anodic Bonding

Anodic bonding is an advanced connection technology, which is invented by Wallis and Pomerantz in 1969. It has been widely used in the field of packaging, such as MEMS, microelectronics, and micro/nanofluidics (Tang et al., 2016). The most commonly used glass substrate in anodic bonding is the borosilicate glass (Pyrex 7740), which has a similar thermal expansion coefficient (i.e., 3.0 ppm/K) to that of Si (i.e., 2.6 ppm/K). In the anodic bonding, surfaces of Si and glass to be bonded are first brought in contact with each other. The schematic diagram of the bonding configuration is shown in Fig. 7.2c. A heating process is necessary to promote ion mobility in the Pyrex glass. The temperature is usually in the range of 200–500 °C. Besides, the applied voltage is also indispensable. The voltage is in the range of 500–1000 V (Hsu et al., 2017).

During the bonding procedure, Na^+ in the glass substrate will be drifted toward the negative electrode when the voltage is applied on both sides of the bonded pairs. A depletion layer with a thickness of hundreds of nanometers adjacent to the Si is formed. The depletion layer is negatively charged, while the Si substrate is positively charged (Tatar et al., 2012). Therefore, a large electrostatic attraction between the bonded pair can drive surfaces into close contact. In this way, the voltage is mainly applied to the depletion layer. Consequently, the bonding interface with a sharp change in the phase composition can be observed by TEM (Fig. 7.2d). Due to the low temperature of anodic bonding, the generated bubbles are mainly originated from the particles adsorbed on the surfaces. To increase the bonding area, a pre-cleaning process on surfaces to be bonded is important (Fig. 7.2e). By observing current changes in the circuit, the bonding process can be studied. When the voltage is just applied, there is a larger current pulse, and then the current decreases. Finally, the current is almost zero, indicating that the bonding process has been completed.

In recent years, the main progress in anodic bonding is focused on new applications and new materials, for example, an electrical feed-through method for the anodic bonding on both sides of silicon-on-insulator (SOI) wafers. Glass/glass anodic bonding using reactive magnetron sputtered TiN_x films as the intermediate layer to achieve fully transparent device packaging. Moreover, the bonding materials have also been changed from silicon to ceramics and metal materials (e.g., Al, Fe, and Ti). To replace the cathode material, several ionic conduction polymer materials have shown excellent potential in alleviating interfacial stress. Among them, one of the breakthroughs in the SiC thin film transfer is the use of anodic bonding (Fig. 7.2f) (Phan et al., 2017). Firstly, single-crystalline 3C-SiC thin films are grown on the Si substrate via a low-pressure chemical vapor deposition process. Then the anode electrode is connected on the side of the SiC/Si wafer, and the cathode electrode is applied on the glass wafer. During the power-up and heating process, the positive charges are attracted to the cathode side, resulting in a depletion layer at the SiC/glass interface. The negative charges in the glass tend to move toward the anode and diffuse into the SiC substrate to generate chemical bonds between the SiC and glass substrate. This method allows the large-scale SiC wafer direct bonding without any buffer layers. Besides, both SiC and glass substrates have good biocompatibility. The excellent transparency of SiC/glass bonded pairs enables the observation of cell growth under a typical inverted optical microscopy (Fig. 7.2g). Therefore, the transferred SiC thin film using the anodic bonding method demonstrates potential applications for high-temperature electronics and biological cell culture.

7.3.3 Ultraviolet (UV) Activated Bonding

UV is an electromagnetic wave with a wavelength ranging from 10 nm to 400 nm. According to the wavelength range, UV can be divided into three types: UVA (315–400 nm), UVB (280–315 nm), and UVC (190–280 nm). Due to the high energy of UV, it can be used for the removal of organic contaminants (Sunkara & Cho, 2012). Currently, the commonly used UV light source is the Hg vapor lamp. It can emit the light with a mixed wavelength of 185 and 254 nm. Because of the photochemical reaction with oxygen (O_2), UV light at a wavelength of 185 nm can be absorbed by O_2. Then O_2 will be decomposed into oxygen-free radicals, and the generated free radicals can combine with O_2 to generate ozone (O_3) (Ahn & Lee, 2015). When the UV light irradiates on the surfaces to be treated, the UV light at 254 nm can be absorbed by the hydrocarbon compounds. With the help of O_3 and oxygen radicals, the organic contaminants will be converted into CO_2 and H_2O and removed from the surfaces (Tang et al., 2009). In addition to cleaning the surface, UV can also introduce an ultrathin dense oxide (SiO_x) film on the silicon surface. This thin film has been considered to be beneficial for direct bonding. Due to the strong surface cleaning and activation capabilities, UV assisted direct bonding has been considered to be a low-cost method (Tran et al., 2013). Therefore, the direct bonding of Si/Si using the UV/O_3 activation has been investigated (Anantha & Tan, 2015). Studies show that the bonding quality is significantly affected by the wafer bow, surface adhesion energy, and nanometer-scale surface topography. With a modified parameter of UV irradiation, Si/Si direct bonding can be achieved with a high bonding strength at low temperatures.

Alternatively, vacuum ultraviolet (VUV) is a kind of short wavelength (100–200 nm) UV light. It can clean surfaces more effectively because of the higher energy than the traditionally used UV light (Shinohara et al., 2011). Due to the lack of ion bombardment, UV/VUV can cause less surface damage than plasma. Similar to the UV/O_3 activation, VUV irradiation can also obtain a hydrophilic surface with an ultrathin hydration layer (Xu et al., 2019c). Besides, surfaces activated by VUV have extremely high surface energies, which is considered to be more beneficial to the direct bonding at low temperatures (Fig. 7.2h). The commonly used VUV light source is the Xe excimer lamp, which can emit a wavelength of 172 nm with a full width at half maximum (FWHM) of 14 nm (Xu et al., 2018b). Up to now, materials that can be bonded by the VUV activation have been extensively explored. The bonding materials including Au/Au, Si/Si, Si/glass, glass/glass, SiC/Si, SiC/SiO_2, SiC/glass, $LiNbO_3$/Si, $LiNbO_3$/glass, $LiNbO_3$/Si, PEEK/Pt, PDMS/PDMS, and

PDMS/Cu have been reported (Fu et al., 2016; Xu et al., 2018c, 2019a, b; Shigetou et al., 2015; Wang et al., 2018b, 2019). All the bonding process is performed at a temperature of less than 300 °C. Interfacial TEM images of the typical SiC bonding are shown in Fig. 7.2i. Because of the ultrahigh energy of VUV light, there is an ultrathin amorphous layer on the activated surface. Therefore, the amorphous bonding interface is usually thicker than that obtained by the other activated bonding methods (e.g., UV/O_3 activation, SAB, wet chemical activation, and plasma activation). Although the bonding of various materials has been realized, the UV or VUV/O_3 activated bonding mechanism is still in the primary stage. More work is needed to develop this emerging and environmentally friendly bonding method.

7.3.4 *Low-Temperature Plasma Activated Bonding*

Plasma activation is the cornerstone of the low-temperature direct bonding method. A wide range of materials including metals, semiconductors, dielectrics, and polymers can be bonded after the plasma activation (Dragoi et al., 2008; Masteika et al., 2014). Due to the plasma activated bonding (PAB) that has excellent compatibility with MEMS and CMOS technology, it is the most widely studied among all bonding methods (Alam et al., 2013; Li et al., 2013; Suni et al., 2005; Howlader et al., 2009). To achieve the PAB for different materials at low temperatures, various bonding methods have been developed.

Single-gas plasma activation is the fundamental method in PAB. The commonly used gases are Ar, O_2, and N_2 (Takagi et al., 1998; Wu et al., 2002). Ar plasma activation belongs to the physical etching with the ion bombardment. It is often used to remove surface oxides for the metal bonding, such as Cu/Cu bonding. O_2 and N_2 plasma activation belong to the chemical reaction because of the high reactivity of oxygen and nitrogen free radicals (Esser et al., 2003). They are often used for the semiconductor homogeneous bonding (e.g., Si/Si, glass/glass, and Al_2O_3/Al_2O_3 bonding) since the plasma-induced ultrathin and dense oxide films are beneficial for the bonding (Fig. 7.3a) (Pasquariello et al., 2000). Compared with the classical fusion bonding, the bonding temperature of PAB has decreased from 800 °C to 300 °C. Although the PAB has been studied intensively over the past few years, there is no conclusive model to explain the low-temperature bonding mechanism. A prominent problem is that there are too many kinds of equipment. It is difficult to compare the bonding results from different groups because of the lack of unified equipment. T. Plach and co-authors make an extension of the classical bonding model and propose a four-step model for the PAB based on their experimental findings and other reported work (Plach et al., 2013). An obvious increase in the bonding strength can be observed when an additional O_2 or N_2 plasma treatment is added at various annealing temperatures (Fig. 7.3b). To investigate the effect of plasma activation on sample surfaces, the comparison between the cross-sectional Si surfaces before and after the plasma activation is conducted (Fig. 7.3c). Due to the change in thickness that is a nanometer-scale, the observation is performed by

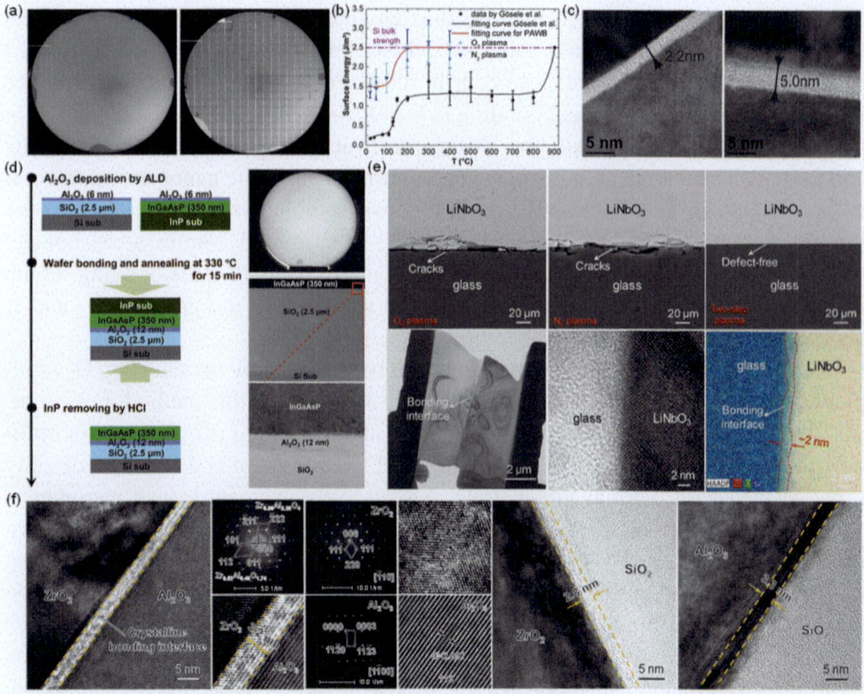

Fig. 7.3 Low-temperature bonding methods: wet chemical activated bonding, anodic bonding, and ultraviolet (UV) activated bonding. (**a**) O_2 plasma activated Si/Si wafer bonding and dicing test (Li et al., 2013). (**b**) Statistical results of the bonding strengths under different kinds of plasma activation (Plach et al., 2013). (**c**) TEM observation for the effect of N_2 plasma on wafer surfaces (Plach et al., 2013). (**d**) Thin film transfer via plasma activated Al_2O_3 bonding (Ikku et al., 2011). (**e**) Comparison of $LiNbO_3$/glass bonding with different kinds of plasma activation (Xu et al., 2018a). (**f**) Plasma activated bonding for YSZ/Al_2O_3, YSZ/SiO_2, and Al_2O_3/SiO_2 at low temperatures (Xu et al., 2020b)

TEM. After the plasma treatment, oxide growth and improved hydrophilicity occur due to the ion injection and reaction during the low-energy bombardment. Accompanied by the changes in the thickness of amorphous layers, a light smoothing also occurs because of the surface swelling. Therefore, in the first stage, the mating surfaces are held together by the hydrogen bonds among the polar –OH groups and H_2O molecules. Because of smoother surfaces, more bonding areas can directly contact. The increased density of –OH groups allows more hydrogen bonds to generate to bridge the interface. The superposition of the aforementioned factors results in the prebonding energies being two to three times that of the classical bonding method (Kikkawa et al., 2000). Besides, more water molecules existed at the bonding interface and in the damaged oxide. In the second stage, part of the water molecules will continue to fill the damaged oxide, and the other part of the water molecules will diffuse out along the interface. At the same time, the diffused molecules will pull the surfaces closer to each other. Once –OH groups are close enough, covalent bonds are formed via the dehydration reaction. In the third stage, the bonding strength becomes saturated when closed –OH groups are consumed completely. The first three stages occur at room temperature. The main difference between the PAB and classical bonding is the closing of the interfacial nanogaps in the fourth stage. For the PAB, the closure starts at a temperature between 100 °C and 200 °C. The damaged oxide plays an important role in the nanogaps' closure. It is like a reservoir that can trap the interfacial water molecules. The oxides become thicker due to this greater amount of water. Consequently, more areas contact directly, leading to a further increase in the bonding strength. This bonding model explains well the importance of the activated wafers with the grown SiO_2 and is accepted by most researchers.

Since the bonding of various materials can be achieved by the one-step O_2 or N_2 plasma activation, the thin film transfer can be achieved by this method. For example, the heterogeneous integration of InGaAsP and thermal Si is important for optoelectronic devices (Fig. 7.3d) (Ikku et al., 2011). However, it is difficult to directly grow InGaAsP on thermal Si substrate. Alternatively, we can deposit nanometer-scale Al_2O_3 on both thermal Si and InGaAsP surfaces, while the InGaAsP is grown on the InP substrate. By using the low-temperature Al_2O_3/Al_2O_3 direct bonding method, InGaAsP/Al_2O_3/SiO_2 multilayers can be fabricated between the InP and Si substrates. Finally, the InP substrate can be removed by soaking in HCl solutions.

In addition to the single-gas plasma activation, sequential plasma activation is a more effective method to modified surfaces with the targeted functional groups. Sequential $O_2 \rightarrow N_2$ plasma activation has been used for the realization of hanging N-related groups on surfaces. Due to the bonding energy of $\equiv Si_2\text{-N-Si} \equiv$ bonds that is higher than that of $\equiv$Si-O-Si$\equiv$ bonds, strong interfaces for Si/Si and quartz/quartz bonded pairs can be achieved at lower temperatures (<200 °C) (Howlader et al., 2006a, 2010b, c, 2011b, 2019; He et al., 2018). Besides the common material bonding, sequential plasma activation can also be used for the heterogeneous integration of optoelectronic materials and third-generation semiconductors (e.g., $LiNbO_3$/glass and GaN/Si) (Xu et al., 2018a; Matsumae et al., 2021). Although the bonding strength does not reach the fracture energy of the substrates, it is sufficient enough to survive back-grinding and other post-fabrication processes (Fig. 7.3e).

Another bonding method is utilizing multiple gas mixture for the plasma activation (Xu et al., 2020b). Multiple functions, such as improvement of hydrophilicity, softening surfaces, and increasing surface energies, can be realized by the one-step process. $O_2/CF_4/H_2O$ plasma activation is a room-temperature bonding method for Si/glass, while $O_2/NH_3/H_2O$ plasma treatment is a low-temperature bonding method for high dielectric oxides (i.e., YSZ/Al_2O_3, YSZ/SiO_2, and Al_2O_3/SiO_2). Some new phenomena also appear because of multiple interactions induced by the mixed plasma. For example, a layered crystalline interface is formed in the YSZ/Al_2O_3 bonding (Fig. 7.3f). It is difficult to explain the reasons based on the current studies, and more work needs to be done to explore its bonding mechanism.

7.4 Room-Temperature Wafer Bonding

To achieve room-temperature or low-temperature bonding, surface cleaning and activation are necessary for the improvement of the bonding ability. Surface cleaning is to remove the adsorbed contaminant, which can prevent the mating surfaces from direct contact. Surface activation is to increase the dangling bonds that are beneficial to strengthen the bonding interface. Room-temperature wafer bonding is an attractive method for the packaging of heterogeneous materials. Without the annealing or heating treatments, problems about the interfacial stress can be effectively solved during the bonding process. Currently, several room-temperature wafer bonding technologies have been developed, i.e., surface activated bonding (SAB), modified SAB, fluorine-contained plasma activated bonding, and Au intermediate-layer bonding. All the bonding methods can be applied to a variety of materials and achieve ultrahigh bonding strength. In this section, we will introduce the advantages and disadvantages of different bonding methods and the range of materials that can be bonded.

7.4.1 Surface Activated Bonding (SAB)

The development of the SAB technique begins in the 1990s; Prof. Tadatomo Suga's group from the University of Tokyo takes the lead in the research on room-temperature SAB. The principle of the SAB method is using the fast argon (Ar) atom or ion beam to bombard surfaces to be bonded in the ultrahigh vacuum environment ($<10^{-5}$ Pa). The oxides and contaminants can be effectively removed. Then the cleaning surfaces come into close contact in the same vacuum environment with appropriate pressure. Atomic bonding can be achieved according to the newly generated bonds across the bonding interface. High bonding strength can be obtained even without subsequent annealing or heating process.

In the early studies, the bonding materials involved are mainly metal-ceramic and metal-semiconductor, such as Al-ceramics, Al-Al, Al-Si_3N_4, Al-stainless steel, Al-Si, Al-Al_2O_3, Cu-Cu, Cu-PbSn solder, etc., at the beginning (Takagi et al., 1999a; Transactions and Society, 2012; Howlader et al., 2010a; Yang et al., 1997). Since the surface activity is ultrahigh after the ion beam bombardment, the problem of secondary oxidation and secondary pollution is easy to occur if this process is carried out in a nonideal state. Therefore, the ion beam bombardment and bonding are usually performed in an ultrahigh vacuum. Although the bonding has been improved, surface oxidation and pollution still exist. Therefore, Prof. Suga specially develops a piece of equipment, which can conduct in situ activation and bonding, for SAB (Fig. 7.4a, b). With the continuous in-depth studies of this technology by Prof. Suga and others, the application of this technique has been extended to the bonding of semiconductor materials. Fast Ar atom or ion beam first removes the surface oxides and contaminants which are harmful to the direct bonding (Fig. 7.4c). Subsequently, atoms at the subsurface layer are destroyed with further bombardment, resulting in an increase of surface energies. Due to the activated surfaces that are in an ultrahigh vacuum environment, surface recovery or secondary oxidation can be effectively prevented. After contacting activated surfaces, covalent or ionic bonds can be formed to bridge the interface.

Eutectic hermetic bonding for Cu metalized substrates needs the deposited layers with a total thickness of several micrometers. However, the bumpless interconnect can be achieved by means of the SAB method (Fig. 7.4d) (Kim et al., 2003; Shigetou et al., 2005). Chemical mechanical polished (CMP) Cu surfaces are bonded without any intermediate layer (Method et al., 2006). A void-free Cu/Cu bonding interface

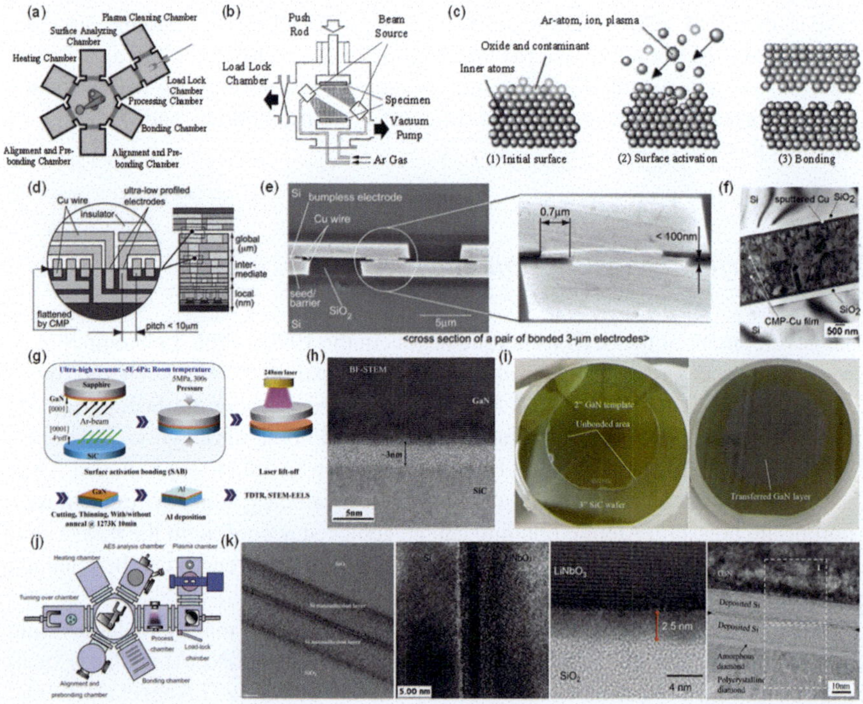

Fig. 7.4 Room-temperature bonding techniques based on surface activated bonding (SAB) method. (**a**) and (**b**) are the schematic diagrams of the robot controlled system and bonding chamber for SAB equipment (Howlader et al., 2006b; Takagi et al., 1998). (**c**) The illustration of the principle of SAB method (Shigetou et al., 2005). (**d**) Schematic diagram of 3D Cu-Cu bumpless interconnect via SAB (Shigetou et al., 2006). (**e**) The optical image of the on-chip integration and SEM image of the bonding interface (Shigetou et al., 2006). (**f**) TEM image of the Cu/Cu bonding interface (Shigetou et al., 2005). (**g**) Schematic diagram of the GaN thin film transfers to the SiC substrate and thermal conductivity test (Mu et al., 2019). (**h**) TEM image of GaN/SiC bonding interface obtained by SAB method (Mu et al., 2019). (**i**) Optical images of the GaN/SiC bonded pair before and after a laser lift-off process (Mu et al., 2019). (**j**) Schematic diagram of the bonding equipment for the modified SAB method (Kondou et al., 2012). (**k**) Typical bonding materials achieved by modified SAB method (Kondou & Suga, 2011; Howlader et al., 2006b; Takigawa et al., 2018; Mu et al., 2018)

is obtained (Fig. 7.4e). Moreover, by embedding Cu into SiO_2 and performing the CMP treatment, a hybrid SAB technique that can realize the bonding of Cu electrodes and SiO_2 at the same time is achieved. With a smooth hybrid surface, a high-quality bonding interface can be confirmed by the TEM observation (Fig. 7.4f) (He et al., 2016). Due to the bonding process that is conducted at room temperature, the interfacial stress caused by the thermal expansion coefficient can be effectively solved. This is important for the high-density 3D bumpless integration. Up to now, the SAB method can be used for the bonding of a variety of homo−/heterogeneous materials, e.g., Si/Si, Si/GaAs, Si/InP, Si/SiC, Si/GaN, Ge/Ge, SiC/SiC, SiC/GaN, SiC/Ga_2O_3, etc. (Xu et al., 2019e; Chung et al., 1997; Mu et al., 2018; Higurashi et al., 2015; Takagi et al., 1996, 2017; Fengwen et al., 2016). Especially in the heterogeneous integration of the third-generation semiconductors, the SAB method shows great advantages because of the high bonding strength, room-temperature bonding, and excellent bonding interfaces. Recently, the wafer-level transfer of the single-crystal GaN thin film to SiC substrates has been developed via the combination of the SAB method and laser irradiation (Fig. 7.4g–i). The GaN thin film grown on a sapphire substrate is bonding to the SiC substrate via the SAB method (Mu et al., 2019). Due to the energy of 248 nm laser that is between the bandgap of GaN and Al_2O_3, sapphire is transparent to the 248 nm laser but GaN has an obvious absorption. Therefore, the GaN thin film will be peeled from the sapphire substrate and remain on the SiC substrate when the laser irradiates the bonded wafer pairs. The GaN/SiC heterogeneous structure has excellent thermal boundary conductance, which is important for high-power electronics.

7.4.2 Modified SAB

Although the SAB method is suitable for the bonding of most materials, it is difficult to realize the bonding of dielectric oxides or ionic compounds with high bonding strength, such as SiO_2 and $LiNbO_3$ (Takagi et al., 1999b; c). To overcome this problem, the modified SAB method using a nanometer-scale Si or Fe layer as the intermediate has been proposed (Suga et al., 2015; Kondou et al., 2012). A process chamber which is used for the Si or metal deposition (Fig. 7.4j) is equipped in the bonding equipment. Before the fast Ar atom or ion beam activation, Si or Fe layers with a thickness of several nanometers are deposited on the dielectric oxide or ionic compound wafers (Takigawa et al., 2018). Then the bonding process proceeds as the SAB method. It has been confirmed that the modified SAB is effective to achieve ultrahigh bonding strength for SiO_2, $LiNbO_3$, and diamond (Fig. 7.4k) (Howlader et al., 2006b). This method expends the range of materials that SAB can bond (Kondou & Suga, 2011; Howlader et al., 2007, 2011a).

7.4.3 Fluorine-Contained Plasma Activated Bonding

Although SAB-based wafer bonding methods have many advantages, they still cannot be widely used in the research. One of the most important reasons is the SAB equipment is very expensive due to the ultrahigh vacuum system. Alternatively, plasma activated bonding is a popular method that has been greatly spread. The generation of plasma and the bonding process can be carried out in a low vacuum environment or the atmosphere (Reiche et al., 2000). Due to the ultrahigh vacuum system that is not required, the operation is convenient and the cost is much lower compared with the SAB method. Therefore, the plasma activated bonding method at room temperature has attracted lots of attention from both academia and industry.

Harbin Institute of Technology and the University of Tokyo have jointly developed a fluorine-contained plasma activated room-temperature wafer direct bonding method (Wang et al., 2017a; Wang & Suga, 2012). The method is mainly to add a very small amount of CF_4 in the traditional O_2 plasma, and surfaces to be bonded are treated with the fluorine-contained plasma (Fig. 7.5a). After 24 h of storage at room temperature, the fracture energy (about 2.5 J/m^2) close to that of the silicon base metal can be obtained without heating. However, due to the fluorine-contained plasma activated surfaces that are hydrophobic, the addition of water vapor in the plasma is essential to improve the bonding ability (Fig. 7.5b) (Wang et al., 2017c). Therefore, the mixed $O_2/CF_4/H_2O$ plasma activation can achieve room-temperature homogeneous Si/Si and heterogeneous Si/glass wafer bonding with ultrahigh bonding strength (Fig. 7.5c). The Si/glass bonding interface is sharp according to the TEM observation. The bonding mechanism is attributed to the softening of surfaces and the introduction of fluorinated bonds that strengthen the bonding interface (Fig. 7.5d). Based on this supposed modified method that uses the PTFE to cover the bottom electrode, fluorine can hang on the activated surfaces due to the sputtering effect caused by the plasma bombardment (Fig. 7.5e) (Wang & Suga, 2016). The bonding energy of the Si/Si bonded pair can be equivalent to the fracture energy of the Si substrate. It also confirms that the fluorine element can play an important role in bonding strength.

Since wafer bonding is often used as a middle process during the device fabrication, subsequent heating or annealing treatments are inevitable for the bonded wafers. However, the void generation in the annealing process is a serious problem in the plasma activated bonding, especially for Si/Si bonded pairs. The origins of voids are mainly divided into two categories: trapped voids and annealing voids. The trapped voids are originated from the particle contaminants adsorbed on wafer surfaces. They are visible after the bonding. Annealing voids are caused by the thermal decomposition of organic contaminants. They are invisible after the bonding at room temperature, but appear during the annealing treatment. The hydrocarbons or micro defects at the bonding interface may act as embryos for the nucleation site of water or hydrogen. There are nearly no visible voids for both O_2 and fluorine-contained plasma activation after annealing at 200 °C (Fig. 7.5f, g) (Wang et al., 2017b). It means that the organic contaminants are mainly removed by the plasma

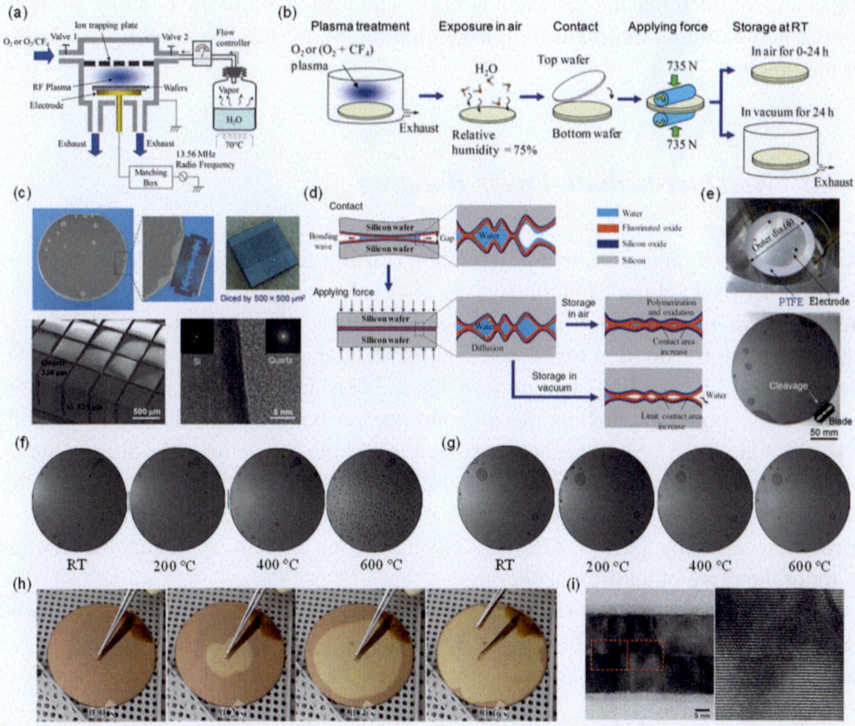

Fig. 7.5 Room-temperature plasma activated bonding: fluorine-contained plasma activated bonding and Au intermediate-layer bonding. (**a**) Schematic diagram of the modified bonding equipment for fluorine-contained plasma activation (Wang et al., 2017c). (**b**) Schematic drawing of the fluorine-contained plasma activated bonding process (Wang et al., 2017a). (**c**) Optical, SEM, and TEM images of Si/glass bonded pairs (Wang et al., 2017c). (**d**) Bonding mechanisms for the fluorine-contained plasma activated bonding (Wang et al., 2017a). (**e**) Realization of room-temperature bonding with fluorine-contained surfaces (Wang & Suga, 2016). Comparison of the (**f**) O_2 and (**g**) fluorine-contained plasma activated Si/Si wafer bonding at different annealing temperatures (Wang et al., 2017b). (**h**) Room-temperature Au/Au intermediate wafer bonding (Higurashi et al., 2017). (**i**) TEM observation for the plasma activated Au/Au bonding interface (Higurashi et al., 2017)

treatment. However, when the annealing temperature is further increased, fewer voids are present for the bonded pairs treated by the fluorine-contained plasma treatment. It is mainly due to the addition of fluorine into the O_2 plasma that can reduce the hydrophilicity of the Si surfaces, allowing the water and hydrogen to diffuse out more easily from the interface.

The material categories for the fluorine-contained plasma activated bonding are limited. This method is more suitable for the bonding of Si-based materials (e.g., Si/Si, Si/SiO_2, and Si/glass). For other materials (e.g., Si_3N_4 and AlN), the increase in the bonding strength is not obvious. Similar to the modified SAB method, fluorine-contained plasma activated bonding can also increase the bonding strength for the non-Si-based materials by introducing the Si or SiO_2 layers with a thickness of nanometers.

7.4.4 Au Intermediate-Layer Bonding

Another room-temperature bonding method is using Au as the intermediate layer to perform the packaging (Yamamoto et al., 2012). Au is a good candidate material for the intermediate-layer bonding due to the high electrical conductivity, good deformability, and high oxidation resistance (Takigawa et al., 2011; Shimatsu et al., 2014). Several Au-Au bonding techniques have been investigated to realize low-temperature bonding, such as thermocompression bonding and diffusion bonding (Okada et al., 2013). Surfaces activated by the plasma can achieve the atomic bonding at room temperature. Besides, the use of Au as the bonding layer enables the bonding performed in ambient air because the gold oxide (Au_2O_3) has a positive formation enthalpy (+19.3 kJ/mol). The oxidation cannot spontaneously occur without the drive of additional energies.

O_2 and Ar are two kinds of commonly used plasma to clean and active surfaces (Yamamoto et al., 2018). The activated mechanisms of two kinds of plasma for Au surfaces are different. Ar plasma is physical etching and O_2 plasma is a chemical reaction (Yamamoto et al., 2019). Studies indicate that Ar plasma treatment can improve the room-temperature wafer-level bonding strength, while O_2 plasma activation cannot strengthen the Au/Au interface. X-ray photoelectron spectroscopy (XPS) analyses show that Au_2O_3 is formed on the Au surface after O_2 plasma activation. Since the oxide has a negative effect on the bonding, the bonding strength is low (0.1–0.2 J/m^2). However, Au_2O_3 will desorb at around 110 °C according to the thermal desorption spectroscopy analysis. The O_2 plasma activated Au/Au bonding strength can increase to 2.5 J/m^2, which is equal to the surface energy of bulk Si. For the Ar plasma-treated Au surfaces, spontaneous bonding can occur when the activated surfaces are contacted with each other (Fig. 7.5h) (Higurashi et al., 2017). The dense bonding interface with high bonding strength is observed by TEM (Fig. 7.5i). The bonding strength of samples that are exposed in the air following Ar plasma treatment decreased more rapidly than that of samples bonded immediately after deposition. It is mainly due to the decrease of surface energies when Ar activated surfaces are exposed in the air condition. Therefore, although Au surfaces that receive Ar plasma treatment are more activated than that of O_2 plasma treatment, the bonding should be conducted immediately after the plasma treatment.

7.5 Application of Wafer Bonding in Plasmonics-Enhanced Micro/Nanofluidic Sensors

In the packaging of MEMS and COMS devices, wafer bonding has played an important role in the sealing, electrical interconnection, and cost control. Although the development of wafer bonding technology relies on the needs of the MEMS and COMS fabrication process, this advanced packaging method also benefits other research fields. One of the most closely related fields is micro/nanofluidics (Mawatari et al., 2014; Xu, 2017). By fabricating micro/nanochannels on the bottom SiO_2 or glass substrate, the top substrate can be integrated with it into one composite via the direct bonding technology (Xu et al., 2013; Wang et al., 2018a). Therefore, the nanometer-scale confined spaces can be used for biological or chemical reactions, molecular diffusion studies, particle manipulation, energy conversion, etc. (Yazici et al., 2020; Shi et al., 2020). In this section, we will introduce an emerging direction in the micro/nanofluidics field: plasmonics-enhanced micro/nanofluidic sensors.

Plasmonic metasurfaces have found a wide range of promising applications in biosensors, chemical sensors, dynamic beam shaping, surface-enhanced Raman spectroscopy (SERS), surface-enhanced IR absorption (SEIRA) spectroscopy, etc. (Wei et al., 2019; Hasan & Lee, 2018; Ren et al., 2020; Chang et al., 2020; Wei & Lee, 2019; Hasan et al., 2017a, b; Dong et al., 2020; Zhou et al., 2020). Due to PDMS that has good bonding abilities, the plasmonics-microfluidic devices can be easily integrated by bonding the substrate with metasurface and PDMS with microchannels (Fig. 7.6a). For the fingerprints of gas and biological molecules (e.g., N-H, O-H, C-H, and C=O), most of them are concentrated in the mid-infrared wavelength beyond 3 μm (Fig. 7.6b) (Rodrigo et al., 2018; Ma et al., 2020a; b). Therefore, the substrate should be transparent infrared materials when using the SEIRA technology for molecular identification. With a small amount of analyte, the plasmonics-microfluidics can achieve dynamic monitoring for chemical and biological reactions with high sensitivity (Fig. 7.6c) (Adato & Altug, 2013). To further improve the sensitivity, the metal-insulator-metal configuration can obtain higher responses in signals because of the quadrupole resonance. However, the enhanced electromagnetic field can only be generated when the distance between metasurfaces and metal mirror is nanometer-scale (Shih et al., 2019). The PDMS is not suitable for the fabrication of nanogaps in MIM due to the collapse. To solve this problem, UV activation has been used for the direct bonding of CaF_2 and SiO_2. The CaF_2-based plasmonics-nanofluidic can quantitatively analyze molecules because of the ultrasmall sensing chamber (Fig. 7.6d). The sensitivity is two orders of magnitude than traditional detected methods (Le & Tanaka, 2017; Le et al., 2018). This sensing platform also shows large responses when the reflective index of analyte changes (Fig. 7.6e) (Le et al., 2020). However, the bonding strength and interface are still not strong enough to resist the water stress corrosion for continuous long-time sensing. To get a strong interface for the infrared material bonding, the mixed gas plasma

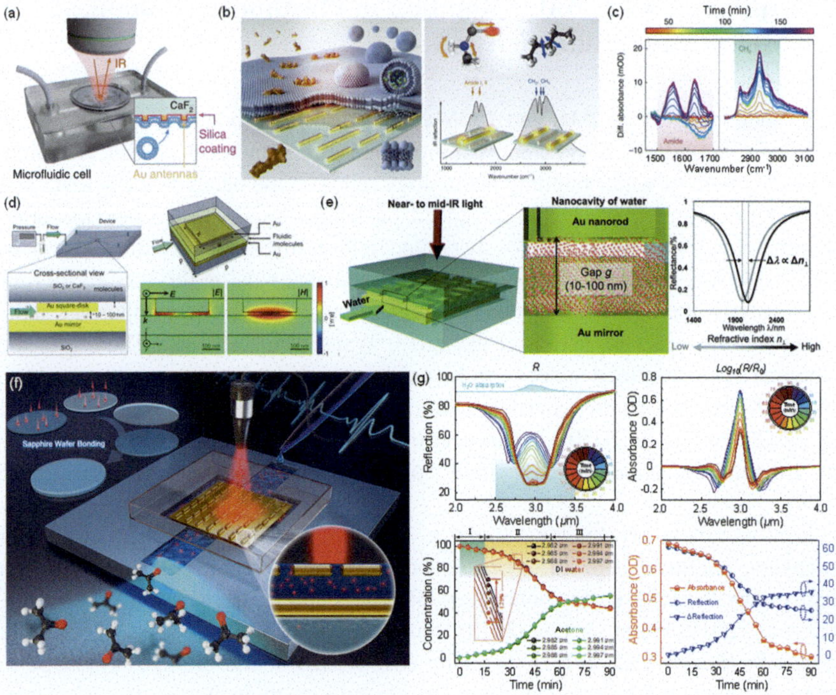

Fig. 7.6 Infrared material bonding for the integration of plasmonics into micro/nanofluidics. (**a**) Schematic diagram of the plasmonics-microfluidic device configuration (Rodrigo et al., 2018). (**b**) Multi-resonant MIR nanoantennas and corresponding measured reflection spectra (Rodrigo et al., 2018). (**c**) Spectral responses during the dynamic test (Rodrigo et al., 2018). (**d**) Device configuration of the CaF_2-based plasmonics-enhanced nanofluidics (Le & Tanaka, 2017). (**e**) Applications of the plasmonic nanofluidic in the recognition of refractive index changes (Le et al., 2020). (**f**) Schematic diagram of the nanometer-scale heterogeneous interfacial sapphire wafer bonding for enabling plasmonics-enhanced nanofluidic MIR spectroscopy (Xu et al., 2020a). (**g**) Dynamic monitoring for the acetone diffusion using the sapphire-based plasmonic nanofluidic (Xu et al., 2020a)

activation has been used for the sapphire bonding (Fig. 7.6f). The sapphire-based plasmonics-nanofluidic platform also demonstrates the ability for the dynamic monitor of molecular diffusion with ultrahigh sensitivity (Fig. 7.6g) (Xu et al., 2020a). Since the transparent window of the sapphire substrate is up to 5.2 μm, it is still limited for the test of macromolecules. Therefore, reliable bonding for infrared materials with larger transparent windows (e.g., MgF_2, CaF_2, and BaF_2) should be developed to satisfy the test requirements.

7.6 Summary and Outlook

Wafer-level packaging based on wafer bonding is an essential process for MEMS device fabrication. It accounts for 50% to 90% in terms of the total cost. Although MEMS technology is originated from the IC process, the development of packaging methods is driven by both of them. In this chapter, we reviewed the technology evolution from the traditional bonding method (i.e., wafer-level eutectic bonding) to the advanced packaging (i.e., low-temperature wafer direct bonding and room-temperature bonding). We introduced the bonding process, using conditions and applicable materials of each method in detail. Differences between traditional bonding methods and advanced packaging technologies were also highlighted in the text. In the final section, we focused on the application of wafer direct bonding in the emerging fields of plasmonics-enhanced nanofluidic sensors. In the past few years, a variety of bonding methods are developed for semiconductors because of the huge market demand. With the booming of the Internet of Things (IoT), sensing technology for personal healthcare and environmental monitoring has become more and more important in our daily life. Therefore, the bonding of infrared materials for the ultrasensitive detection will be more extensive in the future applications. Besides, although all kinds of wafer-level packaging methods have been put into production, the bonding mechanism for the wafer bonding still needs to be further explored. Compared with other mature fabrication technologies, the advanced packaging methods are still in a preliminary stage. More work needs to be invested in basic research to fundamentally solve the problem that occurs in actual production.

Acknowledgments This work is partly supported by research grant of CRP-15th (NRF-CRP15-2015-02) at the National University of Singapore (NUS), Singapore, and partly supported by the National Natural Science Foundation of China (grant no. 51975151) at Harbin Institute of Technology. Jikai Xu also thanks the China Scholarship Council (File No. 201906120176) for supporting this work.

References

Adato, R., & Altug, H. (2013). In-situ ultra-sensitive infrared absorption spectroscopy of biomolecule interactions in real time with plasmonic nanoantennas. *Nature Communications, 4*, 2154.

Ahn, S. Y., & Lee, N. Y. (2015). Solvent-free thermoplastic-poly(dimethylsiloxane) bonding mediated by UV irradiation followed by gas-phase chemical deposition of an adhesion linker. *Journal of Micromechanics and Microengineering, 25*, 75007. https://doi.org/10.1088/0960-1317/25/7/075007

Alam, A. U., Howlader, M. M. R., & Deen, M. J. (2013). Oxygen plasma and humidity dependent surface analysis of silicon, silicon dioxide and glass for direct wafer bonding. *ECS Journal of Solid State Science and Technology, 2*, P515–P523.

Anantha, P., & Tan, C. S. (2015). UV/O_3 assisted InP/Al_2O_3–Al_2O_3/Si low temperature die to wafer bonding. *Microsystem Technologies, 21*, 1015–1020.

Brand, O., & Baltes, H. (2002). Microsensor packaging. *Microsystem Technologies, 7*, 205–208.

Chang, Y., Wei, J., & Lee, C. (2020). Metamaterials-from fundamentals and MEMS tuning mechanisms to applications. *NANO, 9*, 3049–3070.

Chuang, R. W., & Lee, C. C. (2002). Silver-indium joints produced at low temperature for high temperature devices. *IEEE Transactions on Components and Packaging Technologies, 25*, 453–458.

Chuang, T. H., Lin, H. J., & Tsao, C. W. (2006). Intermetallic compounds formed during diffusion soldering of au/cu/Al_2O_3 and cu/Ti/Si with Sn/in interlayer. *Journal of Electronic Materials, 35*, 1566–1570.

Chung, T. R., Yang, L., Hosoda, N., & Suga, T. (1997). Room temperature GaAs-Si and InP-Si wafer direct bonding by the surface activated bonding method. *Nuclear Instruments and Methods in Physics Research Section B: Beam Interactions with Materials and Atoms, 121*, 203–206.

Dong, B., Ma, Y., Ren, Z., & Lee, C. (2020). Recent progress in nanoplasmonics-based integrated optical micro/nano-systems. *Journal of Physics D: Applied Physics, 53*, 213001.

Dragoi, V., Mittendorfer, G., Thanner, C., & Lindner, P. (2008). Wafer-level plasma activated bonding: New technology for MEMS fabrication. *Microsystem Technologies, 14*, 509–515.

Esser, R. H., Hobart, K. D., & Kub, F. J. (2003). Improved low-temperature Si-Si hydrophilic wafer bonding. *Journal of the Electrochemical Society, 150*, G228.

Fengwen, M., Kenichi, I., Haruo, N., et al. (2016). Room-temperature wafer bonding of SiC-Si by modified surface activated bonding with sputtered Si nanolayer. *Japanese Journal of Applied Physics, 55*, 04EC09.

Fu, W., Shigetou, A., Shoji, S., & Mizuno, J. (2016). Low temperature bonding between polyether ether ketone (PEEK) and Pt through vapor assisted VUV surface modification. In *Proceedings—Electron. Components Technol. Conf. 2016-August* (pp. 2541–2546).

Gooch, R., & Schimert, T. (2003). Low-cost wafer-level vacuum packaging for MEMS. *MRS Bulletin, 28*, 55–59.

Hasan, D., & Lee, C. (2018). Hybrid metamaterial absorber platform for sensing of CO_2 gas at mid-IR. *Advancement of Science, 5*, 1700581.

Hasan, D., Pitchappa, P., Wang, J., et al. (2017a). Novel CMOS-compatible Mo-AlN-Mo platform for metamaterial-based mid-IR absorber. *ACS Photonics, 4*, 302–315.

Hasan, D., Pitchappa, P., Pei Ho, C., & Lee, C. (2017b). High temperature coupling of IR inactive C=C mode in complementary metal oxide semiconductor metamaterial structure. *Advanced Optical Materials, 5*, 00778.

He, R., Fujino, M., Yamauchi, A., et al. (2016). Combined surface activated bonding technique for low-temperature cu/dielectric hybrid bonding. *ECS Journal of Solid State Science and Technology, 5*, P419–P424.

He, R., Yamauchi, A., & Suga, T. (2018). Sequential plasma activation methods for hydrophilic direct bonding at sub-200 ° C. *Japanese Journal of Applied Physics, 57*, 02BD03.

Higurashi, E., Sasaki, Y., Kurayama, R., et al. (2015). Room-temperature direct bonding of germanium wafers by surface-activated bonding method. *Japanese Journal of Applied Physics, 54*, 030213.

Higurashi, E., Okumura, K., Kunimune, Y., et al. (2017). Room-temperature bonding of wafers with smooth au thin films in ambient air using a surface-activated bonding method. *IEICE Transactions on Electronics, E100C*, 156–160.

Howlader, M. M. R., Suehara, S., & Suga, T. (2006a). Room temperature wafer level glass/glass bonding. *Sensors and Actuators A: Physical, 127*, 31–36.

Howlader, M. M. R., Suga, T., & Kim, M. J. (2006b). Room temperature bonding of silicon and lithium niobate. *Applied Physics Letters, 89*, 1–4.

Howlader, M. M. R., Suga, T., & Kim, M. J. (2007). A novel bonding method for ionic wafers. *IEEE Transactions on Advanced Packaging, 30*, 598–604.

Howlader, M. M. R., Suga, T., Itoh, H., et al. (2009). Role of heating on plasma-activated silicon wafers bonding. *Journal of the Electrochemical Society, 156*, H846.

Howlader, M. M. R., Kibria, M. G., Zhang, F., & Kim, M. J. (2010a). Hybrid plasma bonding for void-free strong bonded interface of silicon/glass at 200 °C. *Talanta, 82*, 508–515.

Howlader, M. M. R., Zhang, F., & Kibria, M. G. (2010b). Void nucleation at a sequentially plasma-activated silicon/silicon bonded interface. *Journal of Micromechanics and Microengineering, 20*, 065012.

Howlader, M. M. R., Kaga, T., & Suga, T. (2010c). Investigation of bonding strength and sealing behavior of aluminum/stainless steel bonded at room temperature. *Vacuum, 84*, 1334–1340.

Howlader, M. M. R., Zhang, F., & Kim, M. J. (2011a). Annealing temperature-dependent interfacial behavior of sequentially plasma-activated silicon bonded wafers. *Journal of Microelectromechanical Systems, 20*, 17–20.

Howlader, M. M. R., Selvaganapathy, P. R., Deen, M. J., & Suga, T. (2011b). Nanobonding technology toward electronic, fluidic, and photonic systems integration. *IEEE Journal of Selected Topics in Quantum Electronics, 17*, 689–703.

Howlader, M. R., Itoh, H., Suga, T., & Kim, M. (2019). Sequential plasma activated process for silicon direct bonding. *ECS Transactions, 3*, 191–202.

Hsu, C., Chen, G., Lin, Y. H., & Cheng, M. M. C. (2017). Anodic bonding using Gorilla glasses. In *2017 IEEE 12th Int. Conf. Nano/micro Eng. Mol. Syst. NEMS 2017* (pp. 566–569).

Ikku, Y., Yokoyama, M., Iida, R., et al. (2011). ALD Al_2O_3 activated direct wafer bonding for 111-V CMOS photonics platform. In *Conf proc-Int. Conf. Indium phosphide Relat. Mater* (pp. 7–10).

Kikkawa, T., Hirose, K., & Chikaki, S. (2000). Authors would like to thank Dr. M. Kamoshida, Dr. M. Ogawa, Dr. K. Okada, Dr. M. Kikuchi, Dr. A. *Ishitani, 143*, 1773–1779.

Kim, T. H., Howlader, M. M. R., Itoh, T., & Suga, T. (2003). Room temperature cu–cu direct bonding using surface activated bonding method. *Journal of Vacuum Science & Technology, A: Vacuum, Surfaces, and Films, 21*, 449–453.

Kondou, R., & Suga, T. (2011). Si nanoadhesion layer for enhanced SiO_2-SiN wafer bonding. *Scripta Materialia, 65*, 320–322.

Kondou, R., Wang, C., Shigetou, A., & Suga, T. (2012). Nanoadhesion layer for enhanced Si-Si and Si-SiN wafer bonding. *Microelectronics and Reliability, 52*, 342–346.

Le, T. H. H., & Tanaka, T. (2017). Plasmonics–nanofluidics hydrid metamaterial: An ultrasensitive platform for infrared absorption spectroscopy and quantitative measurement of molecules. *ACS Nano, 11*, 9780–9788.

Le, T. H. H., Morita, A., Mawatari, K., et al. (2018). Metamaterials-enhanced infrared spectroscopic study of nanoconfined molecules by plasmonics-nanofluidics hybrid device. *ACS Photonics, 5*, 3179–3188.

Le, T. H. H., Morita, A., & Tanaka, T. (2020). Refractive index of nanoconfined water reveals its anomalous physical properties. *Nanoscale Horizons, 5*, 1016–1024.

Lee, C. C., & Choe, S. (2002). Fluxless in-Sn bonding process at 140 °C. *Materials Science and Engineering A, 333*, 45–50.

Lee, C., & Xie, J. (2009). Design and optimization of wafer bonding packaged microelectromechanical systems thermoelectric power generators with heat dissipation path. *Journal of Vacuum Science & Technology B: Microelectronics and Nanometer Structures, 27*, 1267.

Lee, C. C., Wang, C. Y., & Matijasevic, G. (1993). Au–in bonding below the eutectic temperature. *IEEE Transactions on Components, Hybrids, and Manufacturing Technology, 16*, 311–316.

Lee, C., Huang, W. F., & Shie, J. S. (2000). Wafer bonding by low-temperature soldering. *Sensors and Actuators A: Physical, 85*, 330–334.

Lee, C., Yu, D., Yu, A., et al. (2008). Bonding interface materials evolution of intermediate in/ag layers for low temperature fluxless solder based MEMS wafer level packaging. In *2008 10th Int. Conf. Electron. Mater. Packag. EMAP 2008 2* (pp. 216–219).

Lee, C., Yu, A., Yan, L., et al. (2009). Characterization of intermediate in/ag layers of low temperature fluxless solder based wafer bonding for MEMS packaging. *Sensors and Actuators A: Physical, 154*, 85–91.

Li, D. L., Shang, Z. G., Wang, S. Q., & Wen, Z. Y. (2013). Low temperature Si/Si wafer direct bonding using a plasma activated method. *Journal of Zhejiang University Science C, 14*, 244–251.
Ma, Y., Dong, B., & Lee, C. (2020a). Progress of infrared guided-wave nanophotonic sensors and devices. *Nano Convergence, 7*, 12.
Ma, Y., Dong, B., Wei, J., et al. (2020b). High-responsivity mid-infrared black phosphorus slow light waveguide photodetector. *Advanced Optical Materials, 8*, 1–12.
Made, R. I., Gan, C. L., Yan, L. L., et al. (2009). Study of low-temperature thermocompression bonding in ag-in solder for packaging applications. *Journal of Electronic Materials, 38*, 365–371.
Mai, C., Li, M., & Yang, S. (2015). Low temperature direct bonding of silica glass via wet chemical surface activation. *RSC Advances, 5*, 42721–42727.
Mai, C., Sun, J., Chen, H., et al. (2016). Silicon direct bonding: Via low-temperature wet chemical surface activation. *RSC Advances, 6*, 37079–37084.
Masteika, V., Kowal, J., Braithwaite, N. S. J., & Rogers, T. (2014). A review of hydrophilic silicon wafer bonding. *ECS Journal of Solid State Science and Technology, 3*, Q42–Q54.
Matsumae, T., Fengwen, M., Fukumoto, S., et al. (2021). Heterogeneous GaN-Si integration via plasma activation direct bonding. *Journal of Alloys and Compounds, 852*, 156933.
Mawatari, K., Kazoe, Y., Shimizu, H., et al. (2014). Extended-nanofluidics: Fundamental technologies, unique liquid properties, and application in chemical and bio analysis methods and devices. *Analytical Chemistry, 86*, 4068–4077.
Mu, F., He, R., & Suga, T. (2018). Room temperature GaN-diamond bonding for high-power GaN-on-diamond devices. *Scripta Materialia, 150*, 148–151.
Mu, F., Cheng, Z., Shi, J., et al. (2019). High thermal boundary conductance across bonded heterogeneous GaN-SiC interfaces. *ACS Applied Materials and Interfaces, 11*, 33428–33434.
Okada, A., Shoji, S., Nimura, M., et al. (2013). Vacuum ultraviolet irradiation treatment for reducing gold-gold bonding temperature. *Materials Transactions, 54*, 2139–2143.
Pasquariello, D., Hedlund, C., & Hjort, K. (2000). Oxidation and induced damage in oxygen plasma in situ wafer bonding. *Journal of the Electrochemical Society, 147*, 2699.
Phan, H. P., Cheng, H. H., Dinh, T., et al. (2017). Single-crystalline 3C-SiC anodically bonded onto glass: An excellent platform for high-temperature electronics and bioapplications. *ACS Applied Materials and Interfaces, 9*, 27365–27371.
Plach, T., Hingerl, K., Tollabimazraehno, S., et al. (2013). Mechanisms for room temperature direct wafer bonding. *Journal of Applied Physics, 113*, 4794319.
Qian, W., Kyudong, J., Minseog, C., et al. (2007). Low temperature, wafer level Au-In bonding for ISM packaging. In *2006 7th International Conference on Electronic Packaging Technology ICEPT 06* (p. 7).
Reiche, M., Gösele, U., & Wiegand, M. (2000). Modification of Si(100)-surfaces by SF_6 plasma etching-application to wafer direct bonding. *Crystal Research and Technology, 35*, 807–821.
Ren, Z., Chang, Y., Ma, Y., et al. (2020). Leveraging of MEMS technologies for optical metamaterials applications. *Advanced Optical Materials, 8*, 1–20.
Rodrigo, D., Tittl, A., Ait-bouziad, N., et al. (2018). Resolving molecule-specific information in dynamic lipid membrane processes with multi-resonant infrared metasurfaces. *Nature Communications, 9*, 2160.
Shi, Q., Dong, B., He, T., et al. (2020). Progress in wearable electronics/photonics-moving toward the era of artificial intelligence and internet of things. *InfoMat*, 1–32.
Shigetou, A., Itoh, T., & Suga, T. (2005). Direct bonding of CMP-cu films by surface activated bonding (SAB) method. *Journal of Materials Science, 40*, 3149–3154.
Shigetou, A., Itoh, T., & Suga, T. (2006). Bumpless interconnect of ultrafine cu electrodes by surface activated bonding (SAB) method. *Electron. Commun Japan, part II Electron. (English Transl Denshi Tsushin Gakkai Ronbunshi), 89*, 34–42.
Shigetou, A., Itoh, T., Matsuo, M., et al. (2006). Bumpless interconnect through ultrafine Cu electrodes by means of surface-activated bonding (SAB) method. *IEEE Transactions on Advanced Packaging, 29*, 218–226.

Shigetou, A., Mizuno, J., & Shoji, S. (2015). Vacuum ultraviolet (VUV) and vapor-combined surface modification for hybrid bonding of SiC, GaN, and Si substrates at low temperature and atmospheric pressure. In *Proceedings—Electron. Components Technol. Conf. 2015-July* (pp. 1498–1501).

Shih, K., Ren, Z., Wang, C., & Lee, C. (2019). MIR plasmonic liquid sensing in nano-metric space driven by capillary force. *Journal of Physics D: Applied Physics, 52*, 394001.

Shimatsu, T., Uomoto, M., & Kon, H. (2014). (Invited) room temperature bonding using thin metal films (bonding energy and technical potential). *ECS Transactions, 64*, 317–328.

Shinohara, H., Mizuno, J., & Shoji, S. (2011). Studies on low-temperature direct bonding of VUV, VUV/O_3 and O_2 plasma pretreated cyclo-olefin polymer. *Sensors and Actuators A: Physical, 165*, 124–131.

Suga, T., Takahashi, Y., Takagi, H., et al. (1992). Structure of A1-A1 and A1-Si_3N_4 interfaces bonded at room temperature by means of the surface activation method. *Acta Metallurgica et Materialia, 40*, 133–137.

Suga, T., Mu, F., Fujino, M., et al. (2015). Silicon carbide wafer bonding by modified surface activated bonding method. *Japanese Journal of Applied Physics, 54*, 030214.

Suni, T., Henttinen, K., Lipsanen, A., et al. (2005). Wafer scale packaging of mems by using plasma activated wafer bonding. *Proceedings of the Electrochemical Society PV, 2005-02*, 173–183.

Sunkara, V., & Cho, Y. K. (2012). Investigation on the mechanism of aminosilane-mediated bonding of thermoplastics and poly(dimethylsiloxane). *ACS Applied Materials and Interfaces, 4*, 6537–6544.

Takagi, H., Kikuchi, K., Maeda, R., et al. (1996). Surface activated bonding of silicon wafers at room temperature. *Applied Physics Letters, 68*, 2222–2224.

Takagi, H., Maeda, R., Chung, T. R., et al. (1998). Effect of surface roughness on room-temperature wafer bonding by Ar beam surface activation. *Japanese Journal of Applied Physics, Part 1: Regular Papers & Short Notes, 37*, 4197–4203.

Takagi, H., Maeda, R., Hosoda, N., & Suga, T. (1999a). Transmission electron microscope observations of Si/Si interface bonded at room temperature by Ar beam surface activation. *Japanese Journal of Applied Physics, Part 1: Regular Papers & Short Notes, 38*, 1589–1594.

Takagi, H., Maeda, R., Hosoda, N., & Suga, T. (1999b). Room-temperature bonding of Si wafers to Pt films on SiO_2 or $LiNbO_3$ substrates using Ar-beam surface activation. *Japanese Journal of Applied Physics, Part 2: Letters, 38*, 126–129.

Takagi, H., Maeda, R., Hosoda, N., & Suga, T. (1999c). Room-temperature bonding of lithium niobate and silicon wafers by argon-beam surface activation. *Applied Physics Letters, 74*, 2387–2389.

Takagi, H., Maeda, R., & Suga, T. (2017). Room-temperature wafer bonding of Si to $LiNbO_3$, $LiTaO_3$ and $Gd_3Ga_5O_{12}$ by Ar-beam surface activation. *Journal of Micromechanics and Microengineering, 348*, 11–16.

Takigawa, R., Higurashi, E., Suga, T., & Kawanishi, T. (2011). Air-gap structure between integrated $LiNbO_3$ optical modulators and micromachined Si substrates. *Optics Express, 19*, 15739.

Takigawa, R., Higurashi, E., & Asano, T. (2018). Room-temperature wafer bonding of $LiNbO_3$ and SiO_2 using a modified surface activated bonding method. *Japanese Journal of Applied Physics, 57*, 06HJ12.

Tang, Z., Peng, P., Shi, T., et al. (2009). Effect of nanoscale surface topography on low temperature direct wafer bonding process with UV activation. *Sensors and Actuators A: Physical, 151*, 81–86.

Tang, J., Cai, C., Ming, X., et al. (2016). Morphology and stress at silicon-glass interface in anodic bonding. *Applied Surface Science, 387*, 139–148.

Tatar, E., Torunbalci, M. M., Alper, S. E., & Akin, T. (2012). A method and electrical model for the anodic bonding of SOI and glass wafers. In *Proc IEEE Int. Conf. Micro Electro. Mech. Syst.* (pp. 68–71).

Tran, H. H., Wu, W., & Lee, N. Y. (2013). Ethanol and UV-assisted instantaneous bonding of PMMA assemblies and tuning in bonding reversibility. *Sensors and Actuators B: Chemical, 181*, 955–962.

Transactions ECS, Society TE. (2012). Room temperature wafer bonding by surface activated ALD- Al_2O_3. *ECS Meeting Abstracts, 50*, 303–311.

Wang, C., & Suga, T. (2012). Investigation of fluorine containing plasma activation for room-temperature bonding of Si-based materials. *Microelectronics and Reliability, 52*, 347–351.

Wang, C., & Suga, T. (2016). Communication-fluorinated plasma treatments using PTFE substrates for room-temperature silicon wafer direct bonding. *ECS Journal of Solid State Science and Technology, 5*, P393–P395.

Wang, Q., Choa, S. H., Kim, W. B., et al. (2006). Reliability of hermetic RF MEMS wafer level packaging using au-Sn eutectic bonding. *Key Engineering Materials, 326–328*(I), 609–612.

Wang, C., Liu, Y., Li, Y., et al. (2017a). Mechanisms for room-temperature fluorine containing plasma activated bonding. *ECS Journal of Solid State Science and Technology, 6*, P373–P378.

Wang, C., Wang, Y., Tian, Y., et al. (2017b). Room-temperature direct bonding of silicon and quartz glass wafers room-temperature direct bonding of silicon and quartz glass wafers. *Applied Physics Letters, 110*, 221602.

Wang, C., Liu, Y., & Suga, T. (2017c). A comparative study: Void formation in silicon wafer direct bonding by oxygen plasma activation with and without fluorine. *ECS Journal of Solid State Science and Technology, 6*, P7–P13.

Wang, C., Xu, J., Zeng, X., et al. (2018a). Low-temperature wafer direct bonding of silicon and quartz glass by a two-step wet chemical surface cleaning. *Japanese Journal of Applied Physics, 57*, 02BD02.

Wang, C., Xu, J., Qi, X., et al. (2018b). Direct homo/heterogeneous bonding of silicon and glass using vacuum ultraviolet irradiation in air. *Journal of the Electrochemical Society, 165*, H3093–H3098.

Wang, C., Wang, T., Mawatari, K., & Kitamori, T. (2018c). Communication—Evaporation driven micro/Nanofluidic pumping device. *Journal of the Electrochemical Society, 165*, B184–B186.

Wang, C., Xu, J., Guo, S., et al. (2019). A facile method for direct bonding of single-crystalline SiC to Si, SiO_2, and glass using VUV irradiation. *Applied Surface Science, 471*, 196–204.

Wei, J., & Lee, C. (2019). Anomalous plasmon hybridization in nanoantennas near interfaces. *Optics Letters, 44*, 6041.

Wei, J., Li, Y., Chang, Y., et al. (2019). Ultrasensitive transmissive infrared spectroscopy via loss engineering of metallic nanoantennas for compact devices. *ACS Applied Materials and Interfaces, 11*, 47270–47278.

Welch, W. C., Chae, J., & Najafi, K. (2005). Transfer of metal MEMS packages using a wafer-level solder transfer technique. *IEEE Transactions on Advanced Packaging, 28*, 643–649.

Wu, Z., Xanthopoulos, N., Reymond, F., et al. (2002). Polymer microchips bonded by O_2-plasma activation. *Electrophoresis, 23*, 782–790.

Xu, Y. (2017). Nanofluidics: A new arena for materials science. *Advanced Materials, 1702419*, 1–17.

Xu, Y., Wang, C., Li, L., et al. (2013). Bonding of glass nanofluidic chips at room temperature by a one-step surface activation using an O_2/CF_4 plasma treatment. *Lab on a Chip, 13*, 1048–1052.

Xu, J., Wang, C., Wang, T., et al. (2018a). Mechanisms for low-temperature direct bonding of Si/Si and quartz/quartz: Via VUV/O_3 activation. *RSC Advances, 8*, 11528–11535.

Xu, J., Wang, C., Wu, B., et al. (2018b). Communication-defect-free direct bonding for high-performance glass-on-$LiNbO_3$ devices. *Journal of the Electrochemical Society, 165*, B727–B729.

Xu, J., Wang, C., Tian, Y., et al. (2018c). Glass-on-$LiNbO_3$ heterostructure formed via a two-step plasma activated low-temperature direct bonding method. *Applied Surface Science, 459*, 621–629.

Xu, J., Wang, C., Li, D., et al. (2019a). Fabrication of SiC/Si, SiC/SiO_2, and SiC/glass heterostructures via VUV/O_3 activated direct bonding at low temperature. *Ceramics International, 45*, 4094–4098.

Xu, J., Wang, C., Zhou, S., et al. (2019b). Low-temperature direct bonding of Si and quartz glass using the APTES modification. *Ceramics International, 45*, 16670–16675.

Xu, J., Wang, C., Zhang, R., et al. (2019c). VUV/O_3 activated direct heterogeneous bonding towards high-performance $LiNbO_3$-based optical devices. *Applied Surface Science, 495*, 1–11.

Xu, J., Wang, C., Kang, Q., et al. (2019d). Direct heterogeneous bonding of SiC to Si, SiO_2, and glass for high-performance power electronics and bio-MEMS. In *Proceedings—Electron. Components Technol. Conf. 2019-May* (pp. 1266–1271).

Xu, Y., Mu, F., Wang, Y., et al. (2019e). Direct wafer bonding of Ga_2O_3–SiC at room temperature. *Ceramics International, 45*, 6552–6555.

Xu, J., Wang, C., Ji, X., et al. (2020a). Direct bonding of high dielectric oxides for high-performance transistor applications. *Scripta Materialia, 178*, 307–312.

Xu, J., Ren, Z., Dong, B., et al. (2020b). Nanometer-scale heterogeneous interfacial sapphire wafer-bonding for enabling plasmonic-enhanced nanofluidic mid-infrared spectroscopy. *ACS Nano, 14*, 12159–12172.

Yamamoto, S. I., Higurashi, E., Suga, T., & Sawada, R. (2012). Low-temperature hermetic packaging for microsystems using AuAu surface-activated bonding at atmospheric pressure. *Journal of Micromechanics and Microengineering, 22*, 055026.

Yamamoto, M., Higurashi, E., Suga, T., et al. (2018). Properties of various plasma surface treatments for low-temperature au-au bonding. *Japanese Journal of Applied Physics, 57*, 1–6.

Yamamoto, M., Matsumae, T., Kurashima, Y., et al. (2019). Comparison of argon and oxygen plasma treatments for ambient room-temperature wafer-scale au-au bonding using ultrathin au films. *Micromachines, 10*, 0020119.

Yan, L. L., Lee, C. K., Yu, D. Q., et al. (2009). A hermetic seal using composite thin-film in/Sn solder as an intermediate layer and its interdiffusion reaction with cu. *Journal of Electronic Materials, 38*, 200–207.

Yang, L., Hosoda, N., & Suga, T. (1997). Investigations on the Interface microstructure of stainless steel/aluminum joints created by the surface activated bonding method. *Interdisciplinary Sciences, 5*, 279–286.

Yazici, M. S., Dong, B., Hasan, D., et al. (2020). Integration of MEMS IR detectors with MIR waveguides for sensing applications. *Optics Express, 28*, 11524.

Yu, D., Yan, L., Lee, C., et al. (2008). *Study on high yield wafer to wafer bonding using In/Sn and Cu metallization* (pp. 1242–1245).

Yu, D. Q., Lee, C., Yan, L. L., et al. (2009a). Characterization and reliability study of low temperature hermetic wafer level bonding using in/Sn interlayer and cu/Ni/au metallization. *Journal of Alloys and Compounds, 485*, 444–450.

Yu, D. Q., Yan, L. L., Lee, C., et al. (2009b). Wafer-level hermetic bonding using Sn/in and cu/Ti/au metallization. *IEEE Transactions on Components and Packaging Technologies, 32*, 926–934.

Yu, D. Q., Lee, C., Yan, L. L., et al. (2009c). The role of Ni buffer layer on high yield low temperature hermetic wafer bonding using in/Sn/cu metallization. *Applied Physics Letters, 94*, 3–5.

Zhou, H., Hui, X., Li, D., et al. (2020). Metal–organic framework-surface-enhanced infrared absorption platform enables simultaneous on-Chip sensing of greenhouse gases. *Advancement of Science, 2001173*, 1–11.

Chapter 8
MOEMS-Enabled Miniaturized Biomedical Sensing and Imaging System

Bo Li, Lin Huang, and Zhen Qiu

8.1 Introduction

Micro-opto-electro-mechanical systems (MOEMS) technology has enabled the advances in the miniaturization of an increased variety of biomedical sensing and imaging system. The rapid development of fiber optics, light source, detectors, and electronics for data acquisition has also contributed significantly to the growth of the micro-system field. In the past decades, a large number of bio-sensing and biomedical imaging instruments have been demonstrated with sophisticated MOEMS techniques to scale down to ultrathin form factors. Insertable, wearable, and implantable biomedical devices allow for a broad range of applications in sensing and imaging of both solid and hollow organs, such as the skin, liver, lung, heart, esophagus, stomach, and colon. Furthermore, these tiny devices can be implanted onto freely behaving small animals for longitudinal studies in order to understand the neural activity in real time. The MOEMS enabled miniaturized instrument also allows for visualizing subcellular changes in vivo that have never been observed previously. These will help researchers form new knowledge that may pave the way to revolutionary paradigms in the molecular and cell biology. Cutting-edge opto-electronics, innovative optical design, novel sensing platform, miniaturized

B. Li
Department of Electrical and Computer Engineering, Institute for Quantitative Health Science and Engineering, Michigan State University, East Lansing, MI, USA

L. Huang
Department of Biomedical Engineering, Institute for Quantitative Health Science and Engineering, Michigan State University, East Lansing, MI, USA

Z. Qiu (✉)
Department of Biomedical Engineering, and Electrical and Computer Engineering, Institute for Quantitative Health Science and Engineering, Michigan State University, East Lansing, MI, USA
e-mail: qiuzhen@egr.msu.edu

Z. Yang (ed.), *Advanced MEMS/NEMS Fabrication and Sensors*,
https://doi.org/10.1007/978-3-030-79749-2_8

actuating mechanisms, and packaging strategies are required to advance the MOEMS field. In the near future, these miniaturized instruments will provide quantitative sensing data and images with multiplexed wavelength, with deeper tissue penetration, and in multiple dimensions with high spatial and temporal resolutions. Here, we will introduce a few representative sensing and imaging modalities rather than an exhaustive discussion of the vast number of efforts in this field.

8.2 Emerging MOEMS Technology for Miniaturization

Miniaturized sensing and imaging instruments have played an important role in the basic biological study and translational medicine. For example, the miniaturized insertable devices, endomicroscope, can be used for early detection, image-guided surgery, and therapy in the hollow organs, into which conventional or bulky devices are not able to get access. Endomicroscopy can assist the physician to guide tissue biopsies, in particular, in regions where the surface texture alone is not adequate to identify the diseased regions. Moreover, endomicroscopes can provide great clinical utility by differentiating between precancerous or malignant lesions and inflammation in individuals who are at increased risk for developing cancer. To realize the miniaturization, novel MOEMS-based sensing and actuating mechanisms will be essential in the development. The technical challenges in designing this component include generation of a distortion-free scan pattern with sufficient speed to mitigate in vivo motion artifacts with millimeter package dimensions. In addition, a much greater control of the focal volume in the imaging system, including axial scanning for imaging into the tissue, can be achieved with the actuator positioned distally. A summary of performance for several typical MOEMS actuators that are currently being developed for in vivo imaging is shown in Table 8.1. Electrostatic (ES) scanners and actuators have been widely used for different imaging modalities (Jung et al., 2018; Loewke et al., 2019; Piyawattanametha et al., 2012; Qiu et al., 2014a; Yao et al., 2019a; Chen et al., 2012), including optical coherence tomography (OCT), confocal, multiphoton, and photoacoustic. Electrothermal devices that only need very low driving voltage (<10 V) have been demonstrated in several practical sensing and imaging systems (Chen et al., 2018, 2019; Park et al., 2019; Tanguy et al., 2020), such as OCT and photoacoustic. Electromagnetic (EM) devices, based on MOEMS, have been applied (Yao et al., 2019b) for repetitive sensing and imaging, thanks to its robustness. Piezo-tubings (OD < 1.5 mm) have been custom-made (not batch production fashion) for tiny catheters (Park et al., 2020). Currently, thin-film lead zirconate titanate (PZT)- and aluminum nitride (AlN)-based sensors and actuators have attract more attentions (Choi et al., 2014; Qiu et al., 2010, 2014b) because the commercial cluster system (such as ULVAC) has matured. In addition, fiber-optic sensors (Sheng et al., 2020) hold the promise and become popular in the biomedical sensing applications, while it has been widely used for nondestructive test in the industry.

Table 8.1 Typical MOEMS actuators for miniaturization

Parameter	ES	ET	EM	Piezo-tube	Thin-film PZT/ AlN
Footprint	Compact	Compact	Compact	Compact	Compact
Axial displacement	~1 mm in vacuum	>1 mm at ambient	>1 mm at ambient	None	>0.5 mm at ambient
Tilting (mechanical)	±10°	±20°	±20°	Cantilever fashion	> ± 10°
Driving voltage (V)	Medium (20–150)	Low (<10)	Low (<20)	High (>100)	Low (<20)
Driving current (mA)	~0	~15	~50–100	~0	~ 0
Power consumption	Low	~100 mW	~200 mW	Low	Ultralow
Other	Easy mass production	Easy mass production	Good robustness	Not batch fabrication	Excellent sensing and actuation element

A description and summary of performance for MOEMS actuators. *ES* electrostatic, *ET* electrothermal, electromagnetic, *PZT* lead zirconate titanate, *AlN* aluminum nitride

8.3 Optical Coherence Tomography System Based on MOEMS

Optical coherence tomography (OCT) is essentially an interferometric technology that often uses supercontinuum laser source in the near-infrared regime. The weakly focused light beam will be used for penetrating deep into the highly scattered tissue specimens for noninvasive or minimal invasive biomedical imaging (so-called optical biopsy). In the OCT imaging modality, backscattered light is collected coherently for high-resolution cross-sectional images. Swept-source OCT (SS-OCT) operates at frequency-domain and provides superior imaging performance with enhanced signal-to-noise ratio (SNR) and fast frame rate. Using such kind of OCT endomicroscopy, hollow organ (such as the colon) morphology can be visualized. Enabled by the MOEMS technology, the OCT endomicroscope has been developed into a thin catheter that fits the tool channel of the medical endoscopes widely used in the clinics (Park et al., 2019).

Based on an electrothermal (ET) MOEMS micro-actuator with ultralow driving voltage (3 V), a novel SS-OCT endomicroscope has been miniaturized by integrating the fiber optics and tiny lens system. With an outer diameter of only 5.5 mm, the fully assembled probe can potentially be inserted into the tool channel of the therapeutic medical endoscope. The ET MOEMS device is based on the bimorph actuator, shown in Fig. 8.1c, d. The A scan rate of the SS-OCT system can be up to 100 kHz. To demonstrate the imaging performance, in vivo images on nail and ex vivo images on rat intestine have been acquired at a speed up to 200 frames per second (Fig. 8.2).

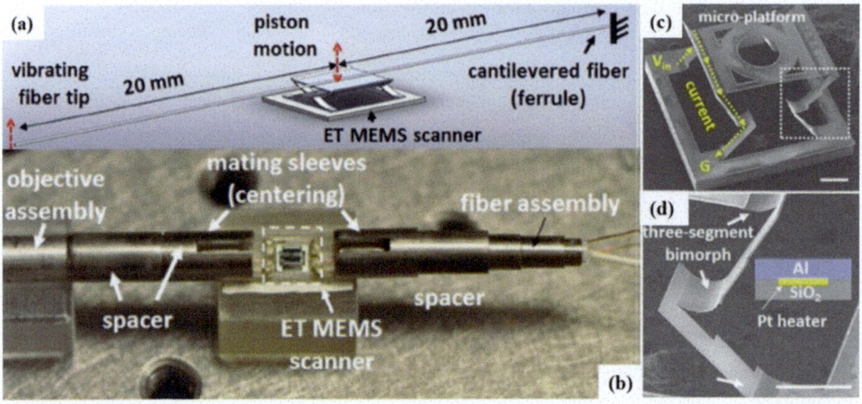

Fig. 8.1 (**a**) Schematics of the fiber scanning mechanism and (**b**) the probe assembly, consisting of an ET MEMS actuator. (**c**) SEM images of an integrated ET MEMS bimorph actuator, scale bar: 500 μm. (**d**) Enlarged view of a three-segmented bimorph actuator [white boxed area in **c**]. ET bimorph actuator is composed of an Al, SiO_2, and Pt layer, scale bar: 500 μm. (Recreated with kind permission from OSA (Park et al., 2019))

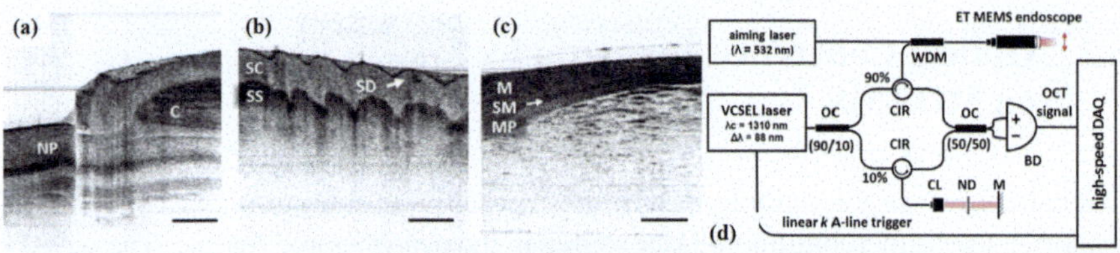

Fig. 8.2 Representative OCT images captured with the ET MEMS-based probe along with an SS-OCT system. (**a**) In vivo human nail fold and (**b**) fingerprint images. Detailed tissue structures, such as the layered nail plate (NP), cuticle (C), stratum corneum (SC), stratum spinosum (SS), and sweat duct (SD) can be clearly identified. (**c**) Ex vivo rat intestine image; the submucosa (SM) layer can be identified between the mucosa (M) and muscularis propria (MP) layers. Scale bars: 500 μm. (**d**) Schematic drawing of a SS-OCT system integrated with the ET MEMS-based scanning probe. A VCSEL with a 100 kHz A-line scan rate was used as the light source. An aiming laser at 532 nm was coupled into the sample arm with a fiber-optic WDM. *OC* fiber-optic coupler, *CIR* fiber-optic circulator, *WDM* wavelength division multiplexer, *BD* balanced detector, *CL* collimating lens, *ND* neutral-density filter, *M* mirror. (Recreated with kind permission from OSA (Park et al., 2019))

The ET MOEMS devices with low driving voltage have been widely applied to multiple imaging modalities, including OCT, photoacoustic, confocal, and wide-field fluorescence imaging. The most advanced two-dimensional scanner (Tanguy et al., 2020), shown in Fig. 8.3, has a small footprint with an effective diameter of 1 mm with tilted mirror surface that is out of plane. The scanner operates at resonant mode on both axes, which enables real-time image with fast frame rate (up to 60 fps) with Lissajous scanning pattern.

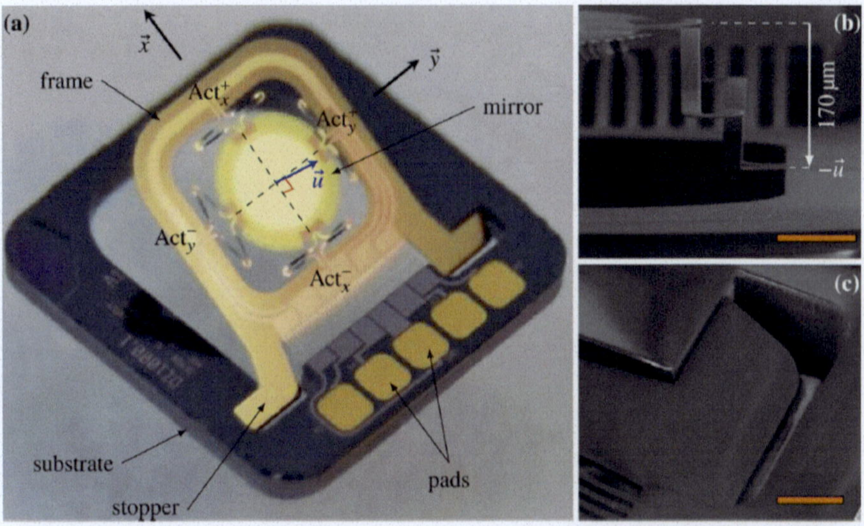

Fig. 8.3 Pictures of the MEMS scanner. (**a**) Overall view. The footprint of the substrate measures 44 mm^2 and the micro-mirror's effective diameter is 1 mm. (**b**) Lateral close-up view of a single actuator. Scale bar: 100 μm. (**c**) Top close-up view of the stopper mechanism. Scale bar: 200 μm. (Recreated with kind permission from OSA (Tanguy et al., 2020))

8.4 Photoacoustic Endomicroscopy

Through detecting ultrasound signal from thermoelastic expansion induced by pulsed laser beam, photoacoustic imaging has become an emerging modality for biomedical applications. Since the seminal photoacoustic microscopic imaging (PAM) study (Chen et al., 2012) based on the electrostatic MOEMS two-dimensional scanner and the microring resonator (MRR), shown in Fig. 8.4 researcher has been making tremendous efforts on developing miniaturized photoacoustic imaging system in the past 10 years.

Using electrothermal two-dimensional scanner, a tiny head-mountable photoacoustic imaging probe (Chen et al., 2019) has been demonstrated for in vivo brain imaging on free-behaving rat. High-resolution vascular imaging in the brain provides real-time functional physiological information. Based on novel low-cost soft nanoimprint lithography (sNIL) technology, researchers have developed a novel disposable ultrasound-sensing chronic cranial window (usCCW) with embedded transparent MRR (Fig. 8.5) for longitudinal imaging and manipulation of various brain activities in live animals (Li et al., 2019).

The newly developed usCCW can be surgically implanted on the skull. This approach will keep maintaining optical access without external ultrasound coupling medium for photoacoustic imaging. Photoacoustic imaging of the cortical vascular network in live mice over 28 days has been demonstrated, as shown in Fig. 8.6. This work establishes the foundation for integrating photoacoustic imaging with modern

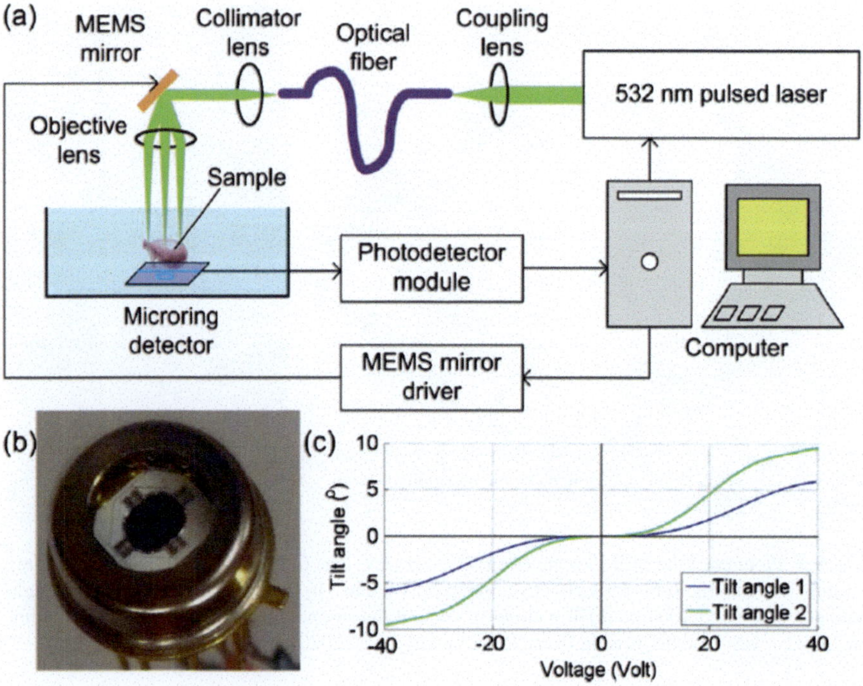

Fig. 8.4 (**a**) Schematic drawing of the MEMS-based PAM system. (**b**) A photograph of the MEMS mirror. (**c**) The calibrated response of mirror tilt angle versus applied voltages. (Recreated with kind permission from OSA (Chen et al., 2012))

brain research. It paves the way to future ultracompact photoacoustic endomicroscopes for clinical applications.

8.5 Confocal Endomicroscopy

Single-axis confocal endomicroscopes based on fiber bundles have been applied for clinical applications, such as the Cellvizio® (Mauna Kea Technologies, FDA approved). The backscattered light from the regions outside of the focal plane can be rejected by the fiber core. To improve the dynamic range on the axial axis and enhance the axial resolution, a new type of dual-axis confocal architecture has been proposed, as shown in Fig. 8.7a. The illumination and collection beam are separated and accurately aligned through crossed optical axes (Fig. 8.7b). Based on novel thin-film PZT actuators with folded beams (Choi et al., 2014; Qiu et al., 2010, 2014b), a novel MOEMS scanner has been developed and used as a monolithic multi-axis scan engine in the distal end of the dual-axis confocal endomicroscope. The outer diameter of the miniaturized probe is around 5.5 mm, which enables wearable and insertable applications.

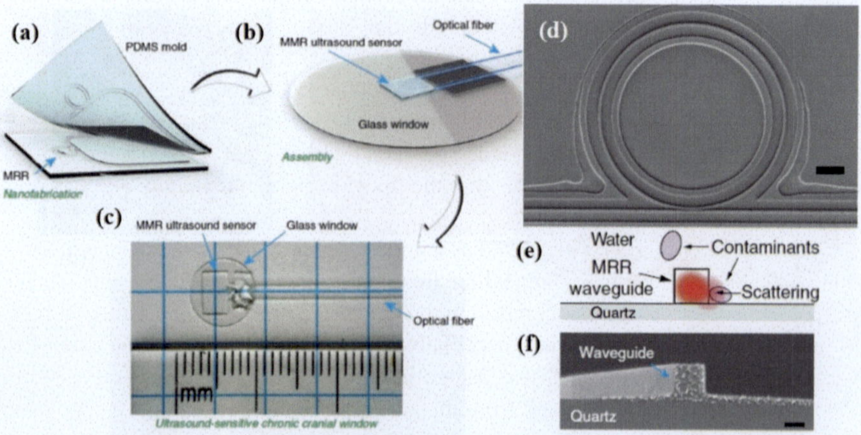

Fig. 8.5 Process to fabricate usCCW: (**a**) Nanofabrication of MRR on transparent substrate using low-cost soft nanoimprint lithography (sNIL) process. (**b**) Assembling an MRR-based ultrasonic detector with matching optical fibers, onto a circular cover slide (8 mm in diameter), and (**c**) finally assembled usCCW device. (**d**) Protecting MRR ultrasonic detector for long-term in vivo imaging. A scanning electron microscope (SEM) image of a MRR ultrasonic detector fabricated by sNIL. Scale bar: 10 μm. (**e**) Schematic and (**f**) SEM image of a cross section of the fabricated MRR. Scale bar: 500 nm. (Recreated with kind permission from Nature Publishing Group (Li et al., 2019))

Using indocyanine green (ICG) (FDA-approved fluorescence dye) topically applied to the tissue specimens, horizontal and vertical cross-sectional imaging of the lower GI tract (colon) has been demonstrated (Jung et al., 2018; Yao et al., 2019a) with subcellular resolution and large field of view (around 1000 μm), as shown in Fig. 8.8a, b.

8.6 Multiphoton System Based on MOEMS Techniques

Based on the nonlinear optics phenomena, multiphoton microscopes perform both fluorescence imaging and label-free autofluorescence imaging with second harmonic and third harmonic generation signals. For example, for two-photon imaging, two lower-energy photons with longer wavelength within the near-infrared regime arrive at tissue biomolecules simultaneously. In general, with 10 to 100 fs pulse width (high peak intensity), ultrafast lasers are usually utilized for excitation on the tissue specimens. The conventional tabletop bulk multiphoton microscopy has been widely used in the research laboratories for all kinds of different applications, including in vivo and ex vivo tissue imaging and *intravital* imaging on small living animals. Most recently, enabled by the advanced fiber optics and MOEMS techniques, fully fiber-based ultracompact endomicroscopes have been developed for a variety of practical applications, such as the in vivo imaging on the brain of living

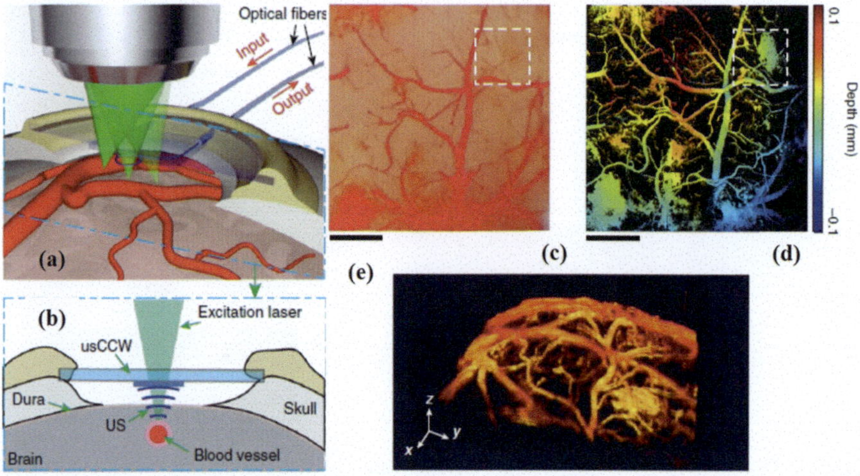

Fig. 8.6 In vivo PAM cortical imaging using a usCCW. (**a**) Illustration of optical scanning through the usCCW, which is surgically implanted on the mouse skull after craniotomy. The inset shows the physical dimension of the usCCW with MRR and fibers attached, which is optically transparent, with a total thickness of 250 μm and a total weight of less than 1 g. The MRR ultrasonic detector is attached on an 8-mm diameter circular substrate and the sensing light is coupled through a pair of 30-cm flexible optical fiber. (**b**) Optical excitation and ultrasonic detection geometry along the cross section highlighted in (**a**). The space between the MRR and the dura is 1 mm and is filled with 0.5% agarose gel. We seal the usCCW with dental cement to prevent infection and leakage. (**c**) Bright-field optical microscopy image of the cortical region through the MRR. (**d**) Depth-encoded maximum-intensity-projection (MIP) PAM image of the same area. The whole image is stitched from nine acquisitions due to the limited laser-scanning field of view. (**e**) Three-dimensional visualization of the vessel orientations and cortical curvature. Scale bars: 200 μm. (Recreated with kind permission from Nature Publishing Group (Li et al., 2019))

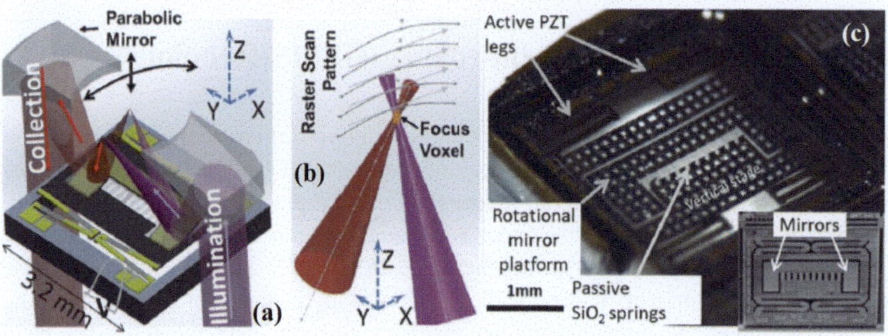

Fig. 8.7 (**a**) Vertical-rotational stage scanning during cross-sectional dual-axis confocal endomicroscopic imaging. (**b**) Detail of raster scanning pattern. (**c**) Vertical-rotational MEMS scanning stage based on active outer vertical displacement and passive inner resonant scanning. Inset: Variant with solid dog-bone mirror surface for dual-axes confocal microscopy

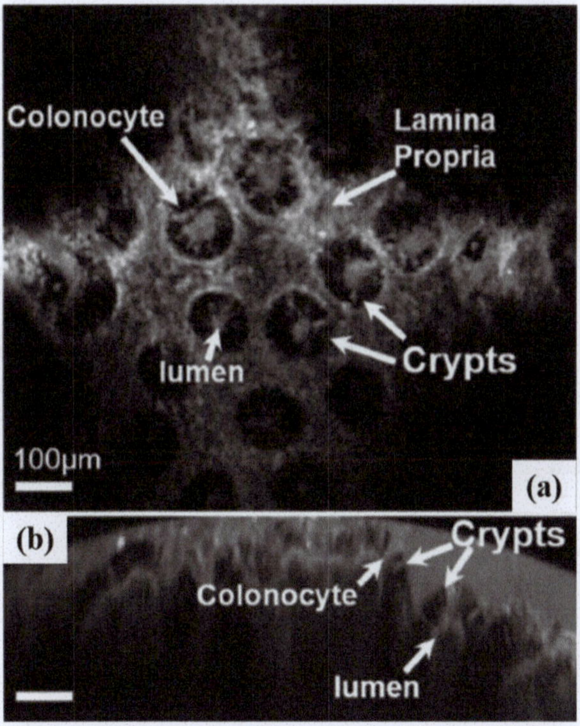

Fig. 8.8 Fluorescence imaging results of the NIR fluorescence intraoperative confocal microscope. Ex vivo fluorescence images of human colon tissue specimens demonstrated "histology-like" imaging with a large field of view (up to 1000 μm), in both the (**a**) en face horizontal cross-sectional image at the 150-m depth and (**b**) vertical cross-sectional image; crypts, colonocytes, and lumen have been visualized with cellular resolutions, scale bar: 100 μm

and free-moving animals. A head-mountable two-photon microscopy has been developed (Zong et al., 2017) using MOEMS electrostatic scanner, custom-made fiber, and tiny gradient index micro lenses. In Johns Hopkins University, an ultra-compact multiphoton endomicroscope has been successfully demonstrated by the team led by Dr. X. Li, a pioneer in the miniaturized multiphoton imaging research field. The new imaging system (Park et al., 2020), shown in Fig. 8.9 is utilizing custom-made double-cladding fiber (DCF) jointly developed by Corning and a tiny piezo-based scanner (outer diameter: 1.5 mm) for lateral scanning at resonant modes.

Made out of PZT-5H, the 8-mm-long piezo-tube-based micro-actuator drives the DCF fiber around resonant frequencies on both *x*- and *y*-axes. The DCF (Corning Inc., NA–0.12 for the core and NA–0.35 for the inner cladding) fiber has been aligned in the center (a cantilever fashion) with a stainless steel (model 304)-based adapter. A superachromatic micro-objective lens (GRINTECH GmbH, NA–0.18 on the fiber side, NA–0.75 on the tissue side, 2.25 mm diameter) has been used at the distal end. The micro-objective lens provides a 300-μm working distance (in air)

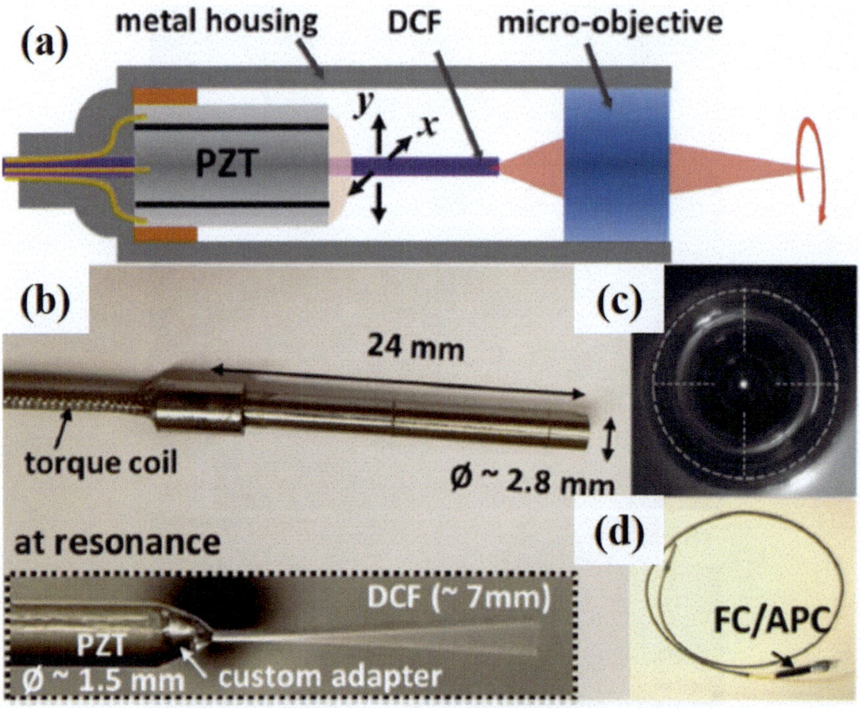

Fig. 8.9 (**a**) The schematic drawing of a resonant fiber-optic scanning endomicroscope. A precision-machined PZT base and a custom adapter were used for centering each component and ensuring effective force transfer from the PZT actuator to the fiber cantilever. (**b**) Photograph of a fully assembled endomicroscope. The fiber scanner and a superachromatic micro-objective are encased inside a hypodermic metal tube of a 2.8-mm diameter and a 24-mm overall rigid length. A 7-mm-long DCF was cantilevered at the distal end of a PZT actuator and resonantly scanned (inset). (**c**) Front view of an endomicroscope showing the DCF was concentrically aligned with the longitudinal axis of the endomicroscope. (**d**) Photograph of an assembled endomicroscope with a protective torque coil over the fiber and PZT drive wires and an FC/APC connector at its proximal end. (Recreated with kind permission from OSA (Park et al., 2020))

with a magnification (~5×). High-resolution imaging has been achieved (0.8 μm lateral and 6.1 μm axial).

For in vivo imaging, the head-mountable endomicroscope has been fully integrated with a custom-made multiphoton imaging system with a femtosecond excitation beam at 920 nm (Ti/sapphire laser with pre-chirp unit) (Fig. 8.10). The potential of the novel endomicroscope to perform in vivo brain imaging on freely behaving mouse has been successfully demonstrated with mouse that has GCaMP6m (in motor cortex) via viral infection. As shown in Fig. 8.11, the imaging system has acquired representative multiphoton fluorescence images of neuronal activities of a freely behaving mouse at two frame rates: 3.3 fps with 1490 × 1024 pixels (circumferential × radial) and 26.4 fps with 1490 × 128 pixels.

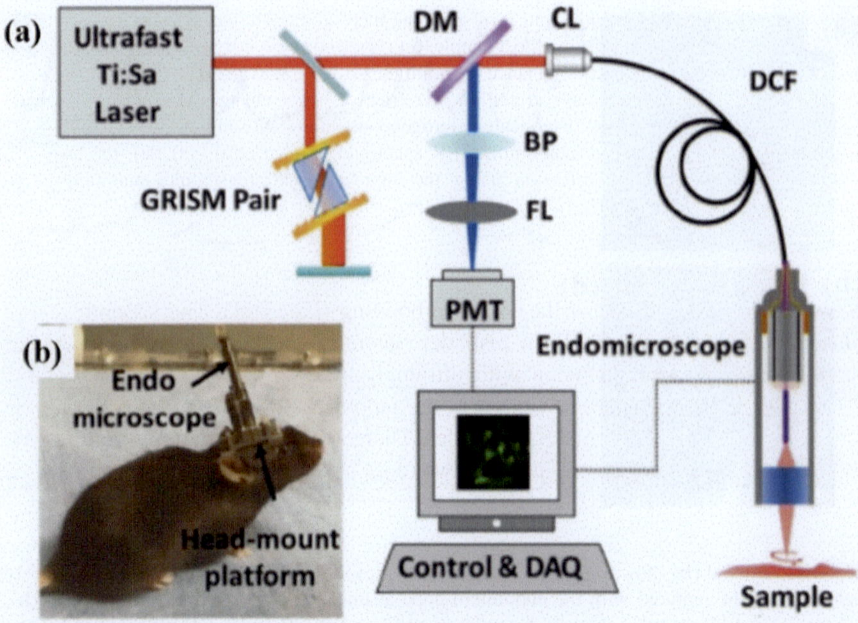

Fig. 8.10 (**a**) Schematic drawing of the MPF endomicroscopy system. (**b**) Photograph of a brain imaging setup for a freely behaving mouse. The endomicroscope was mounted on top of a cranial window over the mouse cortex. *DM* dichroic mirror, *BP* band-pass filter, *FL* focusing lens, *PMT* photomultiplier tube, *CL* coupling lens. (Recreated with kind permission from OSA (Park et al., 2020))

Recently, metalens has become an emerging technology for miniaturized sensing and imaging system design (Arbabi et al., 2018; Pahlevaninezhad et al., 2018), such as spectroscopy, multiphoton, and OCT. A novel proof-of-concept double-wavelength metasurface lens (so-called metalens)-based two-photon imaging system has demonstrated the fluorescent imaging with polyethylene beads (Arbabi et al., 2018). The researchers from Caltech have developed tiny dielectric metasurfaces-based lenses that potentially enable a variety of new applications, including the objective lens, as shown in Fig. 8.12a, b. The new metalens focuses on both 820 and 605 nm light, which are excitation and emission wavelengths, respectively. With the custom-made metasurface lens, the ultracompact two-photon microscopes' fluorescence imaging performance is comparable to the one by conventional imaging system based on traditional and bulk graded-index lenses (Fig. 8.12c, d). In the future, tunable metalenses will potentially enable fast axial scanning for 3D volumetric imaging. In the future, based on the ultrathin metasurface lens system fabricated using nanolithography, miniaturized two-photon microscopes hold the promise for future in vivo imaging of the neural circuit in the brain.

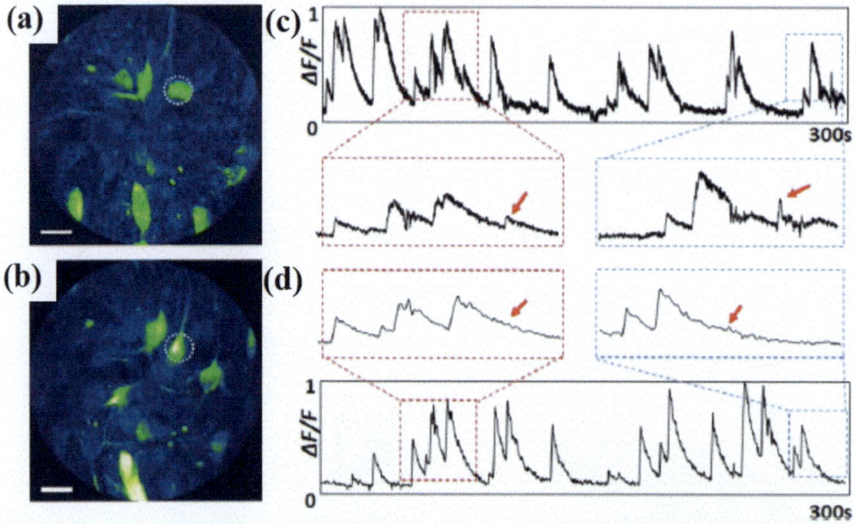

Fig. 8.11 (**a**) and (**b**) show images of neuronal activities in the primary motor cortex of a freely behaving rodent captured with the endomicroscope at imaging speeds of (**a**), 26.4 fps and (**b**), 3.3 fps. (**c**) and (**d**) recorded calcium dynamics of a neuron indicated with a dotted circle in (**a**) and (**b**) over 5 min: (**c**) at 26.4 fps, (**d**) at 3.3 fps. Scale bars: 20 μm. (Recreated with kind permission from OSA (Park et al., 2020))

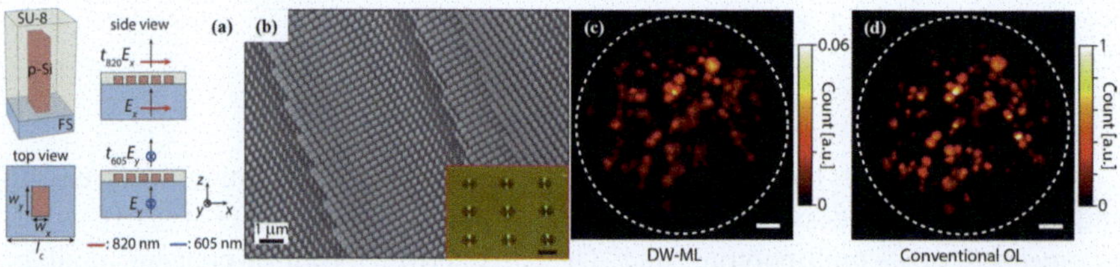

Fig. 8.12 Design and characterization of the metasurface lens. (**a**) Schematic illustration of the meta-atom, consisting of a p-Si nanopost with a rectangular cross-sectional sitting on a fused silica substrate and covered by an SU-8 layer. Top and side views of the nanoposts showing the dimensions and illumination conditions. Tuning the nanopost dimensions, w_x and w_y, allows for independent control of transmission phase of x-polarized light at 820 nm and y-polarized light at 605 nm. (**b**) Scanning electron micrograph of a portion of the fabricated double-wavelength metalens. Inset: Optical image of a few fabricated lenses. Scale bar: 2 mm. (**c**) Two-photon fluorescent microscope image of the microsphere captured using the DW-ML and (**d**) captured using a conventional refractive objective. Scale bars: 10 μm

8.7 Conclusion and Future Direction

Enabled by the most advanced MOEMS techniques, the miniaturized sensing and imaging devices have become more important tools for monitoring the live organs, such as the visualization of the neural activities in the brain and the cellular changes in the hollow organs in vivo. Advanced insertable, wearable, and implantable medical devices, such as endomicroscopes, have been realized by frontier technologies, including light sources, micro-optics, custom-made fibers, tiny sensor, and actuators. Multimodality sensing and imaging system, such as OCT, confocal, photoacoustic, and multiphoton, can perform subcellular optical sectioning with deep penetration in the biological tissue specimens. In the near future, rapid progress in the MOEMS field will further expand the capabilities of the miniaturized biomedical instrumentation through ultrafast speed, ultrasensitive detection, and multiplexed and multidimensional imaging for precision medicine.

References

Arbabi, E., Li, J., Hutchins, R. J., Kamali, S. M., Arbabi, A., Horie, Y., Van Dorpe, P., Gradinaru, V., Wagenaar, D. A., & Faraon, A. (2018). Two-photon microscopy with a double-wavelength metasurface objective lens. *Nano Letters, 18*(8), 4943–4948.

Chen, S.-L., Xie, Z., Ling, T., Guo, L. J., Wei, X., & Wang, X. (2012). Miniaturized all-optical photoacoustic microscopy based on microelectromechanical systems mirror scanning. *Optics Letters, 37*(20), 4263–4265.

Chen, Q., Guo, H., Jin, T., Qi, W., Xie, H., & Xi, L. (2018). Ultracompact high-resolution photoacoustic microscopy. *Optics Letters, 43*(7), 1615–1618.

Chen, Q., Xie, H., & Xi, L. (2019). Wearable optical resolution photoacoustic microscopy. *Journal of Biophotonics, 12*(8), e201900066.

Choi, J., Qiu, Z., Rhee, C.-H., Wang, T., & Oldham, K. (2014). A three-degree-of-freedom thin-film PZT-actuated microactuator with large out-of-plane displacement. *Journal of Micromechanics and Microengineering, 24*(7), 075017.

Jung, I. W., López, D., Qiu, Z., & Piyawattanametha, W. (2018). 2-D MEMS scanner for handheld multispectral dual-axis confocal microscopes. *Journal of Microelectromechanical Systems, 27*(4), 605–612.

Li, H., Dong, B., Zhang, X., Shu, X., Chen, X., Hai, R., Czaplewski, D. A., Zhang, H. F., & Sun, C. (2019). Disposable ultrasound-sensing chronic cranial window by soft nanoimprinting lithography. *Nature Communications, 10*(1), 1–9.

Loewke, N. O., Qiu, Z., Mandella, M. J., Ertsey, R., Loewke, A., Gunaydin, L. A., Rosenthal, E. L., Contag, C. H., & Solgaard, O. (2019). Software-based phase control, video-rate imaging, and real-time Mosaicing with a Lissajous-scanned confocal microscope. *IEEE Transactions on Medical Imaging, 39*(4), 1127–1137.

Pahlevaninezhad, H., Khorasaninejad, M., Huang, Y.-W., Shi, Z., Hariri, L. P., Adams, D. C., Ding, V., Zhu, A., Qiu, C.-W., & Capasso, F. (2018). Nano-optic endoscope for high-resolution optical coherence tomography in vivo. *Nature Photonics, 12*(9), 540–547.

Park, H.-C., Zhang, X., Yuan, W., Zhou, L., Xie, H., & Li, X. (2019). Ultralow-voltage electrothermal MEMS based fiber-optic scanning probe for forward-viewing endoscopic OCT. *Optics Letters, 44*(9), 2232–2235.

Park, H.-C., Guan, H., Li, A., Yue, Y., Li, M.-J., Lu, H., & Li, X. (2020). High-speed fiber-optic scanning nonlinear endomicroscopy for imaging neuron dynamics in vivo. *Optics Letters, 45*(13), 3605–3608.

Piyawattanametha, W., Ra, H., Loewke, K. E., Mandella, M. J., Contag, C. H., Qiu, Z., Wang, T. D., Friedland, S., Liu, J. T., & Kino, G. S. (2012). In vivo near-infrared dual-axis confocal microendoscopy in the human lower gastrointestinal tract. *Journal of Biomedical Optics, 17*(2), 021102.

Qiu, Z., Pulskamp, J. S., Lin, X., Rhee, C.-H., Wang, T., Polcawich, R. G., & Oldham, K. (2010). Large displacement vertical translational actuator based on piezoelectric thin films. *Journal of Micromechanics and Microengineering, 20*(7), 075016.

Qiu, Z., Khondee, S., Duan, X., Li, H., Mandella, M. J., Joshi, B. P., Zhou, Q., Owens, S. R., Kurabayashi, K., & Oldham, K. R. (2014a). Vertical cross-sectional imaging of colonic dysplasia in vivo with multi-spectral dual axes confocal endomicroscopy. *Gastroenterology, 146*(3), 615–617.

Qiu, Z., Rhee, C.-H., Choi, J., Wang, T. D., & Oldham, K. R. (2014b). Large stroke vertical PZT microactuator with high-speed rotational scanning. *Journal of Microelectromechanical Systems, 23*(2), 256–258.

Sheng, Q., Li, B., Uddin, N., Mitul, A. F., Zhu, Y., Qiu, Z., & Han, M. (2020). High resolution, fast response fiber-optic temperature sensor with reduced end conduction effect. *Optics Letters, 45*(21), 6094–6097.

Tanguy, Q. A., Gaiffe, O., Passilly, N., Cote, J.-M., Cabodevila, G., Bargiel, S., Lutz, P., Xie, H., & Gorecki, C. (2020). Real-time Lissajous imaging with a low-voltage 2-axis MEMS scanner based on electrothermal actuation. *Optics Express, 28*(6), 8512–8527.

Yao, C.-Y., Li, B., & Qiu, Z. (2019a). 2D au-coated resonant MEMS scanner for NIR fluorescence intraoperative confocal microscope. *Micromachines, 10*(5), 295.

Yao, J., Peng, T., Sun, B., Zhang, H., Zhao, M., Dai, B., Liu, H., Ding, G., Sawada, R., & Yang, Z. (2019b). A single-fiber endoscope scanner probe utilizing two-degrees-of-freedom (2DOF) high-order resonance to realize larger scanning angle. *IEEE Transactions on Components, Packaging and Manufacturing Technology, 9*(12), 2332–2340.

Zong, W., Wu, R., Li, M., Hu, Y., Li, Y., Li, J., Rong, H., Wu, H., Xu, Y., & Lu, Y. (2017). Fast high-resolution miniature two-photon microscopy for brain imaging in freely behaving mice. *Nature Methods, 14*(7), 713–719.

Chapter 9
Bio-Inspired Flexible Sensors for Flow Field Detection

Yonggang Jiang, Zhiqiang Ma, and Dawei Shen

9.1 Introduction

As flow field greatly influences the altitude and motion of the flying or swimming robots, there is a strong demand to develop pressure and flow sensors with large detection range, fast response, and conformability to curved surfaces for active flow control, disturbance rejection, and flow-assisted navigation. Various detection methods including acoustic Doppler-shift velocimetry, silicon-based sensors, and hot-wire sensors have been developed. Over millennia of evolution, animals have evolved various efficient biological flow sensory systems, in order to survive in complicated environments (Bleckmann et al., 2004; Barth, 2004). Fish lateral line and sensory hairs are excellent biological sensory systems for flow field detection. Inspired by the morphologies and functions of biological receptors, researchers have developed various flow sensors based on biomimetic design principles, achieving excellent sensory capabilities (Jiang et al., 2019a; Han et al., 2018).

Compared to the biological receptors with flexible or stretchable capabilities, most of engineered flow sensors are fabricated on silicon or glass substrates, not suitable for the robotic application with curved surface. Herein, we introduce bio-inspired flexible pressure and flow sensors based on polymers in this chapter with an emphasis on the biomechanical mechanisms for sensitivity enhancement or noise reduction. Section 9.2 describes fish lateral line inspired flexible sensors for hydrodynamic perception, and Sect. 9.3 introduces several examples on flexible hair-like sensors for aerodynamic detection.

Y. Jiang (✉) · Z. Ma · D. Shen
School of Mechanical Engineering and Automation, Beihang University, Beijing, China
e-mail: jiangyg@buaa.edu.cn

Z. Yang (ed.), *Advanced MEMS/NEMS Fabrication and Sensors*,
https://doi.org/10.1007/978-3-030-79749-2_9

9.2 Bio-Inspired Flexible Sensors for Hydrodynamic Detection

9.2.1 Fish Lateral Line

Mechanosensory lateral line systems enable fish and aquatic amphibians to detect local water motions, which play crucial roles in flow-related behaviors, such as rheotaxis, prey, and predators (Coombs et al., 2014). As functional units, the neuromasts in most fish species are distributed across the head, trunk, and tail, as shown in Fig. 9.1a (Schmitz et al., 2008; Ma et al., 2020a). Figure 9.1b schematically illustrates a neuromast containing hair cells covered by a gelatinous cupula into which the hair cell ciliary bundles project. According to the locations of neuromasts, two subsystems of the lateral line system in most teleosts can be distinguished, i.e., superficial neuromast (SN) and canal neuromast (CN) system (Fig. 9.1c) (Coombs et al., 2014). SNs are freestanding on the skin and detect the flow velocity of the surrounding water. CNs are located in subepidermal fluid-filled lateral line canals that are connected to the environment via canal pores, which respond to fluid motion caused by pressure difference between two adjacent canal pores (Coombs et al., 1989). The SN system applies low-pass filtering to the stimulus velocity, whereas the CN system functions as a band-pass filter for various noise sources (Montgomery et al., 2009; Jiang et al., 2019a).

Caves are extreme harsh environments for animals due to a lack of sunlight and often shortage of food sources. The lateral line system of cavefish is probably more sensitive than that of surface fish living in rivers or lakes, to compensate reduced

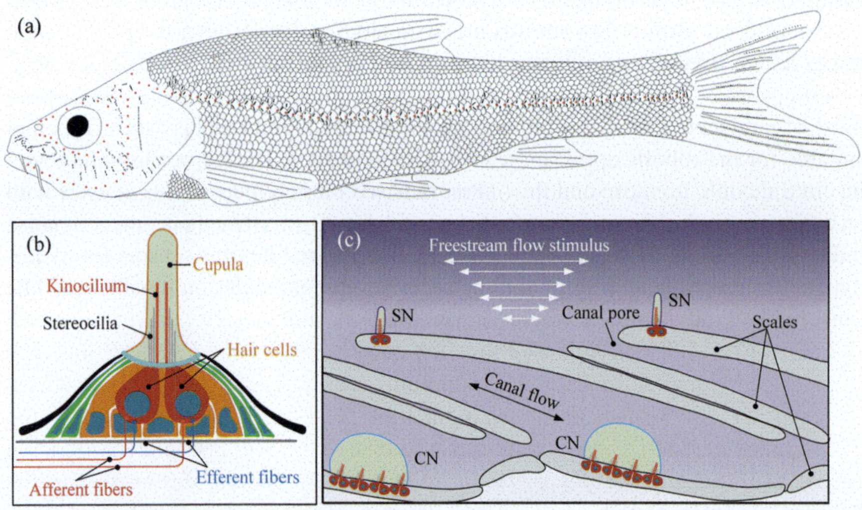

Fig. 9.1 Fish lateral line system. (**a**) Schematic illustration of the neuromast distribution of fish lateral line. Large red dots illustrate CNs, and small black dots represent SNs. (**b**) Schematic structure of a neuromast. (**c**) The locations and detection principles of SN and CN

visual sense and acquire extra flow information to survive in complete dark caves (Yoshizawa et al., 2010). Several anatomical specializations, such as enlarged SNs, constriction structures within lateral line canal nearby CNs, and more sensory neuromasts, endow cavefish lateral line system with enhanced perception ability (Teyke, 1990; Jiang et al., 2016; Ma et al., 2020a).

As the biomechanical interface between neuromast and hydrodynamic environment, the cupula plays many important roles in the sensing performance of the lateral line system. The cupula is able to enhance the sensitivity through the following several aspects: (1) increased cross-sectional area and porous structure in cupula can increase the drag force acting on neuromast and (2) the hydrophilicity and the permeability of the hydrogel-like cupula may also enhance the signal absorption through an enhanced friction factor associated with the material (McConney et al., 2009a). In addition, the SNs in some fish species have cupula with embedded fibers, which probably function as a structural support network for the cupula, allowing the cupula to grow to greater distances away from the stagnant boundary layer of the fish surface.

In some cavefish species, such as *Sinocyclocheilus macrophthalmus*, the lateral line canal gets narrower nearby the CN, as shown in Fig. 9.2a. Denton and Gray (1988) proposed a mathematical model to reveal the canal mechanics within lateral line canal. They measured and modeled the fluid motions inside the actual lateral line canals and canal-like structures (such as capillary tubes) as functions of the frequency of sinusoidal water motions outside the canal. The lateral line canal was represented by a canal consisting of a circular section with diameter d_1, which contained a narrower circular section with diameter d_2 (their corresponding lengths were equal to l_1 and l_2, respectively). If X_{in}/X_{out} designates the ratio between the

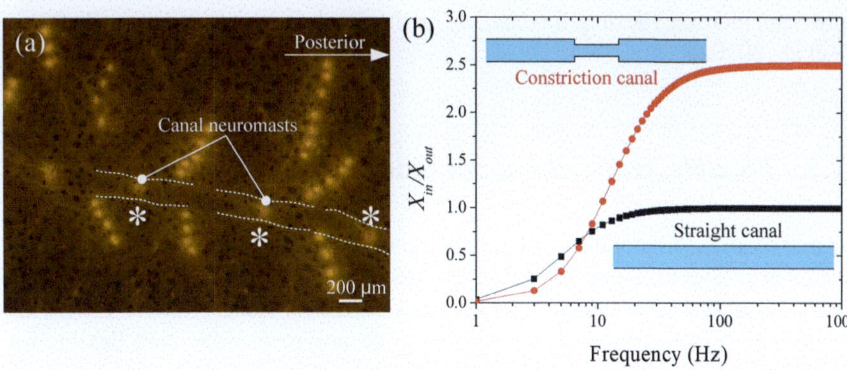

Fig. 9.2 Enhanced sensitivity of canal lateral line system induced by constriction structures. (**a**) The constriction structure in *Sinocyclocheilus macrophthalmus*. The white dash lines represent the canal profiles. The white stars indicate the constriction structures. (**b**) The mechanical sensitivity of lateral line canal with different canal types, including constriction canal and straight canal. The canal length and canal diameter are 5 mm and 0.5 mm, respectively. The length and diameter of the constriction structure are 1 mm and 0.25 mm, respectively. The inserted illustrations show the profiles of the constriction canal and straight canal

relative amplitudes of displacements in the narrower part of the canal and in the liquid adjacent to the canal, it can be found that:

$$\frac{X_{\text{in}}}{X_{\text{out}}} = \frac{j\omega I_{\text{out}}}{j\omega I_{\text{in}} + R_{\text{in}}} \tag{9.1}$$

Here I_{out} and I_{in} are the inertances outside and inside the canal, respectively, which are calculated as follows:

$$I_{\text{out}} = \frac{\rho\left(l_1 + l_2\right)}{a_2} \tag{9.2}$$

$$I_{\text{in}} = \rho\left(\frac{l_1}{a_1} + \frac{l_2}{a_2}\right) \tag{9.3}$$

R_{in} is the frictional (viscous) resistance under the steady (Hagen-Poiseuille) flow conditions, which can be computed as:

$$R_{\text{in}} = 8\pi\eta\left(\frac{l_1}{a_1^2} + \frac{l_2}{a_2^2}\right) \tag{9.4}$$

where $\omega = 2\pi f$, j is the complex operator, ρ is the density of the surrounding fluid ($\rho = 1000$ kg/m^3), and η is the dynamic viscosity ($\eta = 0.001$ Pa s).

As shown in Fig. 9.2b, the constriction structure of the lateral line canal can result in (1) further attenuation of the response at low frequencies, (2) an upward shift in the cutoff frequency, and (3) higher sensitivity at higher frequencies. In addition, decreasing the lateral line canal diameter in the vicinity of the canal neuromast can increase the drag force acting on the canal neuromast (Jiang et al., 2016; Ma et al., 2020a).

9.2.2 *Flexible Artificial Lateral Line Sensors*

Inspired by the fish lateral line system, various artificial lateral line (ALL) devices have been proposed, based on different transduction mechanisms, such as piezoelectric, piezoresistive, optical, etc. (Bora et al., 2018; Jiang et al., 2019a). This section mainly describes ALL devices based on polymers, which are flexible for curved surface.

Mimicking the SN system in fish, Miao and Triantafyllou from the Singapore-MIT Alliance for Research and Technology (SMART) developed flexible ALL velocity sensors based on the piezoelectric and piezoresistive sensing mechanisms. In 2016, they reported a new class of miniature all-polymer flow sensors that closely mimicked the intricate morphology of the mechanosensory ciliary bundles in biological hair cells, as shown in Fig. 9.3a (Asadnia et al., 2016). The artificial ciliary

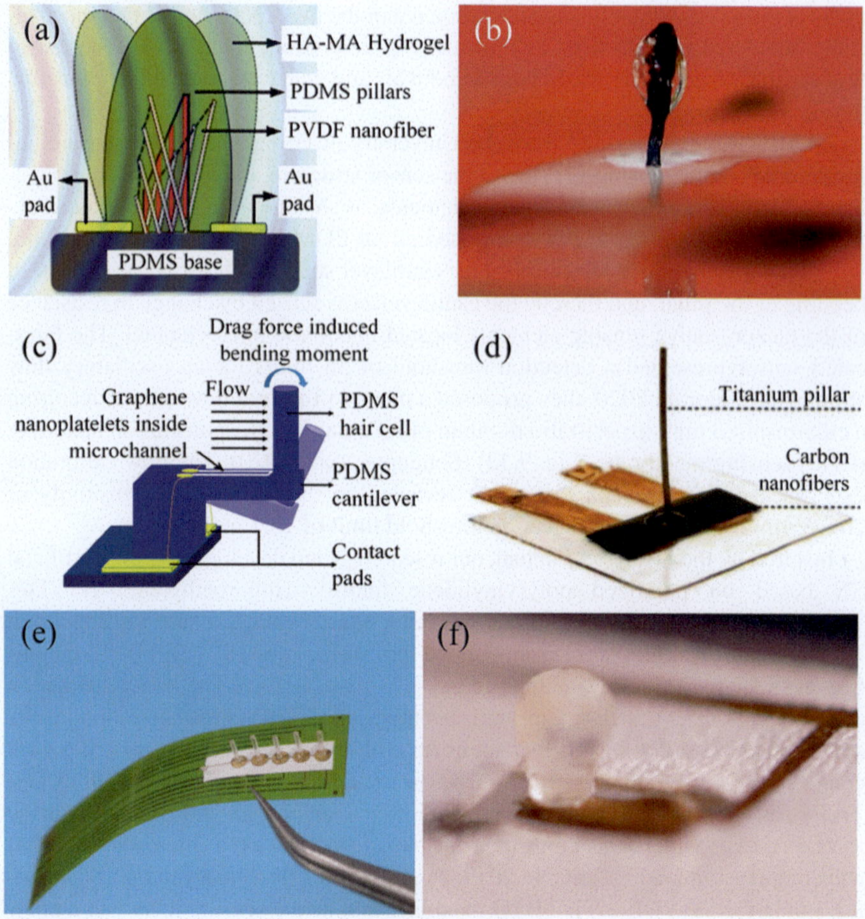

Fig. 9.3 Bio-inspired flexible ALL velocity sensors. (**a**) A flow sensor mimicking the morphology of stereocilia bundles of biological neuromasts (Asadnia et al., 2016). (**b**) A biomimetic flow sensor featuring a piezoelectric membrane with a hydrogel-infused vertically aligned carbon nanotube bundle that mechanically interacts with the flow (Bora et al., 2017). (**c**) An ALL velocity sensor based on a PDMS cantilever with a graphene piezoresistors and a PDMS hair cell (Kamat et al., 2019). (**d**) A bio-inspired flow sensor, featuring a high-aspect-ratio titanium pillar on an electrospun carbon nanofiber (CNF) sensing membrane (Sengupta et al., 2020). (**e**) A flexible artificial SN, featuring a polarized P(VDF-TrFE)/BTO electrospinning nanofiber mat as the sensing layer and a polyimide (PI) film with arrays of circular cavities as the substrate (Hu et al., 2019a). (**f**) A cantilever-based ALL sensor using a BTO/P(VDF-TrFE) nanofiber mat and a hydrogel cupula (Ma et al., 2020c)

bundle was composed of polydimethylsiloxane (PDMS) micro-pillars with graded heights and PVDF piezoelectric nanofiber tip links. The sensors achieved the directional perception ability and a threshold detection limit of 8 μm/s. In 2017, they presented a biomimetic nanoelectromechanical system (NEMS) flow sensor featuring a polyvinylidene fluoride (PVDF) nanofiber sensing membrane with a hydrogel-infused vertically aligned carbon nanotube (VACNT) bundle that mechanically interacted with the flow, as shown in Fig. 9.3b (Bora et al., 2017). The sensor was capable of detecting weak flows in both air and water with a velocity detection threshold of 5 mm/s.

In 2019, Kottapalli group from the University of Groningen developed a cilia-inspired flow sensor using PDMS for the sensor structure and graphene nanoplatelets (GN) as the piezoresistive sensing elements, as shown in Fig. 9.3c (Kamat et al., 2019). The bio-inspired sensor comprised an all-PDMS cantilever-pillar structure with a GN piezoresistor deposited on the cantilever surface. The drag force-induced bending of the pillar, and thereby the cantilever, was sensed by change in resistance of the piezoresistive sensing elements located inside the microchannel. The fabricated sensor presented a detection threshold of 58 mm/s for an oscillatory flow stimulus in water. In 2020, they proposed a new bio-inspired flow sensor, featuring a cilia-inspired high-aspect-ratio titanium pillar on an electrospun carbon nanofiber (CNF) sensing membrane (Fig. 9.3d) (Sengupta et al., 2020). The flow calibration experiments conducted on the sensor revealed an oscillatory flow sensitivity of 26 mV/(m/s), with a lower detection threshold limit of 12.1 mm/s.

Inspired by the SN system in fish, our research group developed flexible artificial SN based on polarized poly(vinylidene fluoride-trifluoroethylene) [P(VDF-TrFE)]/barium titanate (BTO) electrospinning nanofiber mat as the sensing layer. The BTO nanoparticles were compounded into well-aligned P(VDF-TrFE) nanofibers via a far-field electrospinning process (Hu et al., 2019b). The X-ray diffraction (XRD) and Fourier transform infrared spectroscopy (FTIR) characterization results showed that BTO/P(VDF-TrFE) nanofibers with a BTO weight fraction of 5 wt% could improve β-phase crystallinity. The piezoelectric output of the BTO/P(VDF-TrFE) nanofiber mat was significantly improved compared with that of pristine P(VDF-TrFE), which was confirmed by piezoresponse force microscopy (PFM) and compression loading tests. In 2019, we proposed a flexible artificial SN, featuring a polarized P(VDF-TrFE)/ BTO nanofiber mat as the sensing layer and a polyimide (PI) film with arrays of circular cavities as the substrate, as shown in Fig. 9.3e (Hu et al., 2019a). The hydrodynamic experimental results illustrated that the proposed sensor was able to perceive the flow direction within a plane, presenting an "8" figure pattern responding to various flow directions. And the flow velocity calibration results showed that the sensor achieved a detection limitation of 0.23 mm/s. In 2020, we presented a cantilever-based ALL velocity sensor, featuring a BTO/P(VDF-TrFE) nanofiber mat and a hydrogel cupula, as shown in Fig. 9.3f (Ma et al., 2020c). The hydrogel cupula imitating the cupula in a fish lateral line system enhances the sensitivity of ALL velocity sensor through a material-induced drag enhancement mechanism. Through experimental flow characterization, the contributions of the material property of the hydrogel were investigated. The hydrodynamic experimental results showed that the proposed ALL velocity sensor can detect a velocity of 0.42 mm/s.

Inspired by the canal lateral line system in fish, researchers have developed various ALL pressure sensors. In contrast to the conventional diaphragm-based pressure sensors, the canal-type ALL (CALL) pressure sensors exhibit a canal structure with aligned pores on the canal surface and flow-sensing elements inside the canal. Therefore, they are able to detect pressure gradients along the surface. The pressure difference between the canal pores drives the canal fluid, which is opposed by the friction generated due to the canal walls. In 2017, Jiang et al. firstly proposed a fully flexible CALL pressure sensor, featuring a PVDF-/polypropylene-based sensing cantilever and a microfluid canal made from PDMS (Jiang et al., 2017). The proposed sensor illustrated a high-pass filtering capabilities to attenuate low-frequency flow stimulus, similar to its biological prototype, which was resulted from viscous resistance to canal fluid motion at low frequencies. The experimental results illustrated that the proposed device was able to detect a pressure gradient of 11 Pa/m between two adjacent canal pores. In 2019, they fabricated a miniaturized sensor with similar structures, with the assistance of microelectromechanical systems (MEMS) fabrication techniques (Fig. 9.4) (Ma et al., 2019). The designed CALL device consisted of a P(VDF-TrFE)/PI cantilever as a sensing element and a PDMS microfluid canal (Fig. 9.4a–c). The structures and dimensions of the sensor were optimized with the COMSOL software. Inspired by the constriction structure within canal nearby canal neuromast in fish canal lateral line system, artificial constriction structure was introduced in the canal to increase the local flow velocity, thereby

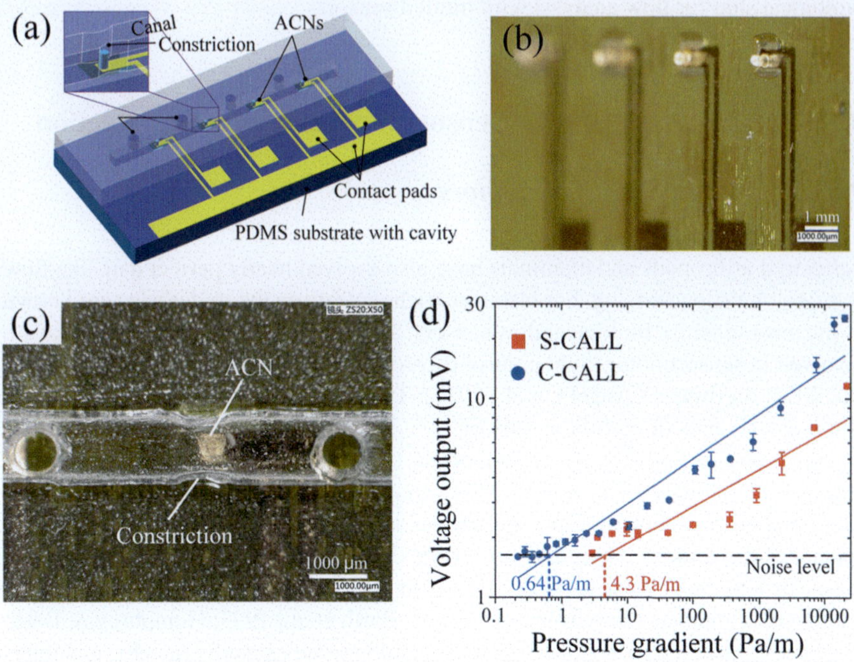

Fig. 9.4 Bio-inspired flexible canal-type ALL (CALL) pressure sensor with a constriction structure. (**a**) A schematic of the proposed device. (**b**) An array of developed artificial neuromasts. (**c**) An optical image of fabricated CALL pressure sensor. (**d**) The comparative hydrodynamic characterization of CALL devices with different canals, i.e., CALL device with straight canal (S-CALL) and CALL device with constriction canal (C-CALL) (Ma et al., 2019)

enhancing the sensory performance of the CALL device. As shown in Fig. 9.4d, the proposed CALL device with constriction canal (C-CALL) achieved a pressure gradient detection limit of 0.64 Pa/m, which was much lower than that of the CALL device with straight canal (S-CALL) (4.3 Pa/m). In addition, the constriction structure could modify the frequency response of the CALL device, i.e., the C-CALL device achieved higher voltage output than S-CALL in high-frequency domain. In 2020, they performed systematic analyses and structural optimization of the constriction canal (Ma et al., 2020b). The hydrodynamic characterization results revealed that the sensitivity of the CALL device increases with the decrease in the width (from 1 mm to 0.6 mm) and length (from 3 mm to 1 mm) of the constriction canal, which is in accordance with the modeling results of canal mechanics. The CALL device with an optimized constriction canal was able to identify the diameter of the cylinder, with a mean identification error of approximately 2.5%. It also demonstrated the identification ability of wake width using the CALL device, indicating the potential for application in hydrodynamic perception.

With the rapid progress of flexible electronics, lateral line-inspired flow sensors with better sensory performance can be expected. However, the perception schemes with multi-model (flow/pressure) sensors have not been well studied. We engineers can still learn sensing fusion and optimization principles of sensor layout from the biological side for flow analysis with limited sensors.

9.3 Bio-Inspired Flexible Sensors for Aerodynamic Detection

9.3.1 Biological Hair Receptors

Terrestrial arthropods and mammals have also evolved nearly perfect hair-like flow sensors for airflow sensing. For instance, trichobothria on the spider's leg are known as the most sensitive biological flow sensors, and wing hairs of the big brown bat are involved in sensorimotor flight control by providing aerodynamic feedback (Tao & Yu, 2012; Sterbing-D'Angelo et al., 2011). This section illustrates the morphology and biomechanics of such flow sensors.

As shown in Fig. 9.5, the trichobothria of spiders show a few morphological features: (1) abundance in number, (2) variation in length, and (3) limited range. The number of trichobothria on spider legs is different depending on the size and species of the spiders; it ranges from several (7–11) to about 100 on each of spider's legs. The lengths of these hairs vary between 100 and 1400 μm in *Cupiennius salei* and the base diameters are from 5 to 15 μm (Barth et al., 1993; Humphrey & Barth 2007). The variation in length can be explained to some extent by the physical interactions between the air and hairs, which is the easiest way for arthropods to tune themselves to a certain frequency range of stimuli by varying the length. The limited range, in fact, is related to the boundary layer thickness near the platform where the hairs are positioned (Casas & Dangles, 2010).

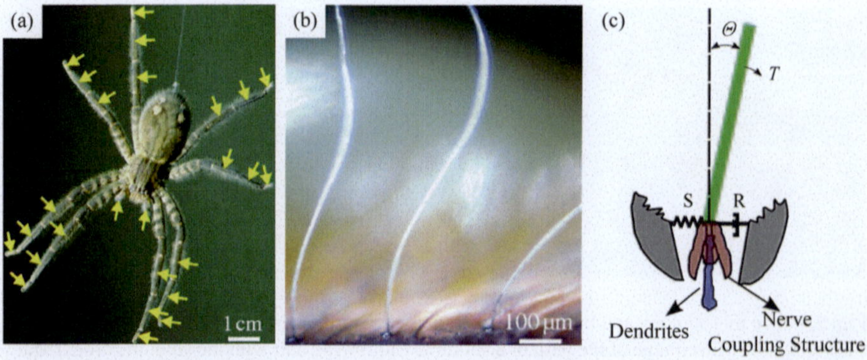

Fig. 9.5 The morphology and biomechanics of the trichobothria in spider legs. (**a**) An adult female spider *Cupiennius salei*. The yellow arrows point to the leg segments carrying trichobothria. (**b**) A group of trichobothria on the spider leg tarsus (McConney et al., 2009b). (**c**) Schematic illustration of a single trichobothria modeling (Tao & Yu, 2012)

As shown in Fig. 9.5c, all the flow-sensing hairs are located in a cuplike cuticular socket, which can be expressed as a hair model with a pin support about which the hair may rotate. The hair shaft forms a simple lever with the long arm exposed to the motion of airflow and the short arm embedded in the exoskeleton coupling with the sensory cells. The length ratio of the outer to inner arm is more than 1000/1, which indicates that the displacement of the hair's tip is scaled down while the force is scaled up. The energy that causes the hair shaft to deflect enough to elicit an action potential is only on the order of 10^{-20} J for a trichobothrium (Barth, 2002). The inner end of the hair shaft couples with the dendrites, and the viscoelastic suspension behaves as a combination of a spring and a dashpot, which resists the deflection of the hair shaft (Fig. 9.5c). As for the trichobothria in spider *Cupiennius salei*, the spring stiffness is on the order of 10^{-11}–10^{-12} Nm/rad and the damping coefficient is on the order of 10^{-14}–10^{-15} Nm/rad (Barth et al., 1993). These extremely low values result in the pure tilt without bending when suffering from the viscous force of airflow. In addition, for some species, such as *C. salei*, the trichobothria surface is conspicuously feathery with 4–14 min branches for each 10 μm hair length, which can increase the drag force acting on the sensilla and thereby enhance the mechanical sensitivity, as it is dominated by viscous forces at *such* a low Reynolds number (McConney et al., 2009b).

Bat wing hair receptors provide flow-related feedback information for enhanced stability during flight. The wing hairs of big brown bat show some distinctive features in hair distribution, morphology, and biomechanics. Figure 9.6a shows the wing schematic of a big brown bat. Though the average density of the hairs is low, the leading edge being more densely populated and median length of the hairs is longer along the leading edge. There are two populations of hairs on the wing: (a) short (<800 μm) and regularly distributed hairs and (b) long hairs. In contrast to the short hair with a linearly tapered cross section, the long hair type is approximately three to ten times longer, larger in diameter, and less tapered (Sterbing-D'Angelo

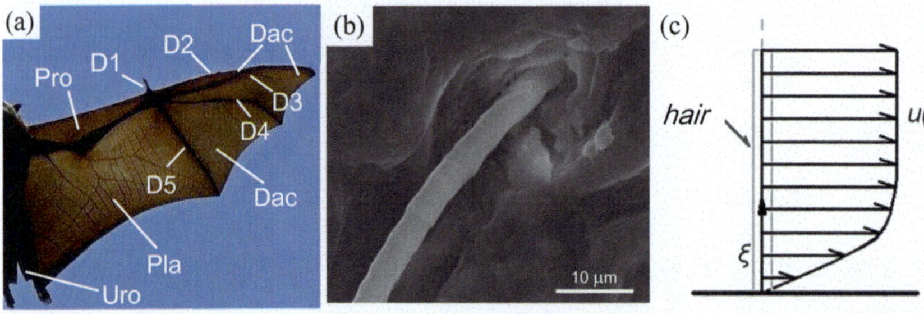

Fig. 9.6 The morphology and biomechanics of wing hairs on the big brown bat. (**a**) Schematic of a bat wing. (**b**) SEM image of the base of a short (331 μm) wing hair with the smooth coronal scale pattern (Sterbing-D'Angelo et al., 2016). (**c**) Nonuniform flow velocity profile incident on the wing hair receptor model (Dickinson, 2010)

et al., 2016). It should be noted that the shaft of bat wing hair appears tightly enveloped by the dome complex and airflow causes deflection of the shaft that in turn stimulate the hair-associated Merkel cells, while the shaft of spider trichobothria is supported by a membrane (Sterbing-D'Angelo et al., 2011).

Mathematical models of hair receptor have received comprehensive exploration in the past few decades. For instance, Dickinson (2010) proposed a hair receptor model to develop a relationship between the transverse forces due to the flow velocity profile incident on the longitudinal axis of the hair (Fig. 9.6c). The calculation results indicate that the hair length is a critical geometric parameter for boundary layer detection and the linearly tapered hair shows higher sensitivities than hairs with uniform diameter, which provide inspiration and guidelines for artificial hair sensor design.

9.3.2 Flexible Artificial Hair-like Sensors

Artificial hair-like (AHL) airflow sensors inspired by flow-sensing hairs have been extensively studied. Various transduction mechanisms, such as piezoresistive (Shi et al., 2013; Liu et al., 2019; Aiyar et al., 2009; Song et al., 2011; Ranganathan et al., 2015; Jiang et al., 2019b), magnetic (Alfadhel et al., 2014), piezoelectric (Li et al., 2010), optical (Paek & Kim, 2014; Bruecker, 2016), and capacitive (Dagamseh et al., 2013), have been used in the development of airflow sensors. In this section, we introduce the research status on the development of flexible artificial hair-like sensors.

Various piezoresistive flexible AHL sensors have been developed by using polymers as structural materials, such as polydimethylsiloxane (PDMS) and polyimide (PI). In 2013, Shi et al. proposed an AHL sensor by using liquid-metal alloy as the piezoresistor, and polydimethylsiloxane (PDMS) membrane as the substrate layer. The liquid-metal piezoresistors deform with the elastomeric substrate, and normal and shear force can be detected with resistance changes of the piezoresistors. In 2019, Liu et al. designed biomimetic electronic hair sensor which consists of nylon fibers as hairs for mechanical signal amplification, PDMS resin as the substrate, and two carbonized papers as piezoresistive mechanoreceptors (Fig. 9.7). The nylon fibers have a diameter and Young's modulus close to those of hairs, and PDMS has a Young's modulus close to that of the human skin. The electronic hair sensor mimicking the human skin is capable of detecting the airflow rate and is effective in identifying the airflow direction. However, the sensitivity and detection ranges were not evaluated in detail.

Polyimide (PI) is another kind polymer material that has been widely used to fabricate flexible AHL sensors. Song et al., (2011) proposed an all-polymer flexible micromachined flow sensor with a polyimide film as the main structure material. The fabricated sensor composed of a carbon-black-based conductive composite elastomer as a piezoresistor and an out-of-plane polyimide-based flow-sensing element. Tensile testing of the composite elastomer yields a gage factor of 7.3, and wind tunnel testing demonstrated a sensitivity as high as 66 Ω/(m/s). Although the fabricated sensors show excellent sensitivity, they suffer from a hysteresis and drift of the baseline resistance. By mimicking the hair cell finding in biological organisms, Ranganathan et al. (2015) proposed the fabrication procedure for small-scale, lightweight, low-cost bio-inspired flow sensors using 13 μm Kapton film to measure flow speeds up to 3.75 m/s.

In 2019, our group reported the fabrication and characterization of a novel flexible airflow sensor (Jiang et al., 2019b). As shown in Fig. 9.8, the hair-like airflow sensor is an out-of-plane cantilever that self-bends from the double-deck PI film substrate. The existence of a stress gradient between the drying polydiallyldimethylammonium chloride (pDADMAC) layer and the PI film was used to cause the hair-like sensor to curve. Gr/PI sensing elements are inkjet-printed at the root of the curved cantilever. The upwardly curved cantilever converts the drag force caused by the external flow to a bending moment. The bending moment introduces a strain that changes the resistance of the piezoresistor as a function of the airflow velocity.

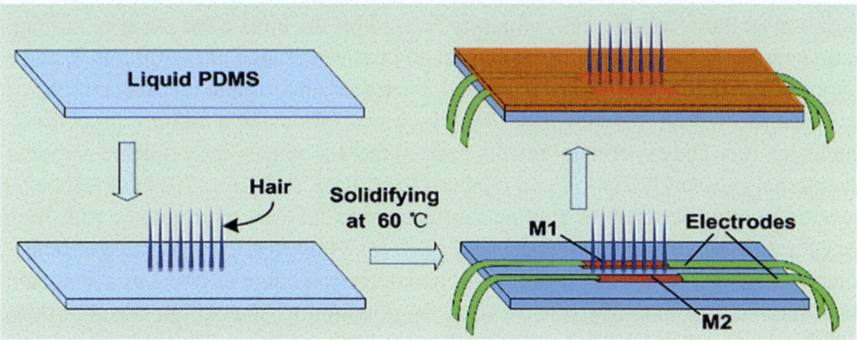

Fig. 9.7 Schematics of an electrical hair sensor consisting of nylon fibers as hairs, carbonized paper as piezoresistors, and PDMS as the substrate (Liu et al., 2019)

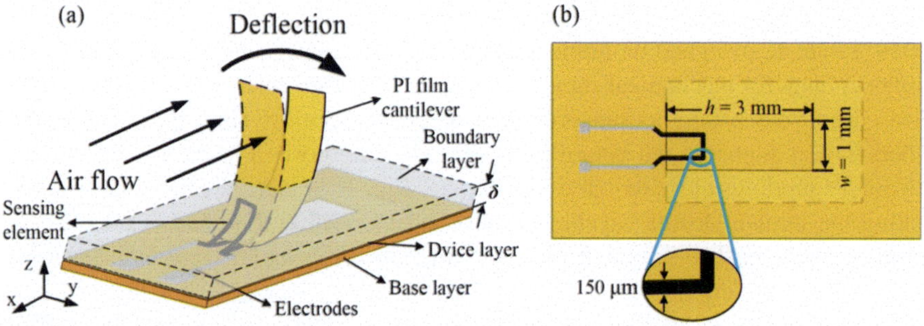

Fig. 9.8 Schematics of an all-polyimide AHL fabricated via an electrical breakdown process (Jiang et al., 2019b)

To attain a sensing element with a high gauge factor, an electrical breakdown (EBD) process was introduced to modify the electrical conductivity of the Gr/PI nanocomposites, thus endowing the sensing element with increased piezoresistivity. Once the applied current exceeded a given threshold, EBD occurred and caused the formation of cavity structures in the polymers, which was essential for the formation of piezoresistivity. A customized tensile setup was used to characterize the gauge factor of the Gr/PI sensing elements, and the sensing element attained a gauge factor of as high as 95. A systematic investigation was performed to characterize the fabricated airflow sensor using a benchtop wind tunnel. The experimental results showed that the fabricated flow sensor achieved a sensitivity of 37.5 kΩ/(m/s), a high resolution of 0.5 m/s, and a large detection range of 0–20 m/s.

Besides the sensors developed based on piezoresistive principles that are mentioned above, giant magnetoimpedance (GMI), piezoelectric, and optical mechanisms are also employed in fabrication of biomimetic airflow sensors. Inspired by the hair-like flow receptors, a magnetic polymer hair-like flow sensor with high resolution and low power consumption was developed (Alfadhel et al., 2014). As shown in Fig. 9.9a, the sensor is comprised of a hair-shaped PDMS pillar embedded with iron nanowires, a GMI sensor, and a substrate layer. When external airflow blows, the tip of the magnetic PDMS pillar will be deflected, and further generate variation in the magnetic field which is created by the embedded permanent magnetic nanowires. The flow sensor reached an operating range of 0–190 mm/s, and a resolution of 0.56 mm/s. In 2010, Li et al. presented an artificial hair receptor based on suspended PVDF fibers (Fig. 9.9b), which was fabricated a thermo-direct drawing technique. The experiment results showed that the sensor has a reliable response to external pressure variation and medium flow. Paek and Kim (2014) developed a new technique that enables routine fabrications of PDMS micropillars with high aspect ratios ranging from 40 and 112 (Fig. 9.9c). However, the fabricated optically interrogated airflow sensor showed a limited detection range of 0–0.2 m/s. It should mention that hair-like flow sensors have the potential to detect high-speed airflow, which is out of the detection range by hot-wire-type anemometers due to thermal saturation.

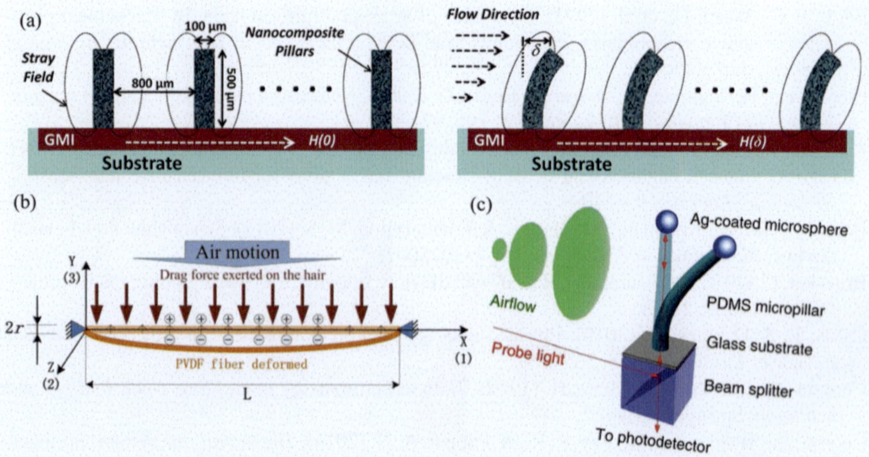

Fig. 9.9 Typical flexible artificial hair-like sensors based on other transduction mechanisms. (**a**) Schematic illustrations of the magnetic artificial hair-like sensor structure and its operation principle (Alfadhel et al., 2014). (**b**) Schematic views of the piezoelectric artificial hair-like flow sensors (Li et al., 2010), (**c**) optically interrogated hair-like flow sensors (Paek & Kim, 2014)

9.4 Concluding Remarks

Through evolution for 1000 years, creatures possess excellent sensory systems for flow field perception, such as fish lateral line system, sensory filiform hairs in arthropods, etc. Inspired by these, researchers have developed various flexible artificial flow sensors which are potential candidates for curved surface, based on different sensing principles. Some of them present excellent sensing performance, comparative to their biological prototypes. However, the understanding of biological sensing principles and the development of flow sensors are still in their baby phase, and there are still many technological challenges that should be resolved for practical applications.

References

Aiyar, A. R., Kim, S. H., & Allen, M. G. (2009). An all polymer air-flow sensor array using a piezoresistive composite elastomer. *Smart Materials and Structures, 18*, 115002.

Alfadhel, A., Li, B., Zaher, A., Yassine, O., & Kosel, J. (2014). A magnetic nanocomposite for biomimetic flow sensing. *Lab on a Chip, 14*, 4362–4369.

Asadnia, M., Kottapalli, A. G. P., Karavitaki, K. D., Warkiani, M. E., Miao, J., Corey, D. P., & Triantafyllou, M. S. (2016). From biological cilia to artificial flow sensors: Biomimetic soft polymer nanosensors with high sensing performance. *Scientific Reports, 6*, 32955.

Barth, F. G. (2002). Spider senses—Technical perfection and biology. *Zoology (Jena, Germany), 105*, 271–285.

Barth, F. G. (2004). Spider mechanoreceptors. *Current Opinion in Neurobiology, 14*, 415–422.

Barth, F. G., Wastl, U., et al. (1993). Dynamics of arthropod filiform hairs. Ii. Mechanical properties of spider trichobothria (cupiennius salei keys). *Philosophical Transactions: Biological Sciences, 340*, 445–461.

Bleckmann, H., Schmitz, H., & von der Emde, G. (2004). Nature as a model for technical sensors. *Journal of Comparative Physiology. A, 190*, 971–981.

Bora, M., Kottapalli, A. G. P., Miao, J., & Triantafyllou, M. S. (2017). Biomimetic hydrogel-CNT network induced enhancement of fluid-structure interactions for ultrasensitive nanosensors. *NPG Asia Materials, 9*, e440.

Bora, M., Giri, A., Kottapalli, P., Miao, J., & Triantafyllou, M. S. (2018). Sensing the flow beneath the fins. *Bioinspiration & Biomimetics, 13*, 025002.

Bruecker, C. (2016). Measurement of near-wall 3D flow velocity from wave-guiding micro-pillars. *Optics Express, 24*, 19.

Casas, J., & Dangles, O. (2010). Physical ecology of fluid flow sensing in arthropods. *Annual Review of Entomology, 55*, 505–520.

Coombs, S., Görner, P., & Münz, H. (1989). *The mechanosensory lateral line. Neurobiology and evolution*. Springer Verlag.

Coombs, S., Bleckmann, H., Fay, R. R., & Popper, A. N. (2014). *The lateral line system*. Springer.

Dagamseh, A. M. K., Bruinink, C. M., Wiegerink, R. J., Lammerink, T. S. J., Droogendijk, H., & Krijnen, G. J. M. (2013). Interfacing of differential-capacitive biomimetic hair flow-sensors for optimal sensitivity. *Journal of Micromechanics and Microengineering, 23*, 035010.

Denton, E. J., & Gray, J. A. B. (1988). Mechanical factors in the excitation of the lateral line of fishes. In Sensory Biology of Aquatic Animals (eds. Atema, J., Fay, R. R., Popper, A. N., Tavolga, W. N.), pp. 595–617. Springer, New York, USA.

Dickinson, B. T. (2010). Hair receptor sensitivity to changes in laminar boundary layer shape. *Bioinspiration & Biomimetics, 5*, 1–11.

Han, Z., Liu, L., Wang, K., Song, H., Chen, D., Wang, Z., Niu, S., Zhang, J., & Ren, L. (2018). Artificial hair-like sensors inspired from nature: A review. *Journal of Bionic Engineering, 15*, 409–434.

Hu, X., Jiang, Y., Ma, Z., Xu, Y., & Zhang, D. (2019a). Bio-inspired flexible lateral line sensor based on P(VDF-TrFE)/BTO nanofiber mat for hydrodynamic perception. *Sensors, 19*, 5384.

Hu, X. H., Yan, X., Gong, L. L., Wang, F. F., Xu, Y. H., Feng, L., Zhang, D. Y., & Jiang, Y. G. (2019b). Improved piezoelectric sensing performance of P(VDF-TrFE) nanofibers by utilizing BTO nanoparticles and penetrated electrodes. *ACS Applied Materials & Interfaces, 11*, 7379–7386.

Humphrey, J. A. C., & Barth, F. G. (2007). Medium flow-sensing hairs: Biomechanics and models. *Advances in Insect Physiology, 34*, 1–80.

Jiang, Y. G., Fu, J. C., Zhang, D. Y., & Zhao, Y. H. (2016). Investigation on the lateral line systems of two cave fish: *Sinocyclocheilus macrophthalmus* and *S. microphthalmus* (Cypriniformes: Cyprinidae). *Journal of Bionic Engineering, 13*, 108–114.

Jiang, Y., Ma, Z., Fu, J., & Zhang, D. (2017). Development of a flexible artificial lateral line canal system for hydrodynamic pressure detection. *Sensors, 17*, 1220.

Jiang, Y., Ma, Z., & Zhang, D. (2019a). Flow field perception based on the fish lateral line system. *Bioinspiration & Biomimetics, 14*, 041001.

Jiang, Y., Shen, D., et al. (2019b). Fabrication of graphene/polyimide nanocomposite-based hair-like airflow sensor via direct inkjet printing and electrical breakdown. *Smart Materials and Structures, 28*, 065028.

Kamat, A. M., Pei, Y., & Kottapalli, A. G. P. (2019). Bioinspired cilia sensors with graphene sensing elements fabricated using 3D printing and casting. *Nanomaterials, 9*, 954.

Li, F., Liu, W., et al. (2010). A novel bioinspired PVDF micro/Nano hair receptor for a robot sensing system. *Sensors, 10*, 994–1011.

Liu, Y. F., Huang, P., Li, Y. Q., Liu, Q., Tao, J. K., Xiong, D. J., et al. (2019). A biomimetic multifunctional electronic hair sensor. *Journal of Materials Chemistry A, 7*, 1889.

Ma, Z., Jiang, Y., Wu, P., Xu, Y., Hu, X., Gong, Z., & Zhang, D. (2019). Constriction canal assisted artificial lateral line system for enhanced hydrodynamic pressure sensing. *Bioinspiration & Biomimetics, 14*, 066004.

Ma, Z., Herzog, H., Jiang, Y., Zhao, Y., & Zhang, D. (2020a). Exquisite structure of the lateral line system in eyeless cavefish *Sinocyclocheilus tianlinensis* contrast to eyed *Sinocyclocheilus macrophthalmus* (Cypriniformes: Cyprinidae). *Integrative Zoology, 15*, 314–328.

Ma, Z., Jiang, Y., Dong, Z., Han, Z., & Zhang, D. (2020b). Hydrodynamic perception using an artificial lateral line device with an optimized constriction canal. *Journal of Bionic Engineering, 17*, 1–11.

Ma, Z., Xu, Y., Jiang, Y., Hu, X., & Zhang, D. (2020c). BTO/P(VDF-TrFE) nanofiber-based artificial lateral line sensor with drag enhancement structures. *Journal of Bionic Engineering, 17*, 64–75.

Mcconney, M. E., Anderson, K. D., Brott, L. L., Naik, R. R., & Tsukruk, V. V. (2009a). Bioinspired material approaches to sensing. *Advanced Functional Materials, 19*, 2527–2544.

McConney, M. E., Schaber, C. F., Julian, M. D., Eberhardt, W. C., Humphrey, J. A. C., Barth, F. G., et al. (2009b). Surface force spectroscopic point load measurements and viscoelastic modelling of the micromechanical properties of air flow sensitive hairs of a spider (cupiennius salei). *Journal of the Royal Society Interface, 6*, 681–694.

Montgomery, J. C., Windsor, S., & Bassett, D. (2009). Behavior and physiology of mechanoreception: Separating signal and noise. *Integrative Zoology, 4*, 3–12.

Paek, J., & Kim, J. (2014). Microsphere-assisted fabrication of high aspect-ratio elastomeric micropillars and waveguides. *Nature Communications, 5*, 3324.

Ranganathan, B. N., Penskiy, I., & Dean, W. (2015). Bio-inspired wind frame state sensing and estimation for MAV applications. In *Proceedings of the IEEE/RSJ International Conference on Intelligent Robots and Systems (IROS)* (pp. 2729–2735).

Schmitz, A., Bleckmann, H., & Mogdans, J. (2008). Organization of the superficial neuromast system in goldfish, *Carassius auratus*. *Journal of Morphology, 269*, 751–761.

Sengupta, D., Trap, D., & Kottapalli, A. G. P. (2020). Piezoresistive carbon nanofiber-based cilia-inspired flow sensor. *Nanomaterials, 10*, 211.

Shi, X., Cheng, C., et al. (2013). *Artificial hair cell sensors using liquid metal alloy as piezoresistors*. IEEE NEMS.

Song, C., Aiyar, A. R., & Kim, S. H. (2011). Exploitation of aeroelastic effects for drift reduction in an all-polymer air flow sensor. *Sensors and Actuators A: Physical, 165*, 66–72.

Sterbing-D'Angelo, S., Chadha, M., & Chen, C. (2011). Bat wing sensors support flight control. *Proceedings of the National Academy of Sciences of the United States of America, 108*, 11291–11296.

Sterbing-D'Angelo, S., Liu, H., Yu, M., & Moss, C. F. (2016). Morphology and deflection properties of bat wing sensory hairs: Scanning electron microscopy, laser scanning vibrometry, and mechanics model. *Bioinspiration & Biomimetics, 11*, 056008.

Tao, J., & Yu, X. (2012). Hair flow sensors: From bio-inspiration to bio-mimicking—A review. *Smart Materials and Structures, 21*, 113001.

Teyke, T. (1990). Morphological differences in neuromasts of the blind cave fish *Astyanax hubbsi* and the sighted river fish *Astyanax mexicanus*. *Brain, Behavior and Evolution, 35*, 23–30.

Yoshizawa, M., Goricki, S., Soares, D., & Jeffery, W. R. (2010). Evolution of a behavioral shift mediated by superficial neuromasts helps cavefish find food in darkness. *Current Biology, 20*, 1631–1636.

Chapter 10
Optofluidic Devices for Bioanalytical Applications

Hui Yang and Martin A. M. Gijs

10.1 Introduction

Today, we are experiencing fast changes in personal health management: from present one-fits-all post-medical care to targeted therapy based on genetics or molecular mechanisms in different time frames (Hamburg & Collins, 2010). Accordingly, the major task of analytical techniques in biological research now relies on detecting, diagnosing, and monitoring molecular dynamics in terms of pathology and physiology, which will benefit early diagnosis, prognosis, and disease monitoring (Jung & Lee, 2015). Therefore, the most pivotal step is generating a customized system that can track the temporal/spatial distribution of targeted molecules or interrogate their behavior in cellular/subcellular level.

To meet with such challenges, in recent years, optofluidic technology has drawn much focuses. Optofluidics, namely, bridging "optics" and "microfluidics," is capable of manipulating light with on-chip fluidic processes, or using light to control fluidic entities. However, recent efforts on optofluidics have already pushed the technology forward and exceeded its original definition, and either (1) using fluidic and optical elements synergistically or (2) employing advanced optical/photonic methods on microfluidics platform to concretize enhanced deployment, reliability, accuracy, throughput, especially in the on-chip aspect, can be defined as optofluidics (Minzioni et al., 2017; Chen et al., 2019; Yang et al., 2019). Over the last two decades, a fast growing interest was noticed for developing microdevices that

---

H. Yang (✉)
Bionic Sensing and Intelligence Center, Institute of Biomedical and Health Engineering, Shenzhen Institutes of Advanced Technology, Chinese Academy of Sciences, Shenzhen, China
e-mail: hui.yang@siat.ac.cn

M. A. M. Gijs
Laboratory of Microsystems, École Polytechnique Fédérale de Lausanne, Lausanne, Switzerland

Z. Yang (ed.), *Advanced MEMS/NEMS Fabrication and Sensors*,
https://doi.org/10.1007/978-3-030-79749-2_10

integrate one or several functionalities (e.g., sample preparation, reaction, separation, and detection) onto a single chip, the latter being of only millimeter-size up to a few square centimeters in size (Haeberle & Zengerle, 2007; Yang & Gijs, 2018; Manz et al., 1990). These microdevices can be used either to perform a certain type of chemical analysis or, more generally, for other nonanalytical lab processes like sample pre-treatment, and they are known as "micro total analysis systems (μTAS)" or "lab-on-a-chip (LOC)" devices. The concept of the integrated μTAS or LOC microfluidic device was introduced by Manz et al. in the early 1990s (Mark et al., 2010). Since then, microfluidic devices have led to great advances in many research fields. Benefiting from microfluidics such as low consumption, micro−/nanoscale sample manipulation, compatibility with consumer electronics, and optofluidic devices that integrate microfluidics with high-end micro-optical components and methods offers various scenarios in bioanalytical applications and personalized medicine (Jung & Lee, 2015).

In this chapter, we focus on recent developments in the realization and use of micro-optical components that are directly integrated onto a microfluidic device. We introduce basic working principles of micro-optics (Sect. 10.2), and optofluidic components and integrated systems (Sect. 10.3). In Sect. 10.4, we will give a brief overview of the microfabrication technologies for realization of microfluidic devices and micro-optical components. Subsequently, optofluidic devices that eventually have been used for a full bioanalytical application (Sect. 10.5) will be presented. We will conclude by explaining the advantages of optofluidic systems for tackling analytical problems in general and highlighting the advantages of different micro-optical components specifically. We hope that the chapter provides the reader with some orientation in the field and enables selecting platforms with appropriate characteristics for his/her application-specific requirement.

10.2 Fundamentals on Micro-Optics

Based on basic optical principles, one can classify micro-optical components into (1) refractive optical components that rely on the change of the refractive index at an interface, such as lenses, prisms, and mirrors; (2) diffractive optical structures that enable shaping of an optical beam by diffractive/interference effects, such as diffraction gratings; and (3) hybrid (refractive/diffractive) structures. Refractive and diffractive optical components share many similarities when they are used to manipulate monochromatic light, but their response to broadband light is very different. For a material with normal dispersion, refractive lenses have larger focal distances for red light than for blue light, and prisms deflect longer wavelengths by a smaller angle; the contrary occurs for diffractive lenses and gratings. This contrasting behavior arises because two different principles are used to shape the light: refractive optics relies on the phase that is gradually accumulated through propagation, while diffractive optics operates by means of interference of light transmitted through an amplitude or phase mask. The decision to use diffractive or refractive optics for a specific optical problem depends on many parameters, e.g., the spectrum of the light source, the aimed optical application (beam shaping, imaging, etc.), the efficiency required, the acceptable stray light, etc. Arbitrary wavefronts can be generated very accurately by diffractive optics. A drawback for many applications is the strong wavelength dependence. Diffractive optics is therefore mostly used with laser light and for nonconventional imaging tasks, like beam shaping, diffusers, filters, and detectors. Refractive optical elements have in general higher efficiency and less stray light, even though in some cases it is more difficult to make refractive lenses with precise focal lengths or aspheric shapes. Moreover, for broadband applications, diffractive optical elements (DOEs) can be combined with refractive optics to correct for the chromatic aberration. This combination allows systems with low weight or which consist of only one material. We summarize in the following fundamental optical principles like light reflection and refraction, light wave interference, diffraction, and polarization.

From a physics point of view, a *light wave* is an oscillatory electric and magnetic field that co-propagate through space. The electric and magnetic field components of this electromagnetic radiation are oscillating in phase perpendicular to the propagation direction and with respect to each other. Optical methods make use of electromagnetic radiation with wavelengths from ~200 nm to a few μm. This interval can be subdivided into the ultraviolet (UV) region (~200–400 nm), the visible (Vis) region (~400–700 nm), and the near-infrared (NIR) region (~700 nm to a few μm). The UV/Vis region is used most widely for microsensing systems, especially in fluorescence detection methods, but also in absorbance and chemiluminescence techniques. When the amplitude of electromagnetic waves composing a light beam oscillates in the same plane, the light is said to be polarized. Polarized light is generally classified into two groups depending on how the electric waves are aligned with the plane of incidence onto a dielectric surface. The plane of incidence is the plane composing the incident, reflected, and transmitted rays. Transverse magnetic (TM), parallel (//), P, and O are all notations to refer to waves that are polarized such that the electric field is in the plane of incidence. Transverse electric (TE), perpendicular (⊥), S, and E are all ways to refer to polarization perpendicular to the plane of incidence. The polarization of light can affect many aspects of optical systems, and materials can have different properties for different polarizations, such as indices of refraction or reflectivity.

Reflection of light happens when light is incident on an interface between two transparent optical media. How light reflects from a surface is governed by the law of reflection: the angle that the reflected ray makes with the normal at the point of incidence is always equal to the angle the incident ray makes with the same normal; the incident ray, reflected ray, and normal always lie in the same plane. The behavior of reflection can also be shown by Fermat's principle, which states that light will take path with the shortest travel time to go from one point to another. Therefore, depending on the incident angle of the light onto a surface, light will only reflect at an angle that is the same as the incident angle, i.e.,

$$\theta_i = \theta_o \tag{10.1}$$

Scattering can also change the direction of propagation of a light beam. An incident wave can be scattered by a diffusely reflecting surface that generates scattered waves into various directions. A wave can also be scattered while propagating through a diffuse liquid or solid medium with a distribution of refractive indices.

Refraction is the bending of light waves at an interface between two media with different optical density. These two transparent optical media are distinguished from one another by the index of refraction n. The relation between the sines of the angles of incidence and refraction and the indices of refraction of the two media is defined by Snell's law:

$$n_i \sin\theta_i = n_r \sin\theta_r \tag{10.2}$$

Total internal reflection can occur when light in a high-index medium reaches a low-index medium ($n_i > n_r$). When $\theta_r = 90°$, the corresponding incident angle is called the critical angle. Ideally, any incident ray at the critical angle or higher will cause total internal reflection. Although all of the incident energy is reflected, the solution of Maxwell's equations predicts the existence of an electromagnetic field in the less dense medium with intensity decaying exponentially away from the interface. This evanescent wave propagates parallel to the interface and has a decay length, i.e., penetration depth d, on the order of the wavelength of the illumination light λ_i. The evanescent intensity field is described by the following set of equations:

$$I(z) = I_0 e^{-z/d} \tag{10.3}$$

$$d = \frac{\lambda_i}{4\pi}\left[n_i^2 \sin^2\theta_i - n_r^2\right]^{-\frac{1}{2}} \tag{10.4}$$

where z is the transverse dimension, $I(z)$ is the intensity of the evanescent field throughout the medium with lower refractive index, I_0 is the intensity at the interface of two media (at $z = 0$), θ_i is the angle of incidence, n_i is the refractive index of the incident medium, and n_r is the refractive index of the secondary medium. The penetration depth usually ranges between 30 and 300 nm. The polarization of the incident beam does not affect the penetration depth, but it does affect the amplitude of the evanescent field. When the incident angle equals the critical value at which $n_i^2\sin^2\theta_i - n_r^2 = 0$, d goes to infinity and the refracted light propagates along the surface; therefore, the incident angle θ_i must be greater than the critical angle $\theta_c = \arcsin(n_r/n_i)$ for total internal reflection to occur.

It was shown that the penetrating evanescent field could be absorbed or diverted by bringing an absorbent or a high-index material into close proximity to the reflecting surface; hence, the total reflection can be frustrated (Harrick, 1962). A classic experiment to demonstrate this phenomenon is done by using two prisms. The light is incident on the surface of the first prism condition of total internal reflection so as to generate the evanescent field, and the second prism reads this evanescent field and emits it to the far field. As the gap between two prisms decreases, the transmittance increases exponentially, except in the near-field region where the transmittance curve deviates from an exponential decay and exhibits linear characteristics due to the strong interaction between the two prism surfaces via the evanescent field. This special characteristic associated with the frustrated total internal reflection (FTIR) has been exploited in different applications, such as near-field microscopy (Wynne et al., 2012; Carney & Schotland, 2001), optical coupling (Hanumegowda et al., 2005; Gupta & Goddard, 2013), and biochemical detection (Bruls et al., 2009; Dittmer et al., 2010). The method of FTIR was extended into a new technique after the excitation of non-radiative surface plasma waves was found in a thin metal film in contact with the prism base. The surface plasmon resonance (SPR) technique is widely used in gas detection and biosensing. By passing polarized light through a high-refractive index prism, the polarized light becomes TIR and generates an evanescent wave that penetrates the thin metal film. Monitoring the intensity of the reflected light as a function of the angle of incidence, the intensity of reflected light will pass through a minimum, as energy from the evanescent wave is absorbed by the surface plasmons in the thin metal film. The angle at which the minimum reflected light intensity occurs is known as the SPR angle. The SPR angle is sensitive to changes in the dielectric refractive index at close proximity to the metal surface (Helmerhorst et al., 2012). By immobilizing receptor molecules at the metal surface, this technique can be applied to diagnostic applications as the interaction of biomolecules will produce a change in the refractive index near the metal surface, causing a shift in the SPR angle and providing a detectable signal. Usually, the ligand of interest is attached to the surface using well-established surface chemistries, and the analyte is then passed over the ligand-derivatized matrix. Any change in mass following the interaction between the ligand and analyte is detected as a change in the angle of the incident light needed to generate the SPR phenomenon. This is measured as an energy or reflectance dip as a function of pixels, which translates to response units (RU) over time. The RU change is directly proportional to molecular mass change, and so binding kinetics and stoichiometry can be measured in real time without using labeling techniques. Snell's law also determines the imaging properties of lenses. A ray incident on the lens refracts at the front surface (according to Snell's law), propagates through the lens, and refracts again at the rear surface. The principles and working mechanisms of lenses will be introduced later.

Interference means that two or more electromagnetic waves can be combined by adding or subtracting electric and magnetic field vectors with proper attention to phase and amplitude. Two overlapping monochromatic waves are incoherent if the waves have no persistent phase relationship. In this case, the power carried by the waves is additive. On the other hand, if the fluctuations of phase and amplitude of two beams are correlated, the beams are coherent. If two monochromatic beams are coherent, propagate in essentially the same direction, are polarized in the same direction, and are in phase, then in regions of overlap, their electric vectors are added and the waves are said to interfere constructively. The power is proportional to the square of the vector sums of the electric vectors. If the waves are 180° out of phase, the vectors are subtracted, the waves interfere destructively, and if the two waves have equal power, the combined wave has no power.

Diffraction refers to the process by which a beam of light is spread out as a result of passing through a narrow aperture or across an edge. The image of a point object is not an ideal point, even when formed by a perfect optical element. This is due to the diffraction phenomenon. Diffraction sets the limit of resolution of an image. The intensity point spread function is the intensity distribution in the image that is formed from a point object. If an optical detection system is exposed to an object emitting incoherent radiation, the formed image is the superposition of the intensity point spread functions associated with all of the points that make up the object. Two points of a diffraction-limited image can be resolved when the central peak of the point spread function of one falls on the first zero of the other and the resulting pattern has two peaks with a 74% valley between them. This is the *Rayleigh criterion*. The diffraction-limited resolution of a system with a circular aperture is

$$\Delta x = 0.61\lambda / \mathrm{NA} \tag{10.5}$$

where λ is the wavelength of the light beam and NA refers to the numerical aperture of the system. The numerical aperture is an important parameter for a lens, related directly to how much light the lens gathers. The definition of numerical aperture is given as

$$\mathrm{NA} = n \sin \alpha \tag{10.6}$$

where n is the index of refraction of the intervening medium between object and lens and α is the half-angle defined by the limiting rays that are collected by the lens. The "light-gathering" power of the microscope's objective lens is thus increased by increasing the refractive index of the intervening medium. In addition, the numerical aperture is closely related to the acceptance angle of optical waveguides and fibers. For conventional optical microscopes, diffraction sets the ultimate limit of resolution of an image. The evanescent components that carry the fine information about the object decay exponentially in a medium with positive permittivity and permeability and are lost before reaching the image plane; only the propagating wave components emanating from the illuminated object can be collected at the image plane (Pendry, 2000; Fang et al., 2005).

It has been demonstrated that dielectric microspheres can be used as microlenses to achieve near-field focusing and magnification, resulting in the capability to resolve features beyond the diffraction limit (Wang et al., 2011; Yan et al., 2015). High-index microspheres (n ~ 1.9–2.1), fully immersed in liquid, actually allow enhanced imaging with minimum resolved feature sizes of ~$\lambda/7$ using white-light illumination for imaging of nano-features (Darafsheh et al., 2012) and adenoviruses (Li et al., 2013), as well as with fluorescent microscopic setup for resolving the structures of subcellular organelles (Yang et al., 2014). In these works, the microspheres were simply placed on top of the sample object, where they collected the underlying sample's near-field nano-features and subsequently transformed the near-field evanescent waves into far-field propagating waves, creating a magnified image in the far field, which is collected by a conventional optical microscope

(Wang et al., 2011; Yang et al., 2014; Yang & Gijs, 2015). The super-resolution imaging capability (i.e., imaging beyond the classical diffraction limit) of microspheres is linked to the so-called photonic nanojet phenomenon. A substantial literature has developed regarding the existence, properties, and potential applications of the photonic nanojet (Yang et al., 2015, 2016a; Chen et al., 2004; Li et al., 2005; McLeod & Arnold, 2008; Gérard et al., 2008; Heifetz et al., 2009). In principle, a photonic nanojet is a narrow light beam with high optical intensity that can be generated by a transparent dielectric symmetric body, like a microcylinder or microsphere, upon illumination. The nanojet that emerges from a microsphere is located in the immediate vicinity of the rear surface of the sphere. It is a nonresonant phenomenon that appears for microspheres with a diameter of the order of 10–100λ and with a ratio of their refractive index nms to the refractive index of the mostly water-based background medium nw that is smaller than about 2. The nanojet can maintain a subwavelength full-width-at-half-maximum (FWHM) transverse beam width along a path that can extend more than ~2λ beyond the sphere, and the minimum FWHM beam width can be smaller than the classical diffraction limit, in fact as small as ~λ/3. The next section introduces the field of optofluidics, from typical components to its applications.

10.3 Optofluidic Components and Integrated Systems

From an engineering point of view, optical detection in chemical/biochemical microfluidic devices is easily implemented via incorporation of optical components, such as light sources, waveguides, microlenses, and detectors, to achieve a fully operational micro-optical system. Along with the development of microfluidic technologies, researchers also improve lab-on-a-chip functionalities by incorporating optical components into the same system (Monat et al., 2007). Most solid-phase optical components are lack of capability on deformation and position adjustment, as they are usually fixed in an integrated system. On the contrary, fluid-based optical components can achieve self-tunability, adaptivity, and precise alignment in the context of optofluidic circuits. In the following years, the number of papers regarding to "optofluidics" has been booming with a wide variety of applications, such as biochemical sensor (Fan et al., 2008), bio-imaging (Cui et al., 2008), energy production and harvesting (Erickson et al., 2011), on-chip manipulation of cells (Huang et al., 2014; Yang et al., 2016b), etc. Many research groups have presented a large collection of optofluidic components for imaging. These optofluidic components are categorized in terms of either their working mechanisms or their functions similar to their macro-scale counterparts (e.g., waveguide, lens, cavity, etc.) (Song & Tan, 2017; Zhao et al., 2013). We will present the working principles of these optofluidic components, and focus on their timely progresses and challenges in this section.

10.3.1 Optofluidic Light Source

Light sources can be applied in microfluidic systems through free-space coupling or by direct integration on-chip. Commercial light-emitting diodes (LEDs) have been widely used as light sources in hybrid assembled devices. However, an optofluidic chip requires synergetic combination between the microfluidic device and micro-optical system. Organic light-emitting diodes (OLEDs) and dye lasers are currently the two most used integrated light sources on microfluidic chips (Yao et al., 2005; Li & Psaltis, 2008). The advantages of these two devices are that they can be fabricated on the same substrate as the microfluidic system and readily integrated with other fluidic functionalities. They can also benefit from the fluidic scheme: liquid OLEDs can allow realization of truly flexible displays without detachment of the liquid-emitting layer from the electrodes (Xu & Adachi, 2009); the problem of OLED degradation resulting from the decomposition of organic materials no longer exists as the liquid emitters can be refreshed by a flow (Kasahara et al., 2013); and the characteristics of dye lasers are amenable to tuning by changing the dye solution (Li & Psaltis, 2008).

The light sources are usually used together with other micro-optic components to generate an optofluidic laser. The first-generation optofluidic laser based on a Fabry-Pérot resonator was developed by Whitesides' group (Vezenov et al., 2005). In this work, fluorescence was constrained in the core flow of an optofluidic waveguide. Both ends of the waveguide were coated with a thin layer of gold, acting as reflecting mirrors. Therefore, the fluorescence bounced back and forth and escaped from the gold layer, forming a laser. This laser system is configurable in the aspects of emission wavelength, numerical aperture, and absorbance by changing dye/solvent composition and flow rates of core/cladding layer. An optofluidic laser based on distributed-feedback (DFB) gratings was proposed (Song et al., 2009). As shown in Fig. 10.1a, a SU-8 pattern with high-refractive index (i.e., DFB resonator) is sandwiched between a cladding flow and a silicon dioxide layer. The DFB grating has relatively large surface area and allows enhanced interaction of the evanescent tail

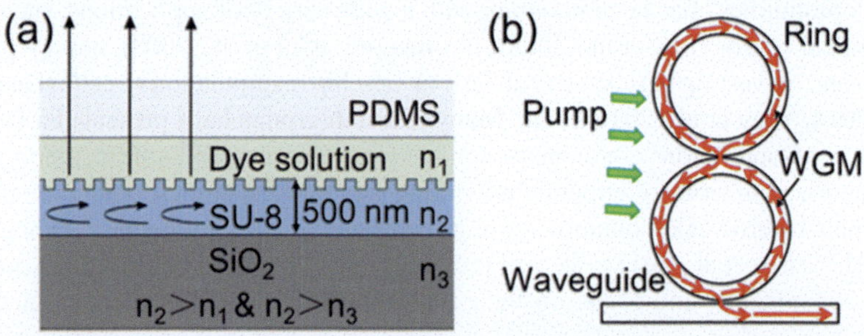

Fig. 10.1 (**a**) DFB resonator-based optofluidic laser. (**b**) Optofluidic ring resonator laser. (Reproduced with permission (Chen et al., 2019), Copyright 2019, Optical Society of America)

with the upper liquid gain medium, forming a surface-emitting optofluidic laser via output coupling.

Though single-mode operation and large tunable wavelength range have been realized with the above two types of optofluidic lasers, the relative low *Q* factors (~1e3) prevent them from achieving low lasing threshold (Fan & White, 2011). As a comparison, an optofluidic ring resonator (OFRR) laser system can achieve high *Q* factor; the schematic is shown in Fig. 10.1b. The system consists of two mismatched ring resonators and a liquid-filled waveguide (Lee et al., 2011). The lower ring is optically coupled with the straight waveguide but physically disconnected. Water-based medium with fluorescent dye is filled inside the structures. Upon excitation, fluorescent light is confined in the gain medium because of the polydimethylsiloxane (PDMS) sidewalls with low-refractive index, and emits through the waveguide. The optical coupling between two OFRRs and the waveguide achieves side mode suppression and a *Q* factor of ~1e8. To date, ring resonators have been exploited in many formats, such as microdroplets (Tang et al., 2009), microknots (Jiang et al., 2007), and microcapillaries (Shopova et al., 2007), providing an excellent optical feedback for low-threshold lasing.

10.3.2 Optofluidic Prism

How to achieve an accurate alignment among multiple optical components is one of the main challenges when designing an optofluidic system. The use of an optofluidic prism to manipulate light path would greatly alleviate such integration issue. A configurable optofluidic prism was reported (Xiong et al., 2011). By converging three laminar flow streams in a triangular-shaped chamber, the apex angle of the prism can be adjusted by altering flow rates. Thus, one can steer optical path in-plane due to light refraction at the interface of two fluids. Further explorations on optofluidic prisms focus on system rotatability and optical scanning robustness. Zappe's group developed a 360° rotatable optofluidic prism which enabled rotational beam scanning in 16 discrete steps (Fig. 10.2) (Kopp et al., 2016). A cylindrical fluidic chamber comprising of two immiscible liquids is wrapped with flexible polymeric foils. Segmented electrodes are embedded in the foils. Upon applying variable voltages over the electrodes, local contact angle of the liquid interface can be changed due to electro-wetting effect. In this device, the total number of beam positions is determined by the number of electrodes, meaning that more meticulous beam rotation can be achieved by using more discrete electrodes.

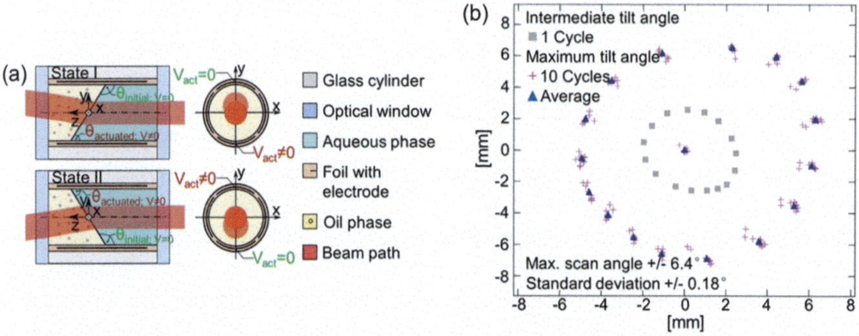

Fig. 10.2 (**a**) Rotatable optofluidic prism with beam positions tuned by electro-wetting effect. (**b**) Beam positions of the tunable prism for ten repeated cycles. Maximum tilt angle (pink crosses), spatial average (blue triangles), and smaller tilt angle (gray squares) are shown in the figure. (Reproduced with permission (Chen et al., 2019), Copyright 2019, Optical Society of America)

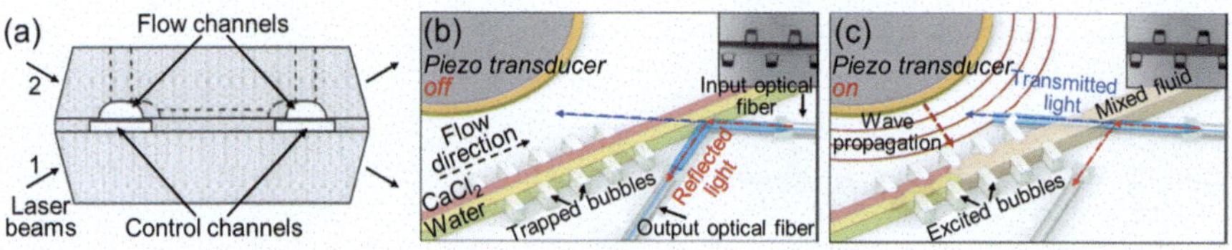

Fig. 10.3 (**a**) Schematic on a three-layer optofluidic switch with four optical facets. The arrows show the directions of both incident and reflected/transmitted laser beams. (**b**, **c**) Working mechanism of an acoustic-driven optofluidic switch without (**b**) and with (**c**) acoustic excitation. (Reproduced with permission (Chen et al., 2019), Copyright 2019, Optical Society of America)

10.3.3 Optofluidic Switch

As an important part of an optofluidic system, optical switches can modify light path and intensity (Pang et al., 2012; Campbell et al., 2004). A 2 × 2 optofluidic switch based on total internal reflection was proposed by Campbell et al. (2004). The device was fabricated using soft lithography, consisting of two distinct layers which are separated by a thin PDMS membrane with low Young's modulus (Fig. 10.3a). The upper layer is a fluidic layer containing a mirror channel with two inlets (one for low-refractive index medium and the other for high-refractive index medium), and the lower layer is a control layer containing five pneumatic valves. These valves are used to control the exchange of sample media in the mirror channel. Upon illumination, the light beam can be transmitted or reflected by tuning the refractive index of the medium in the mirror channel. Huang's group developed an acoustic-driven optofluidic switch (Huang et al., 2012). This device consists of a microfluidic channel that is sandwiched between air-filled cavities in PDMS. Deionized water (n = 1.33) and $CaCl_2$ solution (n = 1.42) are introduced in the main channel. Incident laser beam is reflected at the interface between PDMS (n = 1.41) and DI water when the piezoelectric transducer is turned off (Fig. 10.3b).

On the contrary, the oscillated cavities mix the two media when the transducer is turned on, and the incident laser beam is transmitted through the microchannel since the refractive index of the mixed solution is close to that of PDMS (Fig. 10.3c).

10.3.4 Optofluidic Waveguide

An optical waveguide is a physical structure that transmits light along its axis, which is generally composed of a core with a cladding part. A planar optical waveguide is a waveguide fabricated in a flat format and is particularly interesting to combine with a microfluidic chip. A waveguide is a key optical element for integrated optofluidics, and has the same importance for directing light as the fluidic channel has for directing fluids. The waveguides that have been assembled on microfluidic chips rely on either total internal reflection (TIR) or wave interference for confinement of optical beams (Schmidt & Hawkins, 2008; Hawkins & Schmidt, 2008). TIR-based waveguides are composed of a high-refractive index core material surrounded with a low-refractive index cladding material. A light ray that is incident on the core-cladding interface at an angle above the critical angle of TIR propagates along the longitudinal direction. The requirement for the refractive index of the core to be bigger than that of cladding layer is easy to fulfill in solid-state devices, but poses difficulties for optofluidics. This is because the index of the solid materials used for microfabrication in both polymer- and silicon-based approaches (1.4–3.5) is larger than the index of water and most aqueous solutions (~1.33). Building TIR-based optofluidic systems requires creativity in structural design or using alternate materials. Another approach to waveguiding is the use of wave interferences to confine an optical wave. In this case, multiple layers of material are used as the waveguide cladding. These layers create multiple reflections of the electric field that can interfere constructively or destructively. The key idea is that near-perfect reflection into the original medium can be achieved even if that medium (e.g., water) has a lower refraction index than all of the cladding layer materials (e.g., silicon dioxide, silicon nitride, and titanium dioxide) (Testa et al., 2010).

A typical method for creating integrated optofluidic waveguides with TIR confinement is developed by flowing two or more different laminar liquid streams with different refractive indices inside a fluidic channel, and these streams work as core or cladding. This type of waveguide is referred to as liquid-core/liquid-cladding (L^2) waveguide. The index of the cladding liquid is smaller than that of the core liquid so that the light is guided in the channel by TIR occurring at the interface of the liquids. This concept was first demonstrated using $CaCl_2$ ($n = 1.445$) and water ($n = 1.33$) as core and cladding liquids, respectively, flowing in PDMS ($n = 1.4$) channels (Wolfe et al., 2004, 2005; Brown et al., 2006). Core fluid is first injected into the central channel through a tapered channel. Cladding fluid is then injected on either side of the central channel, resulting in the core fluid being surrounded by cladding fluid on the left and right and PDMS above and below. By changing the flow rates of the liquids, the size and the direction of the waveguides can be reconstructed. Lee et al.

developed a 3D microfluidic L^2 waveguide in which the liquid core was hydrodynamically focused in the center of the PDMS microchannel by the cladding fluids; therefore, the core liquid was not in contact with the channel walls, and this method could be used to decrease the optical loss at the top and bottom interfaces of core and channel walls (Lee et al., 2010). The performance of the L^2 waveguides is affected by the sensitivity of the stream interface to vibrations and diffusion effects. However, compared to solid-state waveguides, the L^2 waveguides show the following advantages: (1) the optical performance is tunable by controlling the fluidic properties and (2) the smooth interfaces between the core and cladding streams minimize optical loss. It has also been shown that the L^2 waveguides can be integrated with biosensing devices. Rosenauer et al. presented a liquid-core/liquid-twin-cladding waveguide for on-chip fluorescence spectroscopy (Rosenauer & Vellekoop, 2009a). The waveguide included an inner and an outer cladding fluid surrounding the core fluid. The analyte was suspended in the inner cladding fluid and excited by the evanescent wave at the fluid interface between the core and the inner cladding fluid (Fig. 10.4). Considerable reduction in sample volume and a low background noise were obtained by tuning the ratios of the flow rates of three different liquids.

10.3.5 Optofluidic Lens

A microlens is an important and key optical component for focusing and collimating light in a microfluidic optical detection system. The motivation for integration of microlenses in microfluidic devices is typically to improve the detection, e.g., by focusing the light in the channel to increase the excitation power for optical measurements (Yang et al., 2015; Schwartz et al., 2010). A microlens can be made of a hard solid surface (e.g., a microsphere) or a deformable interface such as formed

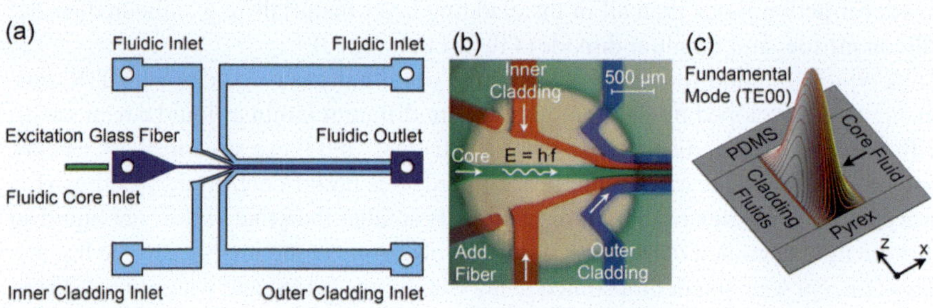

Fig. 10.4 (**a**) Schematic of an optofluidic fluorescence waveguide. (**b**) Micrograph of the L^2 waveguide structure. The hydrodynamic focusing of the core and inner cladding fluids is shown with different stained aqueous ink solutions. (**c**) Simulations from COMSOL show the fundamental mode of the waveguide; the height data expresses the normalized time-average power flow. (Reproduced with permission (Rosenauer & Vellekoop, 2009a), Copyright 2009, American Institute of Physics)

between a soft solid and a liquid, a liquid and a liquid, etc. The solid-state lenses on-chip are usually with fixed focal length, while the optical axis of the lens is vertical to the chip plane, focusing light to the channel from a top or bottom surface (Yang et al., 2015; Schwartz et al., 2010; Yang & Gijs, 2013). This type of lens is most often assembled in hybrid systems. Another type of solid lens has an optical axis parallel to the chip plane, introducing light from the sidewall of the chip (Seo & Lee, 2004; Ro et al., 2005), and can be directly fabricated on-chip. These microlenses are robust, vibration-resistant, and simple to arrange in an array of solid microlenses of specific focal lengths. Other promising and popular microlenses on-chip utilize the curvature formed by the interface between two liquids, between a liquid and a soft solid, or between liquid and air. Such microlenses are also called optofluidic microlenses, and they are dynamic optical elements and their focal lengths can be easily tuned by changing fluid characteristics, such as the type of material, concentration, and flow rate (Mao et al., 2007; Rosenauer & Vellekoop, 2009b). The major advantages of these interface-based microlenses are that they are reconfigurable and they have optically smooth meniscus surfaces. Light can be modulated by simple adjustment of the flow conditions, thus eliminating the need for mechanical or electrical parts for manipulating the lens on the microchip.

Optofluidic lenses can be categorized into in-plane and out-of-plane lenses, depending on the light propagation direction (Song & Tan, 2017; Nguyen, 2010). Yu et al. presented a hydrodynamic out-of-plane optofluidic lens consisting of two layers, which was fabricated by soft lithography, as shown in Fig. 10.5a (Yu et al., 2010). The lower layer contains a circular PDMS aspherical surface that can effectively compensate spherical aberration and improve optical quality, and the upper layer is a PDMS elastic membrane. By controlling injection pressure in the microchannel, the surface contour of the upper layer can be changed, and the focal length of the lens is dynamically adjusted. Other approaches such as using stimuli-responsive hydrogels (Dong et al., 2006), electrical tuning (Mishra et al., 2014), and changing medium's refractive index (Hu et al., 2018) have also been exploited to adjust the focal length of out-of-plane optofluidic lenses.

Since the fabrication process of out-of-plane lenses is relatively complex and the structures should be well-aligned, their applications in optofluidic systems are

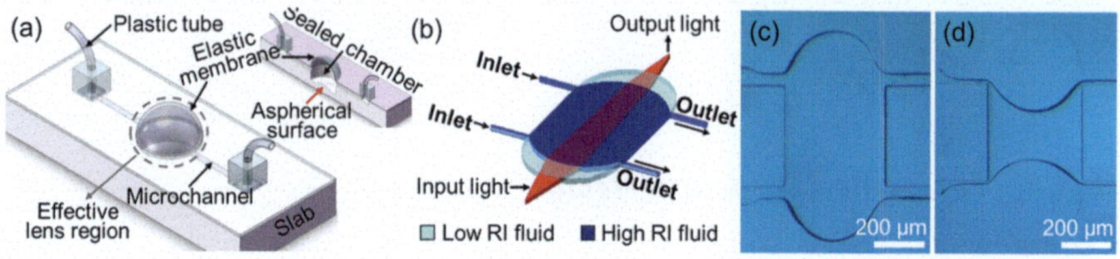

Fig. 10.5 (**a**) Schematic on a liquid-filled out-of-plane lens. The inset illustrates the cross section of the device. (**b**) Schematic on an in-plane optofluidic lens. The high RI fluid medium forms a biconvex microlens that focuses the incident light. (**c, d**) Experimental images on the biconvex microlens (**c**) and the biconcave microlens (**d**), respectively. (Reproduced with permission (Chen et al., 2019), Copyright 2019, Optical Society of America)

limited in many cases (Nguyen, 2010; Mao et al., 2009). In-plane optofluidic lenses, on the other hand, offer more compatibility. The geometry of in-plane optofluidic lens can be changed to adjust focal length by pneumatic tuning (Shi et al., 2010), thermal driving (Liu et al., 2017a), etc. Hydrodynamic tuning is well studied and widely used; this method is based on the formation of a curved liquid/liquid interface, which is controlled by hydrodynamic force. A curved fluid interface can be generated between a core stream and a cladding stream, and this feature can be used as an optofluidic lens and altered by simply changing the flow rate. Mao et al. firstly reported an on-chip hydrodynamically tunable liquid-liquid microlens (Mao et al., 2007). Later, Rosenauer et al. designed a hydrodynamically adjustable convex lens with 3D light focusing ability (Rosenauer & Vellekoop, 2009b). Two miscible fluids of different refractive indices were co-injected into a 90-degree curve in a microfluidic channel. Upon entering the curve, a centrifugal effect caused the fluidic interface to be distorted and one liquid bowed outward into the other. The bowed fluidic interface, coupled with the refractive index contrast between the two fluids, generated the microlens. The curvature of the fluidic interface was modulated by changing the flow rate. Higher flow rates generated a microlens with larger curvature and hence shorter focal length. Hydrodynamically tunable lenses can also be formed in microfluidic channels, in which part of the channel sections are expanded. Fang et al. also presented a hydrodynamic optofluidic lens (Fang et al., 2017). Silicone oil and $CaCl_2$ solution can be injected into an expanded chamber, and used as cladding and core layers of the lens (Fig. 10.5b), respectively, wherein the refractive index of the cladding streams matches that of the device's material to avoid light scattering. By tuning flow rates, the lens geometry can be changed into biconvex (Fig. 10.5c), plano, or biconcave format (Fig. 10.5d). The incident light therefore can be converged, collimated, or diverged, respectively.

Interestingly, it has also been reported that biological materials such as cyanobacteria (Schuergers et al., 2016), diatoms (De Tommasi et al., 2014), red blood cells (RBC) (Miccio et al., 2015), and spider silks (Monks et al., 2016) also show the capability to focus light into a confined region, acting as bio-microlens. Miccio et al. demonstrated using RBC as adaptive optofluidic microlens to perform real-time, fine-resolution measurement on chip (Miccio et al., 2015). Li et al. trapped living yeast and human cell with spherical or disc shapes on a fiber probe to confine excitation light in a sub-wavelength region and achieve fluorescence enhancement (Li et al., 2017).

10.3.6 Integrated Optical Detector

Many optical detectors have been integrated on microfluidic chips, such as silicon photodiodes (Kamei et al., 2003), organic photodiodes (OPDs) (Hofmann et al., 2005), and complementary metal-oxide-semiconductor (CMOS) chips (Cui et al., 2008; Lee et al., 2007). The former two are point detectors without spatial resolution, whereas CMOS detectors can capture a whole image. Silicon photodiodes could render relatively high-sensitivity detection of low analyte concentrations but are too expensive and complicated to fabricate as an integral part of a disposable microfluidic chip. Generally speaking, the optical detectors integrated in point-of-care devices should be inexpensive and easy to fabricate. In contrast to the silicon counterpart, OPDs represent a promising option, as the substrates for their microfabrication can be either a glass or a polymer instead of silicon. For example, when the OPD substrate was printed on a glass slide (Wojciechowski et al., 2009), this system approached the final requirements of a point-of-care system, i.e., it was inexpensive, easy to use, highly sensitive, and robust. Besides realization of photodiodes, CMOS technology allows another type of detector that can be assembled on microfluidic chips and offers the great advantage of the ability to observe the image of the sample object. One of the most widely developed applications of an integrated CMOS detector on-chip is lens-free imaging (Greenbaum et al., 2012). Typically, bio-imaging and biosensing are performed at the microscopic scale, and this usually requires lenses integrated in a microscope system. Lens-free imaging has matured as a modality competitive with traditional lens-based microscopy. In lens-free microscopy, a diffraction pattern resulting from an object is recorded directly on a digital image sensor array without being optically imaged or magnified by any lens elements (Ozcan & McLeod, 2016). Compared to conventional microscopy, this technique is portable and cost-effective, and it provides a field of view of the same size as the sensor size and a numerical aperture close to 1, since the large-area detector is placed very close to the sample. This architecture is advantageous to make the microfluidic system as compact as possible and provides high photon collection efficiency in fluorescence microscopy. Therefore, it is suitable for analysis applications requiring large statistics, such as cytometry. Ozcan's research group has developed different lens-free holographic imaging platforms for on-chip cytometry and diagnostics, with either fluorescent imaging format (Zhu et al., 2011; Arpali et al., 2012) or shadow imaging format (Seo et al., 2009; Wei et al., 2013). Zhu et al. demonstrated a compact optofluidic platform that integrated imaging cytometry and fluorescent microscopy and could be attached to a cell phone (Zhu et al., 2011). The resulting device could be used to rapidly image bodily fluids for cell counts or cell analysis. However, fluorescent labels are unsuitable for a high-resolution lens-free holographic imaging setup because their emission is incoherent with their excitation. Instead, beads with biomolecularly functionalized surfaces can be used to provide specificity, as the scattering from these beads is coherent with their illumination (Ozcan & McLeod, 2016). Wei et al. proposed using gold and silver nanobeads as specific labels to identify and count the number of CD4 and CD8 cells in a cell suspension (Wei et al., 2013). CD4 and CD8 cells are specific types of T lymphocytes, and their relative populations are important for evaluating the stage of human immunodeficiency virus (HIV) infection or AIDS, as well as for evaluating the efficacy of antiretroviral treatment. Counting the relative populations of these cells can be challenging, however, because the only significant difference between these cells is in the types of proteins expressed on their membranes. Under a microscope, both types of cells look virtually identical. In order to sense and count these cells in a lens-free imaging setup, the authors used gold nanoparticles functionalized with anti-CD4 antibodies to label CD4 cells and silver nanoparticles functionalized with anti-CD8 to label CD8 cells. By comparing the spectral response of different cells under lens-free imaging, the investigators were able to discriminate these two types of cells with greater than 95% accuracy using a machine learning algorithm.

To summarize, optofluidic components have shown great potential with respect to various on-chip optical process. It is envisioned that all optical components can be integrated into a single chip in the near future, as well as full functionalities. In terms of highly sensitive bioanalytical analysis in an integrated platform, such optofluidic components and their combinations have to address several critical challenges. Firstly, unlike solid-state optical elements, physical properties of a liquid medium are prone to be instable and temperature dependent, and any disturbance such as heat dissipation would compromise device's performance. Flow control is another concern whenever continuous liquid supply is required. Especially, most structural materials in microfluidic systems have refractive indices very close to the refractive index range of liquids, which hamper the capability to tightly focus a light beam with high numerical aperture. These challenges therefore need more research efforts to develop new approaches.

10.4 Fabrication Technologies of Optofluidic Components

Fabrication of micro-optical devices mainly relied on techniques transferred from the conventional two-dimensional (2D) integrated circuit (IC) and two- or three-dimensional (3D) MEMS processes. Semiconductor fabrication techniques have been adopted to micro-optical structures, including photolithography, thin-film deposition, chemical etching, etc. Silicon-, glass-, glass-silicon-, and glass-polymer-based fabrication techniques were widely studied. However, silicon and glass are hampered from wider applications in micro-optics, because they possess micromachining difficulties and are relatively expensive; moreover, an inconvenience of silicon is the lack of optical transparency at ultraviolet (UV), visible, and near-infrared (IR) wavelengths. Tremendous effort has been made to find alternative materials that are more cost-effective and easier machinable. With the development of related fabrication techniques, polymer–/plastic-based devices have therefore gained increasing interest. Compared with silicon and glasses, polymer materials can avoid high-temperature annealing and stringent cleaning (if they are disposable), and they are more cost-effective and have easier microfabrication, and there exists a wider range of materials to be chosen for characteristics that are required for each specific application, such as good optical transparency, biocompatibility, and chemical or mechanical properties. However, polymer materials usually do not result in strongly bonded layers like glass or silicon, and can exhibit structural deformation during device packaging processes. Each material has therefore both its advantages and disadvantages, and the choice of it will depend on the specific application. New technologies have also been developed in the meanwhile. This section presents current fabrication methodologies to make microfluidic and micro-optic structures.

10.4.1 Microfluidics Technologies

Current methods for fabrication of microfluidic devices can be categorized as replication techniques, such as soft lithography, and direct subtractive fabrication techniques such as laser micromachining (Wu & Gu, 2011).

Soft lithography typically refers to the molding of PDMS, using master templates. PDMS is inexpensive, is biocompatible, and has excellent liquid sealing properties, making it very suitable for microfluidics. In addition, it can be easily bonded to itself, allowing the fabrication of multilayer structures. Thin PDMS diaphragms are elastically extensible so that much larger deflections can be achieved than with hard materials. Fabrication of PDMS structures with micromolding is normally accomplished with four steps: first, a mold (i.e., the master) is fabricated (see Fig. 10.6a) (Duffy et al., 1998). This mold is usually fabricated in photoresist, such as SU-8, by photolithography, or it can also be made by silicon or metal suing chemical etching or laser micromachining techniques. Depending on the mold material and topography, a surface treatment is usually recommended in order to facilitate de-molding. Subsequently, a prepolymer solution (a mixture of PDMS linear polymer and a cross-linking agent) is poured on top of the mold. For high-aspect-ratio mold topographies, this step may need to be performed under vacuum to remove air bubbles. Then, the polymerization is completed. This procedure can be performed for 1–24 h, depending on the curing temperature. Lastly, the PDMS layer is carefully peeled off from the master and then bonded to a substrate material to form a full device. The mold can be reused many times, which reduces the cost considerably. Overall, soft lithography is faster, less expensive, and particularly suitable for prototyping compared to glass and silicon micromachining. Therefore, this method is commonly used for fabricating various microfluidic devices.

Other than soft lithography, *hot embossing* and *injection molding* are widely used to mold plastic materials. *Hot embossing* is a technique that transfers a pattern from a micromachined silicon, quartz, or metal master to a pliable plastic sheet (see Fig. 10.6b) (Li, 2015a). Heat and high pressure allow the plastic sheet to become imprinted. First, the stamp or master is brought in contact with the plastic surface. A well-distributed pressure is applied, while the substrate is heated to above its glass transition temperature. Finally, the substrate is cooled down before removing the stamp. *Injection molding* is another very promising technique for low-cost fabrication of microfluidic devices (see Fig. 10.6c) (Li, 2015b). Thermoplastic polymer materials are heated past their glass transition temperature to make them soft and pliable. The molten plastic is subsequently injected into a cavity that contains the master. Because the cavity is maintained at a lower temperature than the plastic, rapid cooling of the plastic occurs, and the molded part is ready in only a few minutes.

Even though polymer materials are now widely used in microfluidic devices, they still have drawbacks such as poor electrical and mechanical properties, and less resistance to many common solvents. In optofluidics, damage upon tightly focused laser irradiation and fluorescence at certain common wavelengths appear to be

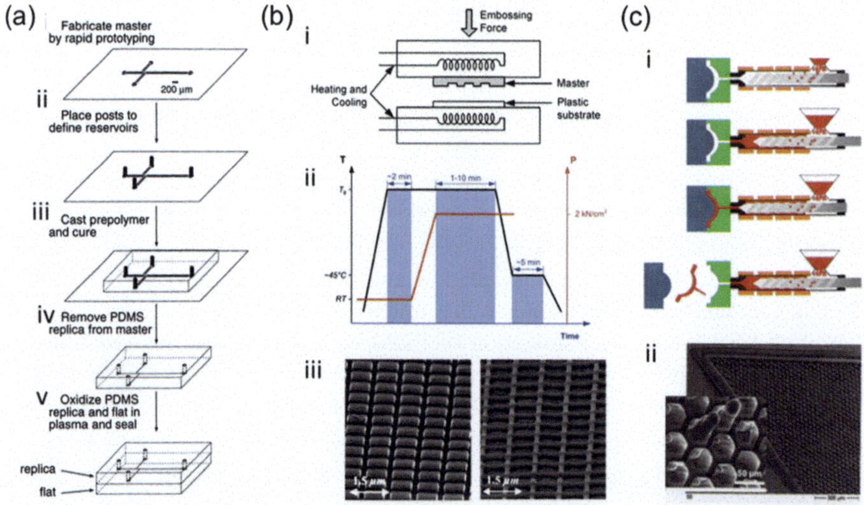

Fig. 10.6 (**a**) Scheme describing the fabrication of enclosed microscopic channels in PDMS. (i) A photoresist master is realized on a silicon wafer. (ii) Glass posts were placed on the wafer to define reservoirs for analytes and buffers. (iii) A prepolymer of PDMS was then cast onto the silicon wafer and cured. (iv) The polymer replica of the master containing a negative relief of channels was peeled away from the silicon wafer, and the glass posts were removed. (v) The PDMS replica and a flat slab of PDMS were oxidized in a plasma discharge and were brought into conformal contact, so that an irreversible seal formed between them. (**b**) (i) Schematic diagram of a hot embossing tool. (ii) Typical temperature (T) and pressure (P) conditions during a hot embossing process. (iii) Silicon master and the embossed PMMA structure. Features are 150 nm wide and 200 nm deep at 1 µm spacing. (**c**) (i) Schematic diagram of an injection molding process. (ii) Injection-molded polymer structures. (Reproduced with permission (Yang & Gijs, 2018), Copyright 2018, Royal Society of Chemistry)

severe when using polymer materials. In contrast, glass-based devices with high optical transparency to visible light and inertia to chemical solvents prove to be very suitable for these applications. Therefore, glass has still been widely used in the commercial biosensing products. Companies such as Micronit, LioniX, or Micralyne market glass microfluidic circuits.

Typical fabrication technique on glass substrate is laser micromachining. *Laser micromachining* is a direct write and etch technique; therefore, it does not require any mask or post-development steps. It is a suitable technique for rapid prototyping of microsystems, and in particular microfluidic devices. Most glasses have poor thermal properties and are brittle, which makes the fabrication of finely machined features a difficult task. In this regard, ultrafast laser microfabrication, which utilizes ultrashort laser pulses of pulse width as narrow as ~100 fs (a femtosecond laser), became a powerful approach for precise and fine processing for optofluidic devices in glasses (Zhang et al., 2008). By exposing a piece of glass to a focused femtosecond laser of a peak intensity well above the damage threshold of the glass, three-dimensional microstructures could be generated in the glass. A unique feature of laser micromachining using a femtosecond laser is that it can deliver high-density energy in a highly controlled and spatially localized way. The nonlinear absorption induced by the extremely high peak intensity in the pulse's focal point ensures that the absorption, and therefore material removal, is confined to the vicinity of the focus, leaving the rest of the substrate unaffected, thus allowing ultraprecision micromachining. In addition to the fabrication of microchannels in glass, microchannels can also be fabricated in polymers by femtosecond lasers (Zheng et al., 2006; Gattass & Mazur, 2008). By direct writing on a glass substrate, a femtosecond laser beam with a small pulse energy (< micro-Joules) can generate a localized increase on the refractive index in the material, therefore generating a waveguide. Moreover, if the peak intensity of the laser pulse is higher than the damage threshold of the glass, the material at the focus is vaporized, and the high pressure pushes the material to both sides of the irradiated region. Consequently, the tracks at the two sides result in a higher refractive index, while the region between the two tracks can guide light. Compared with MEMS techniques, laser microfabrication offers the advantages of simplicity, versatility, and high efficiency. Because there is no need for either a mask or post-development processing, the fabrication rate using femtosecond lasers is approaching that of lithographic techniques, though a femtosecond laser remains a very expensive equipment.

10.4.2 Micro-Optics Technologies

Micro-optical and optofluidic devices have also benefited from microelectronic technology development. This section reviews current fabrication methodologies to make optical structures on a chip, focusing on integrated waveguides and microlenses.

10.4.2.1 Technologies for Waveguide Fabrication

On-chip integrated waveguides can be easily interfaced with external light sources and detectors, while they precisely direct light to the microfluidic channel, both for providing the optical source to locally expose the targets and precisely directing the scattered or fluorescent output signals to the external detectors, without the difficulties of alignment, epoxy-fixing, or breakage problems, occurring when using fibers. Therefore, waveguides have been widely used in integrated microsystems. Integrated waveguides can be constructed using a variety of micromachining procedures. Common processes include lithographic patterning, thin-film deposition, and etching, and these techniques can be used to fabricate a ridged waveguide, i.e., a solid-core waveguide. Usually, a thin layer of the core material is deposited on a planar substrate first. The substrate is coated with photoresist, the latter is then exposed to UV light or X-rays through a lithography mask that defines the waveguide shape, and developed to form a pattern on the surface of the substrate. With the remaining photoresist as a mask for either wet chemical or dry etching to define the ridge waveguide structures. Dry etching methods, for example ion-beam etching, produce smooth edges, particularly on curved sections, but they also cause some lattice damage, which must be removed by annealing if minimum optical losses are desired. Wet chemical etching on the other hand produces less lattice damage, but it is very difficult to control the etch depth and profile. Most chemical etchants are preferential regarding crystal orientation, thus leading to ragged edges on curved sections of waveguides when using Si substrates. After the photolithography and etching procedures to process the core, a cladding layer is deposited. As an example, silicon oxynitride waveguides were fabricated in a standard silicon fabrication line by PECVD and LPCVD deposition processes in combination with 1100 °C annealing treatments to remove light-absorbing hydrogen bonds, and optical lithography for pattern definition and dry RIE for the pattern transfer process (Aparicio et al., 2014). As shown in Fig. 10.7, a 2-μm-thick SiO_2 buffer layer (n = 1.45) and a 1-μm-thick SiO_xN_y core layer (n = 1.85) were both deposited by PECVD; a ~0.5-μm-thick borophosphosilicate glass (BPSG) layer (n = 1.45 nm) was grown by LPCVD in order to optimize the surface planarity after RIE. Finally, the whole system was coated with a thin SiN_x film (50 nm of thickness). This waveguide system was applied to detect low surface concentration (10^{-11} mol cm^{-2}) of a green light-emitting organic dye.

While a solid-core waveguide is assembled into the microchip but fabricated separately from the microchannels, liquid-core waveguides (LCWs) share the same physical volume as the microfluidic channels. Indeed, the microchannel not only transports the sample but also transmits the light, which greatly simplifies the design and reduces light loss through the channel walls. There are many different methods to fabricate LCWs. By using the soft lithography technology to mold PDMS structures, Cho et al. developed a method to construct LCWs in PDMS microchannels by

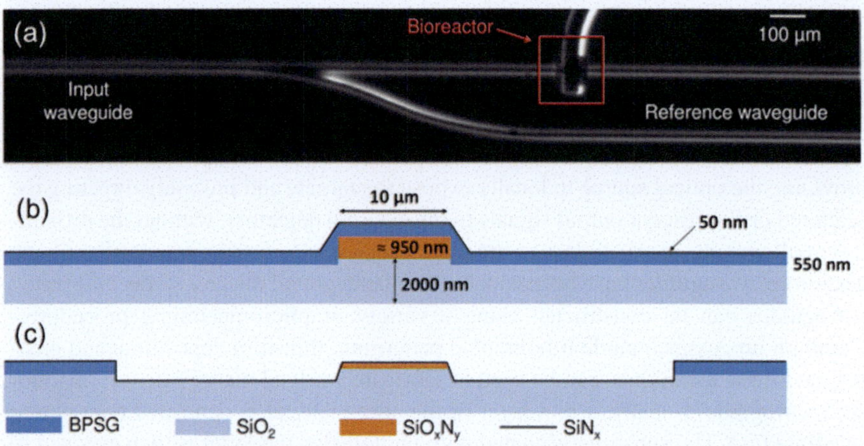

Fig. 10.7 Fabrication procedure of silicon oxynitride waveguides: (**a**) Optical image of the top view of the waveguide; (**b**, **c**) Schematic cross sections of the waveguide structure before (**b**) and at the bioreactor well (**c**). (Reproduced with permission (Yang et al., 2019), Copyright 2019, MDPI)

flowing Teflon® AF solution through the channels (Cho et al., 2010a). This coating approach avoids the difficulty of bonding between two Teflon-coated substrates. The waveguide was integrated into a micro fluorescence-activated cell sorter that, by readout of the fluorescence signal of pre-stained cells, could sort the latter into a split channel positioned downstream.

10.4.2.2 Technologies for Microlens Fabrication

A microlens is an important key component for focusing and collimating light. It can be made as a solid curved surface or as a tunable interface between two deformable materials/media. On-chip hard solid-state lenses usually are fixed-focus lenses, similar to miniaturized traditional lenses that are used in a free-space detection system. Different methods have been developed for building solid microlenses. Thermal reflow processes have been widely used for the fabrication of large arrays of hemispherical microlenses. In this technique, a glass substrate is first coated with a layer of photoresist. Photolithography technique is then used to pattern the photoresist layer and generate a series of cylindrical islands, which are heated above the glass transition temperature of the photoresist. Due to surface tension, the shape of the photoresist cylinders changes to minimize the surface energy and becomes hemispherical, therefore generating a microlens array on the substrate. As the surface tension governs the contact angle of the photoresist and the shape of the microlens eventually, a thin layer of additives can be deposited on the substrate to influence the surface tension between the substrate and the photoresist, tuning the focal length and numerical aperture of the fabricated lenses. Yang et al. utilized a bottom polyimide layer to form a pedestal to sustain the upper photoresist lens after the heat reflow process (Yang et al., 2004). The interactive force between two material interfaces causes the upper photoresist to form a spherical profile and transform the polyimide pedestal into a trapezoid with arc sides (as shown in Fig. 10.8). Advantages of the thermal reflow technique include a low material consumption, low manufacturing costs, the intrinsic simplicity of the technique, and an easy process control. Microlens arrays can be produced over large surface areas. However, the low transparency and thermal instability of photoresist features during the reflow process limit its widespread application (Yang & Gijs, 2018).

Tunable microlenses are based on adjusting the curvature of the interface between two media, such as liquid/liquid, liquid/air, and liquid/soft solid interfaces. These microlenses are dynamic optical elements, and their focal length can be tuned by changing either the refractive index or the geometry of the microlens. The refractive index of the liquid in a micro-optofluidic device can be modulated by changing fluid characteristics, such as species, concentration, and flow patterns of liquid streams. The geometry of the lens can be modulated by a variety of concepts known in microfluidics. If the lens is formed by a liquid/solid interface of a soft material such as PDMS, the shape of a thin flexible membrane spanning a cavity with the fluid inside can be controlled by applying pressure. If the lens is formed by liquid/liquid or liquid/air interfaces, its shape can be controlled by means of hydrodynamics,

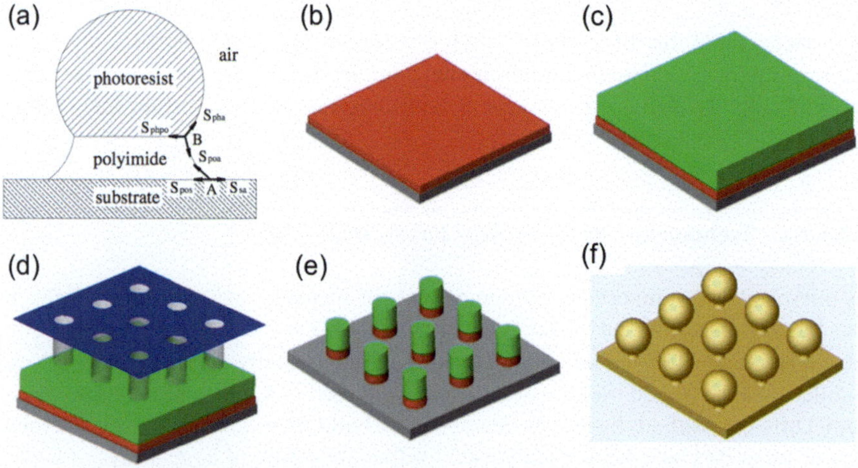

Fig. 10.8 Illustration of a micro-ball lens formation in the thermal reflow process. A polyimide layer is coated on a Si wafer beneath a photoresist layer. Photolithography process is used to pattern these two materials through a mask. Micro-ball lens array with a pedestal can be generated after heat reflow. (Reproduced with permission (Yang et al., 2019), Copyright 2019, MDPI)

electro-wetting, or other physical effects. When the lenses are tuned by manipulating the refractive index of a liquid, the composition of the lens liquid can be changed by controlled diffusion of species. With microfluidic techniques, when a laminar flow condition is maintained in the microfluidic channel, the concentration distribution of species is determined by molecular diffusion. Thus, a concentration gradient or a refractive index gradient can be generated along the flow direction. Mao et al. reported a liquid gradient refractive index (L-GRIN) lens (Mao et al., 2009). The lens was formed by diffusion of $CaCl_2$ from a core stream with a concentration of 3.5 M to cladding streams of deionized water (see Fig. 10.9). The refractive index decreased gradually from $n = 1.41$ to $n = 1.33$. The concentration profile and the corresponding refractive index profile could be controlled by the flow rate of the core stream and the two cladding streams. By adjusting the flow rates, a large gradient of refractive index was achieved, and it allowed a sharper bend of light beams, resulting in a shorter focal length. By adjusting the flow rate of the cladding streams asymmetrically, the concentration profile swung to one side, allowing the light beam to be focused off the optical axis.

10.5 Optofluidic Devices for Bioanalytical Application

This section introduces the main applications of optofluidic systems, and we will focus on cytometry and cell biology studies, as well as nucleic acid and protein detection systems.

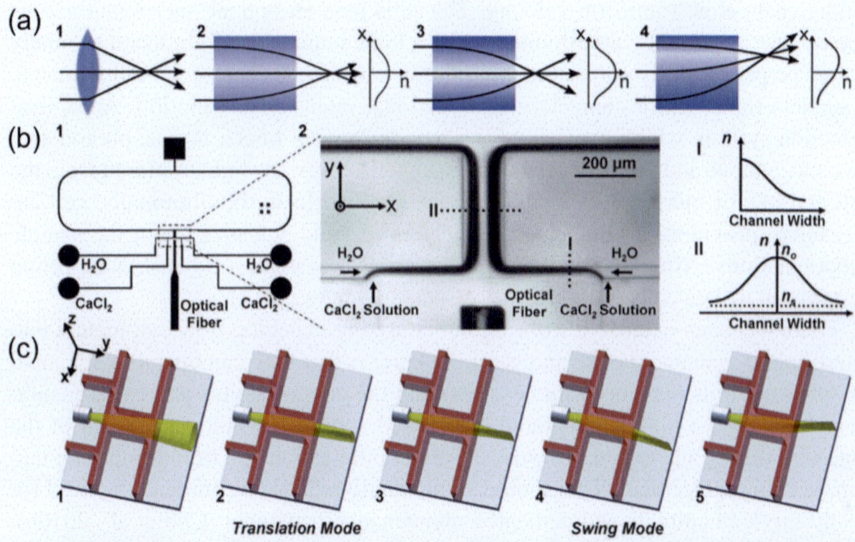

Fig. 10.9 Principle and design of a liquid gradient refractive index (L-GRIN) lens. (**a**) A schematic diagram showing the comparison between the classic refractive lens (a1) and GRIN lens (a2). Change of the refractive index contrast in GRIN lens can result in change of focal distance (a2–a3), and shift of optical axis can result in change of output light direction (a4). (**b**) Schematic of the L-GRIN lens design (b1), microscopic image of the L-GRIN lens in operation (b2, left), and the expected refractive index distribution at two locations (I and II) inside the lens (b2, right). High optical contrast areas (dark streaks) were observed near the fluidic boundaries (b2, left), suggesting significant variation of refractive index due to the $CaCl_2$ diffusion. (**c**) Schematic drawing showing two operation modes of the L-GRIN lens: the translation mode with variable focal length including no-focusing (c1), a large focal distance (c2), and a small focal distance (c3); and the swing mode with variable output light direction (c3–c5). Illustration of a micro-ball lens formation in the thermal reflow process. A polyimide layer is coated on a Si wafer beneath a photoresist layer. Photolithography process is used to pattern these two materials through a mask. Micro-ball lens array with a pedestal can be generated after heat reflow. (Reproduced with permission (Yang & Gijs, 2018), Copyright 2018, Royal Society of Chemistry)

10.5.1 Cytometry and Cell Biology Studies

Flow cytometry or fluorescence-activated cell sorting (FACS) is a commonly used powerful analytical technique in research, in which the properties of individual elements of a population, such as cells or microparticles, are measured. It has a wide range of applications, including immunophenotyping, gene diagnosis, bacteria analysis, and clinical hematology diagnosis. Typically, these applications require the labeling of cells with multiple fluorophores for correlated analysis of cellular characteristics and properties. Fluorescently tagged antibodies are often used to mark and identify cells via immunofluorescence (Godin et al., 2008). After the labeling procedure, the cells are introduced in a fluidic system and hydrodynamically focused by a sheath flow, ensuring that cells travel through the center of the fluidic channel at a uniform velocity. The cells pass an optical interrogation zone one by one, where they are illuminated by a light source. The light beam normally passes perpendicularly across the microfluidic channel. As each cell is illuminated, it scatters light with a characteristic directional intensity distribution. An optical detection system, which consists of waveguides/optical fibers, lenses, photodetectors, etc., guides and collects the detection signals. Thus, the light scattered from the cell and one or more colors of fluorescence emitted from the illuminated cell are measured, providing a number of parameters to yield statistics about the sample subpopulations. After identification by the optical system, a downstream sorting system can isolate cells of interest (Piyasena & Graves, 2014).

There has been extensive work on miniaturizing benchtop flow cytometers into microfluidic systems; the advantage of the latter is that they can operate with small sample (and thus reagent) volumes, lowering the cost of assays and experimentation. Moreover, a highly integrated chip-based approach would turn much of the time-consuming and expensive work of setup optimization and troubleshooting into simple chip replacement. This should significantly reduce or eliminate the need for costly service contracts and dedicated maintenance personnel (Cho et al., 2010b). Following the practice of commercial instruments, many microfluidic cytometers still use traditional bulk optics and typically require a confocal microscope stage for mounting commercial high-magnification objectives. However, when the optical components can be co-integrated onto a single chip, the probability of hardware failure will be reduced. More importantly, an optofluidic cytometer offers the potential for innovative architectures and increased functionality. The high-level integration, along with parallel processing techniques, can be exploited for screening at higher throughput. Many different formats of waveguides and microlenses have been integrated on-chip for improving the optical system performance. For example, Watts et al. presented a functional photonic-microfluidic flow cytometer with integrated optical waveguides and beam shaping components (Watts et al., 2010, 2013). The waveguide cores, optical lenses, and microfluidic channels were simultaneously fabricated in SU-8 on a glass substrate, which also served as a bottom cladding region. PDMS was used to seal the fluid channels and to simultaneously act as the upper cladding of the waveguides. Air regions formed by removing material from either side of a waveguide core region formed the side cladding of the waveguides. The ridge waveguide terminated at a system of vertical cylindrical lenses that were designed to focus the input beam originating from the waveguide into a tunable spot in the middle of the microfluidic channel. Beam waists of different width could be obtained via changing the lens design to detect particles with different sizes (Watts et al., 2010). In order to enhance the detection of the forward scattered light in the microfluidic flow cytometer, a notch was placed in the lens system, which caused a dark spot on the facet of a collection waveguide. This design helped to remove axial rays from the excitation beam and decreased the background noise (Watts et al., 2013).

Other than cytometric analysis, optofluidic systems have also been used for manipulation of cells and for cell biology analysis. The application of optofluidic technologies for various optics-based single-cell analyses, which involved single-cell manipulation, treatment, and physical property detection, was reviewed by Huang et al. (2014). Optical tweezer-based single-cell manipulation is based on the generation of a large electromagnetic field gradient at the focal point of a beam of light formed by a lens. This resulted in an attractive force that pulls the particle to the focused spot, if the particle permittivity is larger than that of the surrounding medium. Another strategy adopted to optically manipulate cells is the use of scattering forces induced by a laser beam to push cells to certain areas on the chip. Moreover, researchers have demonstrated the use of optically induced mechanical and electrical forces to perform single-cell manipulation. For example, a high-power laser pulse generated a microbubble that provided the actuation force to transport a cell to a collection channel (Chen et al., 2013). Micro-optical structures not only have been used for generating forces for on-chip cell manipulation, but also were at the basis of biophysical characterization of cells and cellular phenotyping in general. An overview of the different optical detection methods that have been used for the latter was presented in the review article of Tung et al. (2012). Herein, optofluidic devices were categorized according to the different detection techniques used, namely, fluorescence, surface-enhanced Raman spectroscopy, surface plasmon resonance, or interferometric detection. Rodriguez-Ruiz et al. have discussed the engineering aspects of integrating optical spectroscopy with microfluidic systems, such as making the choice of the appropriate detection mechanism, i.e., colorimetry, fluorimetry, absorbance, and/or dispersion/scattering to be combined with microlenses of different materials and mirrors, depending on the application (Rodríguez-Ruiz et al., 2016).

Ibarlucea et al. demonstrated cell screening in a disposable PDMS chip operating in a multiple internal reflection detection mode configuration (Ibarlucea et al., 2010; Vila-Planas et al., 2011). The chip was tailored such that the interrogation region of the sample allowed reducing the mean flow cell volume by including self-aligned lenses and focusing mirrors on the chip. In particular, light propagating in the system followed a zigzag optical path with the help of integrated air mirrors (see Fig. 10.10a). The system was able to perform either scattering, scattering + absorption, or absorption measurements without any hardware modification. Later the same group demonstrated a monolithically integrated biophotonic lab-on-a-chip for cell culture and simultaneous pH monitoring (Muñoz-Berbel et al., 2013). The device incorporated a filter with 3-μm-high size-exclusion microchannels, capable of efficiently trapping cells in the incubation chamber, as well as optical elements for real-time interrogation of two microfluidic chambers. Both yeast cells and vascular smooth muscle cells as a mammalian cell model were tested. Liang et al. measured the refractive index of single living cells transiting in a microfluidic channel (Liang et al., 2007). Chemical or ingredient changes inside a cell can be detected by its refractive index measurement; one of the possible applications is to distinguish cancer cells from healthy cells. The measurement system integrated an external cavity laser (the microfluidic channel with/without passing cell was part of the optical cavity) and a microlens (see Fig. 10.10b). When a cell was transiting through the cavity, this resulted in a frequency shift and intensity modulation of the power of a broadband laser source, from which one derived the cell's refractive index.

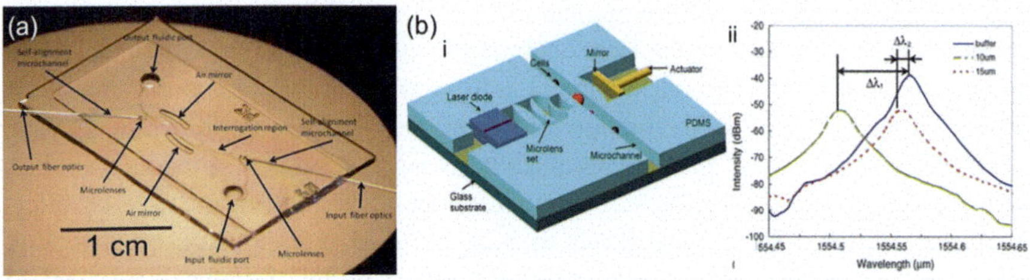

Fig. 10.10 (**a**) Picture of a PDMS-based multiple internal reflection system. It comprises self-alignment microchannels for accurate fiber optics positioning, microlenses, and air mirrors. All these micro-optical elements ensure an adequate zigzag path of the light along the interrogation region, which can be filled through the input fluidic port. (**b**) (i) Schematic diagram of an integrated optical cavity detection chip. Cells or microparticles are delivered into the external cavity of a laser along a microfluidic channel with a width of 100 μm. On account of the different indices between a living cell and the buffer medium, the cell moving through the analysis region changes the effective length of the external cavity and subsequently affects the emission characteristics of the laser diode. (ii) Experimental results for detection of PS beads, showing the emission spectrum obtained for the loaded PBS buffer, a 10 μm bead, and a 15 μm bead, respectively. (Reproduced with permission (Yang & Gijs, 2018), Copyright 2018, Royal Society of Chemistry)

Calibration of the system was based on the use of two standard PS beads of known refractive index. It was found that tested cancerous cells had a refractive index in the 1.392–1.401 range, while this was in the 1.35–1.37 range for normal cells.

10.5.2 Nucleic Acid and Protein Detection

There is a significant demand nowadays for high-throughput biomolecular analysis tools to assist diagnosis in research and clinical applications. Both DNA and proteins are frequently used to correlate biomarkers to specific diseases. Microfluidics offers quantifiable benefits over existing procedures in nucleic acid and protein analysis. PCR and fluorescence detection on microfluidic devices have led to a significant improvement in total processing time. While featuring low limits of detection, microfluidic devices can also improve the sensitivity and speed and reduce costs of an immunoassay over ELISA by overcoming the issues of the large sample and reagent volumes. Furthermore, by integrating different steps into a miniaturized system, microfluidics can also overcome the limitations of the use of multiple instruments with sample loss due to large dead volumes and the risk of contamination.

Detection schemes that rely on integrated optics have successfully demonstrated DNA fragment analysis. Both optical fibers and integrated waveguides have been used to deliver fluorescence excitation light into the system and collect emitted fluorescent signals. Guo et al. presented an optofluidic device to perform detection of DNA and its mutations (Guo et al., 2012). The device was composed of a droplet-generation unit, a mixing and reaction unit, and a fiber-based optical detection unit with two fibers precisely positioned through fiber grooves, one for fluorescence excitation and the other connected to an off-chip photomultiplier tube (PMT) for signal collection (see Fig. 10.11a). By detecting the fluorescence signal changes of the target DNA/molecular beacon complex in single droplets, the device was able to distinguish molecular beacon, wild-type single-strain DNA BRCA1, single-mutation single-strain DNA, and control single-strain DNA in droplets with a detection throughput of 2000 droplets per second. Cai et al. reported a microdevice using hybrid optofluidic integration for detection and quantification of Ebola virus (Cai et al., 2015). In this work, sample preparation and target preconcentration were implemented on a PDMS-based microfluidic chip, and liquid-core optical waveguides on a silicon chip were used for single nucleic acid fluorescence detection. Magnetic microbeads functionalized with a synthetic oligonucleotide were used to capture the target nucleic acids; the bonded nucleic acids were then thermally released from the beads, re-suspended and fluorescently labeled, and sent to the optofluidic chip for detection. Brennan et al. presented an optofluidic platform, which had integrated microfluidic channels for fluidic delivery, a resonant mirror waveguide for optical detection, and surface chemistry for covalent immobilization of DNA probes (Brennan et al., 2008). The integrated device detected DNA binding in two configurations: (1) as an excitation platform for fluorescence measurements with Cy3-tagged target DNA and (2) as a refractive index platform, where changes in effective refractive index due to binding were monitored.

Besides detection of DNA hybridization, various designs that integrated optical waveguides and microlenses with microfluidics have been proposed to improve detection characteristics in immunoassays. Waveguide systems have realized noise rejection improvements by probing a region of interest of only hundreds of nanometers from the surface in heterogeneous assays using evanescent field excitation, preventing emission from unbound bulk material. The main advantage of this technique is to guide the light very close to the microfluidic channel, and then ensure an excellent coupling efficiency between the waveguide and the microfluidic channel. Kemmler et al. developed a point-of-care system to simultaneously measure c-reactive protein, interleukin-6, and procalcitonin (Kemmler et al., 2009). The system consisted of a planar waveguide fluorometer, a fluidic setup, and exploited sandwich bioassays. By microarray technology, one specific type of antibody for each parameter was spotted on the surface of the waveguide, and the detection was fluorescence-based and worked with specific labeled antibodies. The light was guided in the waveguide creating an evanescent wave close to the surface, which excited the surface-immobilized fluorophores very efficiently. The detection limit of CRP, IL-6, and PCT were 0.35, 0.08, and 1 ng/mL, respectively, with using a sample volume of ~120 μL. Liu et al. developed a planar waveguide-based array immunosensor, allowing measurements of up to 24 analytes simultaneously (Liu et al., 2017b). In this system, laser light was coupled into a planar optical waveguide via a beveled angle, forming eight individual TIR spots where excitation and detection were enabled (see Fig. 10.11b). A multichannel microfluidic chip was employed to isolate the parallel TIR lines physically so as to form eight independent flow channels on the same chip, avoiding the cross-reactivity of antibodies and supporting

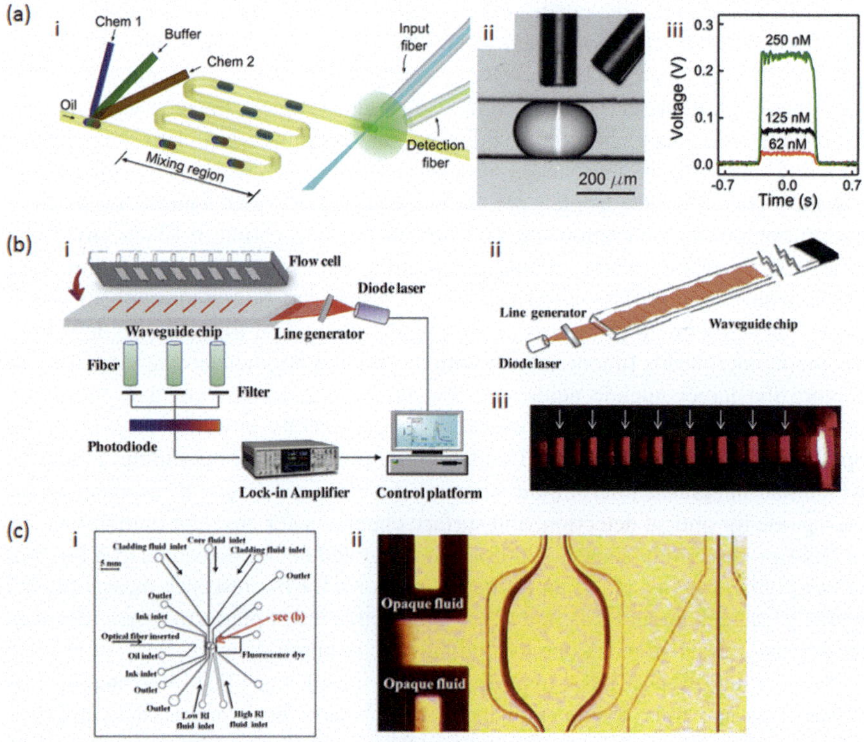

Fig. 10.11 (**a**) (i) Schematic of a droplet-based optofluidic device composed of a droplet-generation unit, a mixing and reaction unit, and a fiber-based optical detection unit. (ii) Fluorescence intensity characterization of the droplet-based with a microscopic snapshot of a droplet excited by a 20-μm-wide laser beam. (iii) Electric signal of three droplets with different fluorescent dye concentrations, as generated by a photomultiplier tube. (**b**) (i) Schematic diagram of a planar waveguide-based array immunosensor with a removable multichannel PDMS flow cell. (ii) Excitation light path inside the planar waveguide transducer. (iii) Photograph of waveguide chip with eight reflection lines for biosensing, as indicated by white arrows. (**c**) (i) Schematic illustration showing main components of a PDMS optofluidic chip comprising an all-liquid tunable in-plane biconvex microlens. (ii) Magnified image of aperture and microlens components of the optofluidic chip. (Reproduced with permission (Yang & Gijs, 2018), Copyright 2018, Royal Society of Chemistry)

various bioassay conditions. By employing fluorescent detection with fluorophore-labeled antibodies binding to the surface of the waveguide with the analyte derivative covalently attached, multi-analyte biosensing was realized.

Besides the aforementioned waveguides, also integrated microlenses have been used in optofluidic fluorescence systems to improve the system compactness and efficiency through more focused excitation of fluorescence volumes. Song et al. presented optofluidic lenses which were hydrodynamically formed with a liquid-core liquid-cladding configuration (Song et al., 2010, 2011). By tuning the flow rate ratio between core and cladding streams, the lens could be changed from focusing mode into diverging mode, and the focal length and divergent angle were also tunable. The focusing effect of the converging lens was clearly observed by using the fluorescent dye rhodamine B. Chaó et al. developed an optofluidic chip, which comprised a tunable in-plane biconvex microlens and micro-prism to control the focal length and deviation angle of a light beam (see Fig. 10.11c) (Chao et al., 2013). The L^2 microlens enabled the focal length to be adjusted in the range of 2.9–7.6 mm, while the microprism chamber filling with two fluids generated a deviation angle range of 6.2°–22.3°. Mohammed et al. demonstrated a capillary action-based autonomous microfluidic device in conjunction with on-chip planar microlenses, for optical detection of the cardiac biomarker troponin I (Mohammed & Desmulliez, 2013). The 2D planar lenses and the microfluidic channels were fabricated on a PMMA substrate using a single CO_2 laser process. The illumination from a single LED light source propagated through the microlenses and was focused onto an area of 0.5 mm^2, which represents the excitation area within the microfluidic channel, where a detection limit of 0.08 ng/mL was achieved, which is of the order of clinically acceptable concentrations. More related works are included in several review articles (Zhao et al., 2010; Threm et al., 2012; Schmidt & Hawkins, 2011).

10.6 Concluding Remarks

The integration of micro-optical elements with microfluidic platforms can advantageously contribute to the automation of many bioanalytical applications and that such miniaturized components are key to the improvement of the performance of an analytical system, especially when the latter has microfluidic features. Especially, the optofluidic integration results in high interaction efficiency between light and fluids, which is very desirable for sensing applications. In this chapter, optofluidic system for bioanalytical applications are exemplified, focusing on two main classes of components: optofluidic lasers and lenses. These elements have shown great potentials in various on-chip optical processes, but several challenges should be addressed such as disturbance on optical properties due to the instability of liquid medium, difficulty on flow control, and similar refractive indices between liquid and most structural materials of microfluidic systems.

In Sect. 10.2, we have concisely introduced basic features and advantages of engineered microfluidic systems that permit the implementation of all manipulation and handling steps necessary for designing an analytical process. We have then reviewed the field of micro-optics after introducing basic properties of electromagnetic waves and dielectric interface phenomena like scattering and refraction in Sect. 10.3. In Sect. 10.4, we have presented basic engineering building blocks of an optofluidic system, such as light sources, prisms, switches, optical detectors, and importantly micro-waveguides and microlenses. We presented also different technologies for realizing microfluidic and optofluidic chips in the next section. Current methods for fabrication of microfluidic devices and micro-optical components have been introduced. As main applications of microfluidic systems having on-board integrated micro-optical components, we have consecutively discussed cytometric applications, cell biology studies, and nucleic acid and protein detection systems.

Summarizing, we believe that the integration of micro-optical components on a microfluidic chip presents key advantages for many large-scale biological studies at the individual cell level and is of benefit to automated analytical screening. We have shown that microfluidics and optofluidics have enormous potentials for replacing the traditional labor-intensive analytical procedures indeed. Moreover, these devices permit accurate spatiotemporal application of compounds, so that dose-response curves and screening data sets with enriched information can be obtained. Overall, optofluidic technologies largely demonstrated potential to dramatically improve many aspects of analytical research by providing superior performance when compared to the classical protocols. The ongoing transition of any modern analytical application from an observation-based discipline to a mechanism-based science suggests that systems capable of analyzing large samples in a well-defined spatiotemporal way and with better sensitivity will benefit from increasing attention in the next years and likely will become invaluable tools in the near future. Micro-optical components and optofluidic devices certainly will play a prominent role in this endeavor.

Acknowledgments This work was supported by the National Natural Science Foundation of China (61805271, 62074155), Key-Area Research and Development Program of Guangdong Province (2019B020226004), and Shenzhen Science and Technology Innovation Commission (JCYJ20170818154035069).

References

Aparicio, F. J., Froner, E., Rigo, E., Gandolfi, D., Scarpa, M., Han, B., Ghulinyan, M., Pucker, G., & Pavesi, L. (2014). Silicon oxynitride waveguides as evanescent-field-based fluorescent biosensors. *Journal of Physics D: Applied Physics, 47*, 405401.

Arpali, S. A., Arpali, C., Coskun, A. F., Chiang, H.-H., & Ozcan, A. (2012). High-throughput screening of large volumes of whole blood using structured illumination and fluorescent on-chip imaging. *Lab on a Chip, 12*, 4968–4971.

Brennan, D., Lambkin, P., Moore, E. J., & Galvin, P. (2008). An integrated optofluidic platform for DNA hybridization and detection. *IEEE Sensors Journal, 8*, 536–542.

Brown, M., Vestad, T., Oakey, J., & Marr, D. W. M. (2006). Optical waveguides via viscosity-mismatched microfluidic flows. *Applied Physics Letters, 88*, 134109.

Bruls, D. M., Evers, T. H., Kahlman, J. A. H., Lankvelt, P. J. W. V., Ovsyanko, M., Pelssers, E. G. M., Schleipen, J. J. H. B., Theije, F. K. D., Verschuren, C. A., Wijk, T. V. D., Zon, J. B. A. V., Dittmer, W. U., Immink, A. H. J., Nieuwenhuis, J. H., & Prins, M. W. J. (2009). Rapid integrated biosensor for multiplexed immunoassays based on actuated magnetic nanoparticles. *Lab on a Chip, 9*, 3504–3510.

Cai, H., Parks, J. W., Wall, T. A., Stott, M. A., Stambaugh, A., Alfson, K., Griffiths, A., Mathies, R. A., Carrion, R., Patterson, J. L., Hawkins, A. R., & Schmidt, H. (2015). Optofluidic analysis system for amplification-free, direct detection of Ebola infection. *Scientific Reports, 5*, 14494.

Campbell, K., Groisman, A., Levy, U., Pang, L., Mookherjea, S., Psaltis, D., & Fainman, Y. (2004). A microfluidic 2 × 2 optical switch. *Applied Physics Letters, 85*, 6119–6121.

Carney, P. S., & Schotland, J. C. (2001). Three-dimensional total internal reflection microscopy. *Optics Letters, 26*, 1072–1074.

Chao, K.-S., Lin, M.-S., & Yang, R.-J. (2013). An in-plane optofluidic microchip for focal point control. *Lab on a Chip, 13*, 3886–3892.

Chen, Z., Taflove, A., & Backman, V. (2004). Photonic nanojet enhancement of backscattering of light by nanoparticles: A potential novel visible-light ultramicroscopy technique. *Optics Express, 12*, 1214–1220.

Chen, Y., Wu, T.-H., Kung, Y.-C., Teitell, M. A., & Chiou, P.-Y. (2013). 3D pulsed laser-triggered high-speed microfluidic fluorescence-activated cell sorter. *Analyst, 138*, 7308–7315.

Chen, S., Hao, R., Zhang, Y., & Yang, H. (2019). Optofluidics in bio-imaging applications. *Photonics Research, 7*, 532–542.

Cho, S. H., Chen, C. H., Tsai, F. S., Godin, J. M., & Lo, Y.-H. (2010a). Human mammalian cell sorting using a highly integrated micro-fabricated fluorescence-activated cell sorter (μFACS). *Lab on a Chip, 10*, 1567–1573.

Cho, S. H., Godin, J. M., Chen, C.-H., Qiao, W., Lee, H., & Lo, Y.-H. (2010b). Recent advancements in optofluidic flow cytometer. *Biomicrofluidics, 4*, 043001.

Cui, X., Lee, L. M., Heng, X., Zhong, W., Sternberg, P. W., Psaltis, D., & Yang, C. (2008). Lensless high-resolution on-chip optofluidic microscopes for Caenorhabditis elegans and cell imaging. *Proceedings of the National Academy of Sciences of the United States of America, 105*, 10670–10675.

Darafsheh, A., Walsh, G. F., Negro, L. D., & Astratov, V. N. (2012). Optical super-resolution by high-index liquid-immersed microspheres. *Applied Physics Letters, 101*, 141128.

De Tommasi, E., De Luca, A. C., Lavanga, L., Dardano, P., De Stefano, M., De Stefano, L., Langella, C., Rendina, I., Dholakia, K., & Mazilu, M. (2014). Biologically enabled subdiffractive focusing. *Optics Express, 22*, 27214–27227.

Dittmer, W. U., Evers, T. H., Hardeman, W. M., Huijnen, W., Kamps, R., Kievit, P. D., Neijzen, J. H. M., Nieuwenhuis, J. H., Sijbers, M. J. J., Dekkers, D. W. C., Hefti, M. H., & Martens, M. F. W. C. (2010). Rapid, high sensitivity, point-of-care test for cardiac troponin based on optomagnetic biosensor. *Clinica Chimica Acta, 411*, 868–873.

Dong, L., Agarwal, A. K., Beebe, D. J., & Jiang, H. (2006). Adaptive liquid microlenses activated by stimuli-responsive hydrogels. *Nature, 442*, 551–554.

Duffy, D. C., McDonald, J. C., Schueller, O. J. A., & Whitesides, G. M. (1998). Rapid prototyping of microfluidic systems in poly(dimethylsiloxane). *Analytical Chemistry, 70*, 4974–4984.

Erickson, D., Sinton, D., & Psaltis, D. (2011). Optofluidics for energy applications. *Nature Photonics, 5*, 583–590.

Fan, X., & White, I. M. (2011). Optofluidic microsystems for chemical and biological analysis. *Nature Photonics, 5*, 591–597.

Fan, X., White, I. M., Shopova, S. I., Zhu, H., Suter, J. D., & Sun, Y. (2008). Sensitive optical biosensors for unlabeled targets: A review. *Analytica Chimica Acta, 620*, 8–26.

Fang, N., Lee, H., Sun, C., & Zhang, X. (2005). Sub-diffraction-limited optical imaging with a silver superlens. *Science, 308*, 534–537.

Fang, C., Dai, B., Xu, Q., Zhuo, R., Wang, Q., Wang, X., & Zhang, D. (2017). Hydrodynamically reconfigurable optofluidic microlens with continuous shape tuning from biconvex to biconcave. *Optics Express, 25*, 888–897.

Gattass, R. R., & Mazur, E. (2008). Femtosecond laser micromachining in transparent materials. *Nature Photonics, 2*, 219–225.

Gérard, D., Wenger, J., Devilez, A., Gachet, D., Stout, B., Bonod, N., Popov, E., & Rigneault, H. (2008). Strong electromagnetic confinement near dielectric microspheres to enhance single-molecule fluorescence. *Optics Express, 16*, 15297–15303.

Godin, J., Chen, C.-H., Cho, S. H., Qiao, W., Tsai, F., & Lo, Y.-H. (2008). Microfluidics and photonics for bio-system-on-a-Chip: A review of advancements in technology towards a microfluidic flow cytometry chip. *Journal of Biophotonics, 1*, 355–376.

Greenbaum, A., Luo, W., Su, T.-W., Göröcs, Z., Xue, L., Isikman, S. O., Coskun, A. F., Mudanyali, O., & Ozcan, A. (2012). Imaging without lenses: Achievements and remaining challenges of wide-field on-chip microscopy. *Nature Methods, 9*, 889–895.

Guo, F., Lapsley, M. I., Nawaz, A. A., Zhao, Y., Lin, S.-C. S., Chen, Y., Yang, S., Zhao, X.-Z., & Huang, T. J. (2012). A droplet-based, optofluidic device for high-throughput, quantitative bioanalysis. *Analytical Chemistry, 84*, 10745–10749.

Gupta, R., & Goddard, N. J. (2013). A novel leaky waveguide grating (LWG) device for evanescent wave broadband absorption spectroscopy in microfluidic flow cells. *Analyst, 138*, 1803–1811.

Haeberle, S., & Zengerle, R. (2007). Microfluidic platforms for lab-on-a-chip applications. *Lab on a Chip, 7*, 1094–1110.

Hamburg, M. A., & Collins, F. S. (2010). The path to personalized medicine. *The New England Journal of Medicine, 363*, 301–304.

Hanumegowda, N. M., Stica, C. J., Patel, B. C., White, I., & Fan, X. (2005). Refractometric sensors based on microsphere resonators. *Applied Physics Letters, 87*, 201107.

Harrick, N. J. (1962). Use of frustrated total internal reflection to measure film thickness and surface reliefs. *Journal of Applied Physics, 33*, 2774–2775.

Hawkins, A. R., & Schmidt, H. (2008). Optofluidic waveguides: II. Fabrication and structures. *Microfluidics and Nanofluidics, 4*, 17–32.

Heifetz, A., Kong, S.-C., Sahakian, A. V., Taflove, A., & Backman, V. (2009). Photonic nanojets. *Journal of Computational and Theoretical Nanoscience, 6*, 1979–1992.

Helmerhorst, E., Chandler, D. J., Nussio, M., & Mamotte, C. D. (2012). Real-time and label-free bio-sensing of molecular interactions by surface plasmon resonance: A laboratory medicine perspective. *Clinical Biochemist Reviews, 33*, 161–173.

Hofmann, O., Miller, P., Sullivan, P., Jones, T. S., de Mello, J. C., Bradley, D. D. C., & de Mello, A. J. (2005). Thin-film organic photodiodes as integrated detectors for microscale chemiluminescence assays. *Sensors and Actuators B: Chemical, 106*, 878–884.

Hu, Y., Rao, S., Wu, S., Wei, P., Qiu, W., Wu, D., Xu, B., Ni, J., Yang, L., Li, J., Chu, J., & Sugioka, K. (2018). All-glass 3D optofluidic microchip with built-in tunable microlens fabricated by femtosecond laser-assisted etching. *Advanced Optical Materials, 6*, 1701299.

Huang, P.-H., Lapsley, M. I., Ahmed, D., Chen, Y., Wang, L., & Huang, T. J. (2012). A single-layer, planar, optofluidic switch powered by acoustically driven, oscillating microbubbles. *Applied Physics Letters, 101*, 141101.

Huang, N.-T., Zhang, H.-L., Chung, M.-T., Seo, J. H., & Kurabayashi, K. (2014). Recent advancements in optofluidics-based single-cell analysis: Optical on-chip cellular manipulation, treatment, and property detection. *Lab on a Chip, 14*, 1230–1245.

Ibarlucea, B., Fernandez-Rosas, E., Vila-Planas, J., Demming, S., Nogúes, C., Plaza, J. A., Büttgenbach, S., & Llobera, A. (2010). Cell screening using disposable photonic lab on a chip systems. *Analytical Chemistry, 82*, 4246–4251.

Jiang, X., Song, Q., Xu, L., Fu, J., & Tong, L. (2007). Microfiber knot dye laser based on the evanescent-wave-coupled gain. *Applied Physics Letters, 90*, 233501.

Jung, K.-H., & Lee, K.-H. (2015). Molecular imaging in the era of personalized medicine. *Journal of Pathology and Translational Medicine, 49*, 5–12.

Kamei, T., Paegel, B. M., Scherer, J. R., Skelley, A. M., Street, R. A., & Mathies, R. A. (2003). Integrated hydrogenated amorphous Si photodiode detector for microfluidic bioanalytical devices. *Analytical Chemistry, 75*, 5300–5305.

Kasahara, T., Matsunami, S., Edura, T., Oshima, J., Adachi, C., Shoji, S., & Mizuno, J. (2013). Fabrication and performance evaluation of microfluidic organic light emitting diode. *Sensors and Actuators A, 195*, 219–223.

Kemmler, M., Koger, B., Sulz, G., Sauer, U., Schleicher, E., Preininger, C., & Brandenburg, A. (2009). Compact point-of-care system for clinical diagnostics. *Sensors and Actuators B: Chemical, 139*, 44–51.

Kopp, D., Lehmann, L., & Zappe, H. (2016). Optofluidic laser scanner based on a rotating liquid prism. *Applied Optics, 55*, 2136–2142.

Lee, H., Liu, Y., Ham, D., & Westervelt, R. M. (2007). Integrated cell manipulation system—CMOS/microfluidic hybrid. *Lab on a Chip, 7*, 331–337.

Lee, K. S., Kim, S. B., Lee, K. H., Sung, H. J., & Kim, S. S. (2010). Three-dimensional microfluidic liquid-core/liquid-cladding waveguide. *Applied Physics Letters, 97*, 021109.

Lee, W., Li, H., Suter, J. D., Reddy, K., Sun, Y., & Fan, X. (2011). Tunable single mode lasing from an on-chip optofluidic ring resonator laser. *Applied Physics Letters, 98*, 061103.

Li, D. (Ed.). (2015a). *Encyclopedia of microfluidics and nanofluidics* (pp. 2109–2114). Springer.

Li, D. (Ed.). (2015b). *Encyclopedia of microfluidics and nanofluidics* (pp. 2089–2090). Springer.

Li, Z., & Psaltis, D. (2008). Optofluidic dye lasers. *Microfluidics and Nanofluidics, 4*, 145–158.

Li, X., Chen, Z., Taflove, A., & Backman, V. (2005). Optical analysis of nanoparticles via enhanced backscattering facilitated by 3-D photonic nanojets. *Optics Express, 13*, 526–533.

Li, L., Guo, W., Yan, Y., Lee, S., & Wang, T. (2013). Label-free super-resolution imaging of adenoviruses by submerged microsphere optical nanoscopy. *Light: Science and Applications, 2*, e104.

Li, Y., Liu, X., Yang, X., Lei, H., Zhang, Y., & Li, B. (2017). Enhancing upconversion fluorescence with a natural bio-microlens. *ACS Nano, 11*, 10672–10680.

Liang, X. J., Liu, A. Q., Lim, C. S., Ayi, T. C., & Yap, P. H. (2007). Determining refractive index of single living cell using an integrated microchip. *Sensors and Actuators A, 133*, 349–354.

Liu, H., Shi, Y., Liang, L., Li, L., Guo, S., Yin, L., & Yang, Y. (2017a). A liquid thermal gradient refractive index lens and using it to trap single living cell in flowing environments. *Lab on a Chip, 17*, 1280–1286.

Liu, L., Zhou, X., Lu, M., Zhang, M., Yang, C., Ma, R., Memon, A. G., Shi, H., & Qian, Y. (2017b). An array fluorescent biosensor based on planar waveguide for multi-analyte determination in water samples. *Sensors and Actuators B: Chemical, 240*, 107–113.

Manz, A., Graber, N., & Widmer, H. M. (1990). Miniaturized total chemical analysis systems: A novel concept for chemical sensing. *Sensors and Actuators B: Chemical, 1*, 244–248.

Mao, X., Waldeisen, J. R., Juluri, B. K., & Huang, T. J. (2007). Hydrodynamically tunable optofluidic cylindrical microlens. *Lab on a Chip, 7*, 1303–1308.

Mao, X., Lin, S.-C. S., Lapsley, M. I., Shi, J., Juluri, B. K., & Huang, T. J. (2009). Tunable liquid gradient refractive index (L-GRIN) lens with two degrees of freedom. *Lab on a Chip, 9*, 2050–2058.

Mark, D., Haeberle, S., Roth, G., Stetten, F. V., & Zengerle, R. (2010). Microfluidic lab-on-a-chip platforms: requirements, characteristics and applications. *Chemical Society Reviews, 39*, 1153–1182.

McLeod, E., & Arnold, C. B. (2008). Subwavelength direct-write nanopatterning using optically trapped microspheres. *Nature Nanotechnology, 3*, 413–417.

Miccio, L., Memmolo, P., Merola, F., Netti, P. A., & Ferraro, P. (2015). Red blood cell as an adaptive optofluidic microlens. *Nature Communications, 6*, 6502.

Minzioni, P., Osellame, R., Sada, C., Zhao, S., Omenetto, F., Gylfason, K. B., Haraldsson, T., Zhang, Y., Ozcan, A., Wax, A., Mugele, F., Schmidt, H., Testa, G., Bernini, R., Guck, J., Liberale, C., Berg-Sørensen, K., Chen, J., Pollnau, M., Xiong, S., Liu, A.-Q., Shiue, C.-C., Fan, S.-K., Erickson, D., & Sinton, D. (2017). Roadmap for optofluidics. *Journal of Optics, 19*, 093003.

Mishra, K., Murade, C., Carreel, B., Roghair, I., Oh, J. M., Manukyan, G., van den Ende, D., & Mugele, F. (2014). Optofluidic lens with tunable focal length and asphericity. *Scientific Reports, 4*, 6378.

Mohammed, M.-I., & Desmulliez, M. P. Y. (2013). Planar lens integrated capillary action microfluidic immunoassay device for the optical detection of troponin I. *Biomicrofluidics, 7*, 064112.

Monat, C., Domachuk, P., & Eggleton, B. J. (2007). Integrated optofluidics: A new river of light. *Nature Photonics, 1*, 106–114.

Monks, J. N., Yan, B., Hawkins, N., Vollrath, F., & Wang, Z. (2016). Spider silk: Mother nature's bio-superlens. *Nano Letters, 16*, 5842–5845.

Muñoz-Berbel, X., Rodríguez-Rodríguez, R., Vigués, N., Demming, S., Mas, J., Büttgenbach, S., Verpoorte, E., Ortiz, P., & Llobera, A. (2013). Monolithically integrated biophotonic lab-on-a-chip for cell culture and simultaneous pH monitoring. *Lab on a Chip, 13*, 4239–4247.

Nguyen, N.-T. (2010). Micro-optofluidic lenses: A review. *Biomicrofluidics, 4*, 031501.

Ozcan, A., & McLeod, E. (2016). Lensless imaging and sensing. *Annual Review of Biomedical Engineering, 18*, 77–102.

Pang, L., Chen, H. M., Freeman, L. M., & Fainman, Y. (2012). Optofluidic devices and applications in photonics, sensing and imaging. *Lab on a Chip, 12*, 3543–3551.

Pendry, J. B. (2000). Negative refraction makes a perfect lens. *Physical Review Letters, 85*, 3966–3969.

Piyasena, M. E., & Graves, S. W. (2014). The intersection of flow cytometry with microfluidics and microfabrication. *Lab on a Chip, 14*, 1044–1059.

Ro, K. W., Lim, K., Shim, B. C., & Hahn, J. H. (2005). Integrated light collimating system for extended optical-path-length absorbance detection in microchip-based capillary electrophoresis. *Analytical Chemistry, 77*, 5160–5166.

Rodríguez-Ruiz, I., Ackermann, T. N., Muñoz-Berbel, X., & Llobera, A. (2016). Photonic lab-on-a-chip: Integration of optical spectroscopy in microfluidic systems. *Analytical Chemistry, 88*, 6630–6637.

Rosenauer, M., & Vellekoop, M. J. (2009a). A versatile liquid-core/liquid-twin-cladding waveguide micro flow cell fabricated by rapid prototyping. *Applied Physics Letters, 95*, 163702.

Rosenauer, M., & Vellekoop, M. J. (2009b). 3D fluidic lens shaping—A multiconvex hydrodynamically adjustable optofluidic microlens. *Lab on a Chip, 9*, 1040–1042.

Schmidt, H., & Hawkins, A. R. (2008). Optofluidic waveguides: I. concepts and implementations. *Microfluidics and Nanofluidics, 4*, 3–16.

Schmidt, H., & Hawkins, A. R. (2011). The photonic integration of non-solid media using optofluidics. *Nature Photonics, 5*, 598–604.

Schuergers, N., Lenn, T., Kampmann, R., Meissner, M. V., Esteves, T., Temerinac-Ott, M., Korvink, J. G., Lowe, A. R., Mullineaux, C. W., & Wilde, A. (2016). Cyanobacteria use micro-optics to sense light direction. *eLife, 5*, e12620.

Schwartz, J. J., Stavrakis, S., & Quake, S. R. (2010). Colloidal lenses allow high-temperature single-molecule imaging and improve fluorophore photostability. *Nature Nanotechnology, 5*, 127–132.

Seo, J., & Lee, L. P. (2004). Disposable integrated microfluidics with self-aligned planar microlenses. *Sensors and Actuators B: Chemical, 99*, 615–622.

Seo, S., Su, T.-W., Tseng, D. K., Erlinger, A., & Ozcan, A. (2009). Lensfree holographic imaging for on-chip cytometry and diagnostics. *Lab on a Chip, 9*, 777–787.

Shi, J., Stratton, Z., Lin, S.-C. S., Huang, H., & Huang, T. J. (2010). Tunable optofluidic microlens through active pressure control of an air–liquid interface. *Microfluidics and Nanofluidics, 9*, 313–318.

Shopova, S. I., Zhou, H., Fan, X., & Zhang, P. (2007). Optofluidic ring resonator based dye laser. *Applied Physics Letters, 90*, 221101.

Song, C., & Tan, S. H. (2017). A perspective on the rise of optofluidics and the future. *Micromachines, 8*, 152.

Song, W., Vasdekis, A. E., Li, Z., & Psaltis, D. (2009). Optofluidic evanescent dye laser based on a distributed feedback circular grating. *Applied Physics Letters, 94*, 161110.

Song, C., Nguyen, N.-T., Tan, S.-H., & Asundi, A. K. (2010). A tuneable micro-optofluidic biconvex lens with mathematically predictable focal length. *Microfluidics and Nanofluidics, 9*, 889–896.

Song, C., Nguyen, N.-T., Yap, Y. F., Luong, T.-D., & Asundi, A. K. (2011). Multi-functional, optofluidic, in-plane, bi-concave lens: Tuning light beam from focused to divergent. *Microfluidics and Nanofluidics, 10*, 671–678.

Tang, S. K., Li, Z., Abate, A. R., Agresti, J. J., Weitz, D. A., Psaltis, D., & Whitesides, G. M. (2009). A multi-color fast-switching microfluidic droplet dye laser. *Lab on a Chip, 9*, 2767–2771.

Testa, G., Huang, Y., Zeni, L., Sarro, P. M., & Bernini, R. (2010). Liquid core ARROW waveguides by atomic layer deposition. *IEEE Photonics Technology Letters, 22*, 616–618.

Threm, D., Nazirizadeh, Y., & Gerken, M. (2012). Photonic crystal biosensors towards on-chip integration. *Journal of Biophotonics, 5*, 601–616.

Tung, Y.-C., Huang, N.-T., Oh, B.-R., Patra, B., Pan, C.-C., Qiu, T., Chu, P. K., Zhang, W., & Kurabayashi, K. (2012). Optofluidic detection for cellular phenotyping. *Lab on a Chip, 12*, 3552–3565.

Vezenov, D. V., Mayers, B. T., Conroy, R. S., Whitesides, G. M., Snee, P. T., Chan, Y., Nocera, D. G., & Bawendi, M. G. (2005). A low-threshold, high-efficiency microfluidic waveguide laser. *Journal of the American Chemical Society, 127*, 8952–8953.

Vila-Planas, J., Fernandez-Rosas, E., Ibarlucea, B., Demming, S., Nogúes, C., Plaza, J. A., Domínguez, C., Büttgenbach, S., & Llobera, A. (2011). Cell analysis using a multiple internal reflection photonic lab-on-a-chip. *Nature Protocols, 6*, 1642–1655.

Wang, Z., Guo, W., Li, L., Luk'yanchuk, B., Khan, A., Liu, Z., Chen, Z., & Hong, M. (2011). Optical virtual imaging at 50 nm lateral resolution with a white-light nanoscope. *Nature Communications, 2*, 218.

Watts, B. R., Kowpak, T., Zhang, Z., Xu, C.-Q., & Zhu, S. (2010). Formation and characterization of an ideal excitation beam geometry in an optofluidic device. *Biomedical Optics Express, 1*, 848–860.

Watts, B. R., Zhang, Z., Xu, C.-Q., Cao, X., & Lin, M. (2013). A method for detecting forward scattering signals on-chip with a photonic-microfluidic integrated device. *Biomedical Optics Express, 4*, 1051–1060.

Wei, Q., McLeod, E., Qi, H., Wan, Z., Sun, R., & Ozcan, A. (2013). On-chip cytometry using plasmonic nanoparticle enhanced lensfree holography. *Scientific Reports, 3*, 1699.

Wojciechowski, J. R., Shriver-Lake, L. C., Yamaguchi, M. Y., Füreder, E., Pieler, R., Schamesberger, M., Winder, C., Prall, H. J., Sonnleitner, M., & Ligler, F. S. (2009). Organic photodiodes for biosensor miniaturization. *Analytical Chemistry, 81*, 3455–3461.

Wolfe, D. B., Conroy, R. S., Garstecki, P., Mayers, B. T., Fischbach, M. A., Paul, K. E., Prentiss, M., & Whitesides, G. M. (2004). Dynamic control of liquid-core/liquid-cladding optical waveguides. *Proceedings of the National Academy of Sciences of the United States of America, 101*, 12434–12438.

Wolfe, D. B., Vezenov, D. V., Mayers, B. T., Whitesides, G. M., Conroy, R. S., & Prentiss, M. G. (2005). Diffusion-controlled optical elements for optofluidics. *Applied Physics Letters, 87*, 181105.

Wu, J., & Gu, M. (2011). Microfluidic sensing: State of the art fabrication and detection techniques. *Journal of Biomedical Optics, 16*, 080901.

Wynne, T. M., Dixon, A. H., & Pennathur, S. (2012). Electrokinetic characterization of individual nanoparticles in nanofluidic channels. *Microfluidics and Nanofluidics, 12*, 411–421.

Xiong, S., Liu, A., Chin, L., & Yang, Y. (2011). An optofluidic prism tuned by two laminar flows. *Lab on a Chip, 11*, 1864–1869.

Xu, D., & Adachi, C. (2009). Organic light-emitting diode with liquid emitting layer. *Applied Physics Letters, 95*, 053304.

Yan, Y., Li, L., Feng, C., Guo, W., Lee, S., & Hong, M. (2015). Microsphere-coupled scanning laser confocal nanoscope for sub-diffraction-limited imaging at 25 nm lateral resolution in the visible spectrum. *ACS Nano, 8*, 1809–1816.

Yang, H., & Gijs, M. A. M. (2013). Microtextured substrates and microparticles used as in situ lenses for on-chip immunofluorescence amplification. *Analytical Chemistry, 85*, 2064–2071.

Yang, H., & Gijs, M. A. M. (2015). Optical microscopy using a glass microsphere for metrology of sub-wavelength nanostructures. *Microelectronic Engineering, 143*, 86–90.

Yang, H., & Gijs, M. A. M. (2018). Micro-optics for microfluidic analytical applications. *Chemical Society Reviews, 47*, 1391–1458.

Yang, H., Chao, C.-K., Lin, C.-P., & Shen, S.-C. (2004). Micro-ball lens array modeling and fabrication using thermal reflow in two polymer layers. *Journal of Micromechanics and Microengineering, 14*, 277–282.

Yang, H., Moullan, N., Auwerx, J., & Gijs, M. A. M. (2014). Super-resolution biological microscopy using virtual imaging by a microsphere nanoscope. *Small, 10*, 1712–1718.

Yang, H., Cornaglia, M., & Gijs, M. A. M. (2015). Photonic nanojet array for fast detection of single nanoparticles in a flow. *Nano Letters, 15*, 1730–1735.

Yang, H., Trouillon, R., Huszka, G., & Gijs, M. A. M. (2016a). Super-resolution imaging of a dielectric microsphere is governed by the waist of its photonic nanojet. *Nano Letters, 16*, 4862–4870.

Yang, T., Bragheri, F., & Minzioni, P. (2016b). A comprehensive review of optical stretcher for cell mechanical characterization at single-cell level. *Micromachines, 7*, 90.

Yang, H., Zhang, Y., Chen, S., & Hao, R. (2019). Micro-optical components for bioimaging on tissues, cells and subcellular structures. *Micromachines, 10*, 405.

Yao, B., Luo, G., Wang, L., Gao, Y., Lei, G., Ren, K., Chen, L., Wang, Y., Hu, Y., & Qiu, Y. (2005). A microfluidic device using a green organic light emitting diode as an integrated excitation source. *Lab on a Chip, 5*, 1041–1047.

Yu, H., Zhou, G., Leung, H. M., & Chau, F. S. (2010). Tunable liquid-filled lens integrated with aspherical surface for spherical aberration compensation. *Optics Express, 18*, 9945–9954.

Zhang, H., Ho, S., Eaton, S. M., Li, J., & Herman, P. R. (2008). Three-dimensional optical sensing network written in fused silica glass with femtosecond laser. *Optics Express, 16*, 14015–14023.

Zhao, Y., Zhao, X., & Gu, Z. (2010). Photonic crystals in bioassays. *Advanced Functional Materials, 20*, 2970–2988.

Zhao, Y., Stratton, Z. S., Guo, F., Lapsley, M. I., Chan, C. Y., Lin, S.-C. S., & Huang, T. J. (2013). Optofluidic imaging: Now and beyond. *Lab on a Chip, 13*, 17–24.

Zheng, H. Y., Liu, H., Wan, S., Lim, G. C., Nikumb, S., & Chen, Q. (2006). Ultrashort pulse laser micromachined microchannels and their application in an optical switch. *International Journal of Advanced Design and Manufacturing Technology, 27*, 925–929.

Zhu, H., Mavandadi, S., Coskun, A. F., Yaglidere, O., & Ozcan, A. (2011). Optofluidic fluorescent imaging cytometry on a cell phone. *Analytical Chemistry, 83*, 6641–6647.

Chapter 11
Wearable MEMS Sensor Nodes for Animal Health Monitoring System

Hirofumi Nogami, Hironao Okada, Seiichi Takamatsu, Narifumi Kawano, Takeshi Kobayashi, Ryutaro Maeda, and Toshihiro Itoh

11.1 Introduction

To achieve the goal of creating a safer and more secure society, wireless sensor network technology has been a promising approach for various applications, such as structural health, human health, and animal health monitoring systems. Structural health monitoring is used to improve the safety and reliability of buildings, bridges, and lifeline systems by detecting damage before it reaches a critical state (Yun & Min, 2011; Yamashita et al., 2016). Damage is detected using many wireless sensor nodes or sensor arrays installed on the structure. Human health monitoring detects sleep disorders, insomnia, cardiac disorder, and heart and lung diseases through the logging of a person's daily life (walking or sleeping) using GPS, ECG, blood pressure sensors, photoplethysmographic sensors, laser Doppler blood flowmeters, temperature sensors, and triaxial and magnetic sensors (Hadjidj et al., 2013; Takeshita et al., 2019; Adil et al., 2020). These technologies have also been introduced into

H. Nogami (✉)
Faculty of Engineering, Kyushu University, Nishi-ku, Fukuoka, Japan
e-mail: nogami@mech.kyushu-u.ac.jp

H. Okada · T. Kobayashi
Sensing System Research Center, National Institute of Advanced Industrial Science and Technology (AIST), Namiki, Tsukuba, Ibaraki, Japan

S. Takamatsu · T. Itoh
Graduate School of Engineering, The University of Tokyo, Bunkyo-ku, Tokyo, Japan

N. Kawano
Omuta City Zoo, Omuta, Fukuoka, Japan

R. Maeda
School of Mechanical Engineering, Xi'an Jiaotong University, Xi'an, People's Republic of China

Device Technology Research Institute, AIST, Tsukuba, Ibaraki, Japan

Z. Yang (ed.), *Advanced MEMS/NEMS Fabrication and Sensors*,
https://doi.org/10.1007/978-3-030-79749-2_11

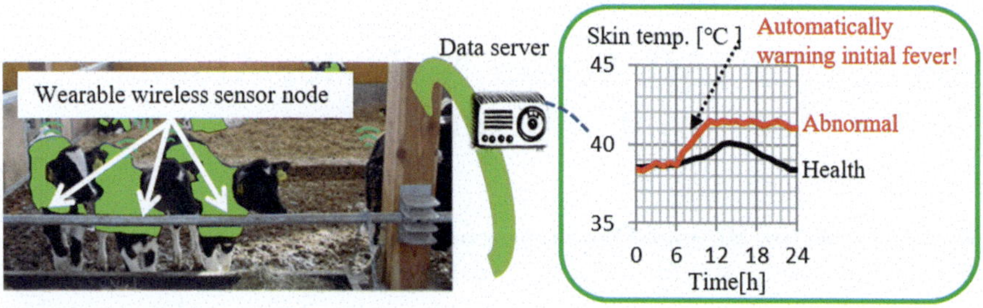

Fig. 11.1 Concept of cow health monitoring system

agriculture, including animal health monitoring (Khem et al., 2018; Nishihara et al., 2013; Okada et al., 2009b). It is believed that wireless sensor nodes attached to animals, in conjunction with a wireless health monitoring system, can achieve the early detection and prevention of diseases and thus reduce economic losses. In this chapter, we introduce wearable wireless sensors for animal health monitoring system.

Figure 11.1 indicates the concept of our cow health monitoring system. In this system, sensor nodes are attached to cows to measure their biological data (body temperature, activity, heart rate, and so on). If one wireless sensor node transmits abnormal data (e.g., fever), the system logs the data on the collected sickness pattern and automatically sends out a warning signal. Furthermore, the system compares the logged data with the expected normal data from a cow and determines the causes of the abnormal reading. Other research groups have a developed wearable wireless estrus detection sensor, a portable estrus intensity detection sensor, and a bolus-type pH monitoring sensor (Andersson et al., 2016; Iwasaki et al., 2015; Zhang et al., 2018).

The following are important. Farmers can use sensors and the system without loss of time and effort. The size and weight of sensor nodes do not interfere with the behavior of the animal. Sensor nodes must have low power consumption to avoid frequent replacement of batteries. They must be durable so that they cannot be broken by animal behavior. In this chapter, we summarized unsuccessful and successful results for sensor nodes actually attached to chickens, cows, and giraffes.

11.2 Chicken Health Monitoring System

11.2.1 Event-Driven Measurement System for Low Power Consumption

Wearable sensor nodes for small animals such as chickens require ultralow power consumption since the weight of the battery should not limit animal behavior. Figures 11.2a, b shows a model of a wireless sensor node and the event-driven

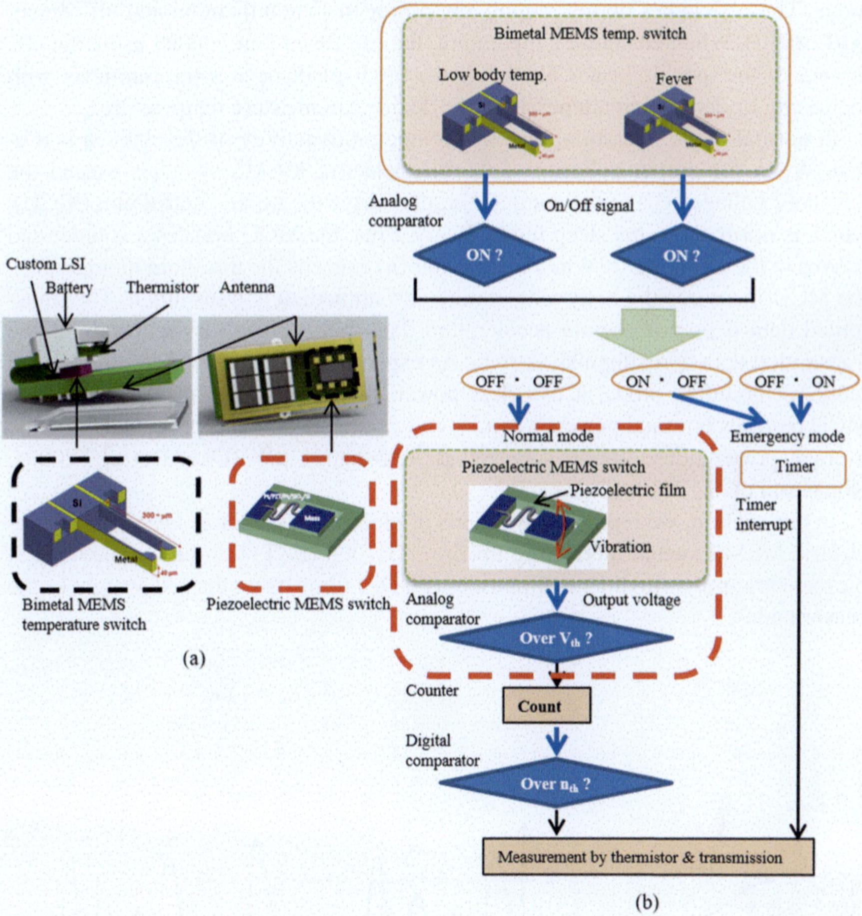

Fig. 11.2 Model of a wireless sensor node (**a**) and the event-driven system for ultralow power consumption (**b**)

system for ultralow power consumption. The sensor node is composed of a thermistor, a custom LSI, an antenna, a battery, a bimetal microelectromechanical system (MEMS) temperature switch, and a piezoelectric MEMS switch. An event-driven system is used to lower the standby energy consumption of the wireless sensor node to approximately 0 mW by using a bimetal MEMS temperature switch and a piezoelectric MEMS switch.

Wireless sensor nodes operate in two modes, the emergency and normal modes, depending on the body temperature of the chickens. Emergency-mode operation is used when the body temperature of the chicken is abnormally high or low, and this data is transmitted to the timer interrupt. The on-off action of two bimetal MEMS temperature switches allows the transition from the normal mode to the emergency mode. These switches do not require any power to switch them on and off (Zhang et al., 2012). When changing temperature, the cantilever pair contact using the difference of the specific heat. Calculating the tip displacement of the cantilever with increasing or decreasing temperature, the device can measure temperature.

In normal-mode operation, the body temperature/activity of the chicken is normal. When the output voltages of the piezoelectric MEMS switches exceed the threshold voltage (V_{th}) of the comparator, they drive the micro control unit (MCU), which is normally in the sleep mode. Every 2 ms, the MCU updates a counter and returns to the sleep mode. When the counter (n) exceeds the threshold number (n_{th}), the MCU measures the body temperature and immediately transmits it. The transmitted data does not contain acceleration data, but we use the frequency of the transmitted data to distinguish activity. As expected, when the MCU is in the sleep mode, or counting mode, it uses less power. In the data transmission mode, the MCU requires minimal power; when V_{th} and n_{th} are set to their optimum values, the average power consumption achieved is less than 1 μW (Okada et al., 2009a; Kobayashi et al., 2010) (Fig. 11.3).

In this section, we report on two kinds of piezoelectric switch. One is the piezoelectric MEMS switch fabricating $Pb(Zr_{0.52},Ti_{0.48})O_3$ (PZT) thin films. The other is piezoelectric poly(vinylidene difluoride) (PVDF) films using the self-weight of the sensor nodes.

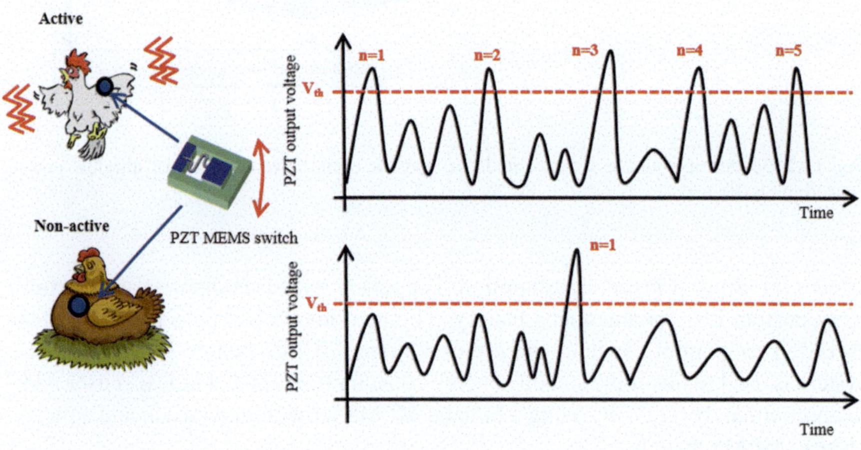

Fig. 11.3 Relationship between threshold voltage (V_{th}) and output voltages of chicken behavior. Suitable V_{th} can distinguish an active chicken from a nonactive chicken

11.2.2 PZT MEMS Switch Version Sensor Node

11.2.2.1 Fabrication of PZT MEMS Switch

The piezoelectric MEMS switch has been fabricated using PZT thin films by microfabrication processing technology. This process is the same as previously presented (Kobayashi et al., 2008). Figure 11.4 shows the process flow. The process began with the deposition of Pt/Ti/PZT/Pt/Ti/SiO_2 multilayers on silicon-on-insulator (SOI) wafers with a 5-μm-thick structural Si layer. The SOI wafers were first oxidized at 1100 °C to form thermal oxide (thickness, 1 μm). After the oxidation, Pt (0.2 μm)/Ti (0.05 μm) thin films were deposited by DC magnetron sputtering to form a bottom electrode. Next, 2.5-μm-thick PZT thin films were formed by sol-gel deposition. The PZT thin films have (100)/(001) preferred orientation and a remnant polarization of 10.5 μC/cm^2. Finally, Pt (0.2 μm)/Ti (0.05 μm) thin films were deposited by DC magnetron sputtering.

The obtained multilayers were etched by dry and wet etching processes to form the piezoelectric accelerometer. The Pt/Ti thin films were etched through mask 1 as a top electrode using Ar ions. The PZT thin films were wet-etched through mask 2 for patterning the piezoelectric thin film, using a mixture of HF, HNO_3, and HCl. The Pt/Ti thin films through mask 3 were etched using Ar ions, and thermal SiO_2 thin films were etched by reactive ion etching (RIE) with CHF_3 gas to pattern the bottom electrodes. Structural Si and buried oxide (BOX) layers were etched by RIE with SF_6 (for Si) and CHF_3 (for BOX) gases through mask 4 to pattern the cantilever. Finally, Si substrates and BOX were etched from their back surface.

Figure 11.5 shows the photographs and design of the PZT MEMS switch. S-shaped PZT cantilevers with proof masses have a resonance frequency of approximately 24 Hz. We have utilized their superharmonic resonance to detect chicken movements with frequencies as low as 5–15 Hz (Nogami et al., 2013b).

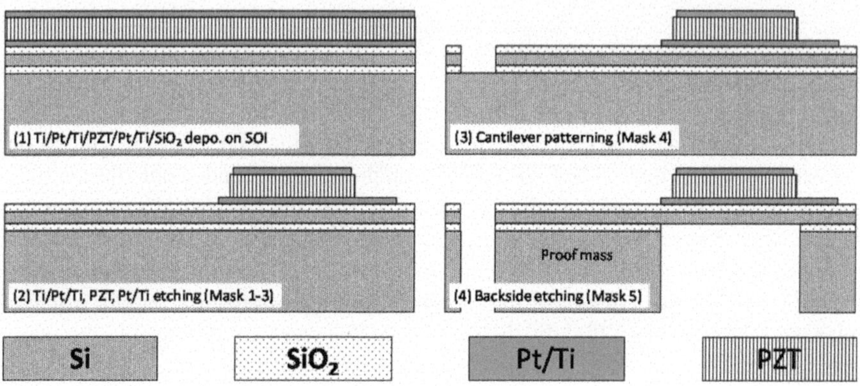

Fig. 11.4 MEMS process for PZT thin film MEMS switches

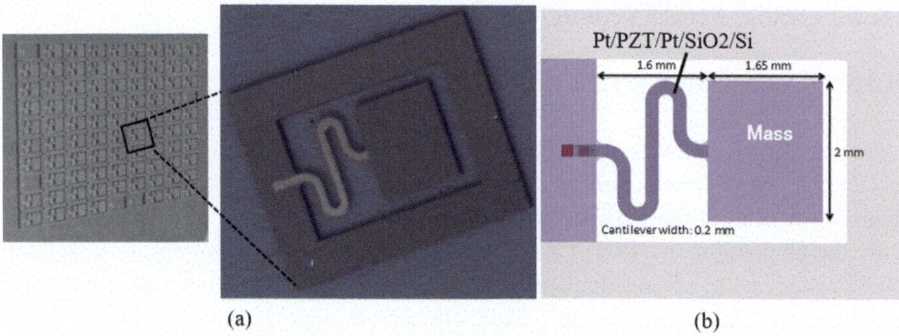

Fig. 11.5 Photographs (**a**) and design (**b**) of PZT MEMS switches

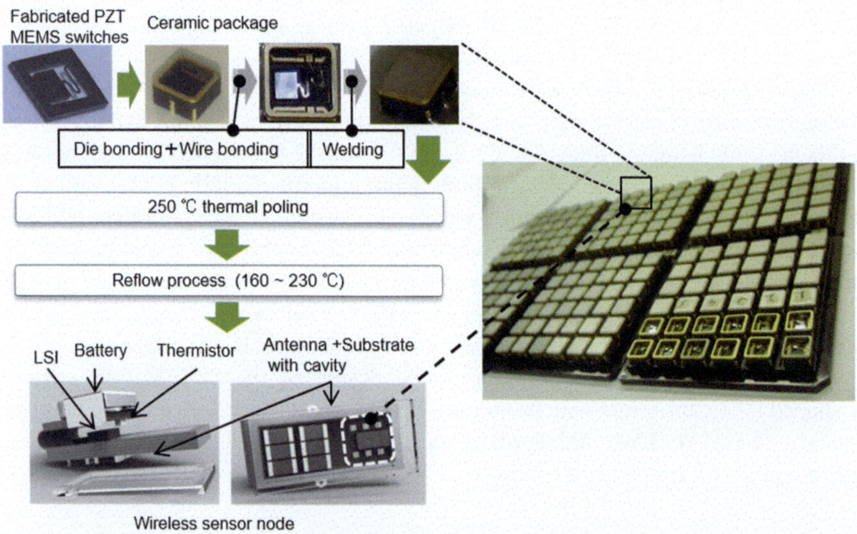

Fig. 11.6 Procedure for implementing the PZT MEMS switches on the wireless sensor node

11.2.2.2 Procedure for Implementing PZT MEMS Switches on Wireless Sensor Node

Figure 11.6 illustrates the procedure for implementing the PZT MEMS switches on the wireless sensor node. The PZT MEMS switches are die-bonded onto the ceramic package and wire-bonded onto the electric pads. Next, the PZT MEMS switches are sealed. To improve piezoelectric output voltage characteristics, the PZT MEMS switches are poled at 250 °C-18 V-5 min. After the 250 °C thermal poling treatment, the PZT MEMS switches have only slightly lower piezoelectric output voltages after the reflow process. Finally, the PZT MEMS devices are reflowed to the wireless sensor nodes. In the reflow process, we used a lead-free solder (TB46-M741 Japan, with the recommended reflow temperature of 160–230 °C).

11.2.2.3 Experimental Procedure

Figure 11.7 shows the fabricated wireless sensor nodes and the experimental environment. Sensor nodes are 14.0 × 55.0 mm^2. The weight of the sensor node is 4.3 g including the battery, which is light enough to allow chickens to move freely. The battery is Sr726W (Sony), whose capacity is 35 mAh, which is sufficient to drive a 1 μW sensor node for 2 years. These sensor nodes were attached to the wings of more than 100 chicken wings in chicken coops.

11.2.2.4 Successful Result and Failure

The measurement result of one sensor is shown in Fig. 11.8a, where the *y*-axis indicates the number of receiving times per 30 min. This indicates the amount of activity. The number of receiving times is large in the daytime, and small in the nighttime. The difference between nighttime and daytime indicates the difference in the amount of activity between night and daytime.

There is the problem that more than 90% of the sensor nodes could not provide measurements. After the test, we detached the sensor nodes from chickens and examined the PZT switches. More than 90% of the sensor nodes had broken PZT switches (Fig. 11.8b). The broken switches were caused by many of chickens pecking at them as soon as sensor nodes were attached to chickens.

S-shaped PZT MEMS switches were adequate for low vibrations (below 15 Hz, 0.5 m/s^2), but they were beset by the structural problem of fragmentation upon impact (chicken pecking).

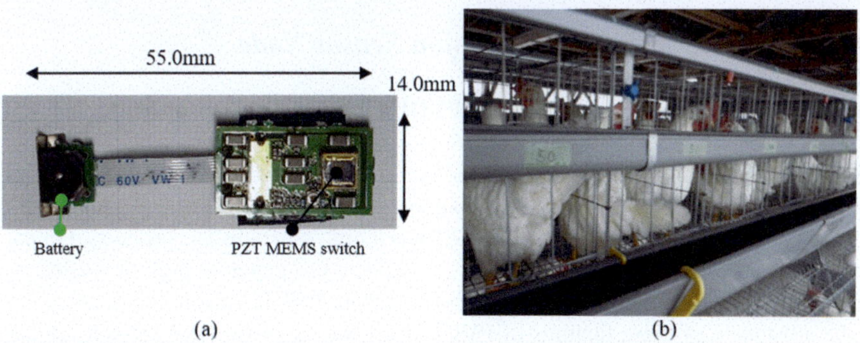

Fig. 11.7 Fabricated wireless sensor nodes (**a**) and the experimental environment (**b**)

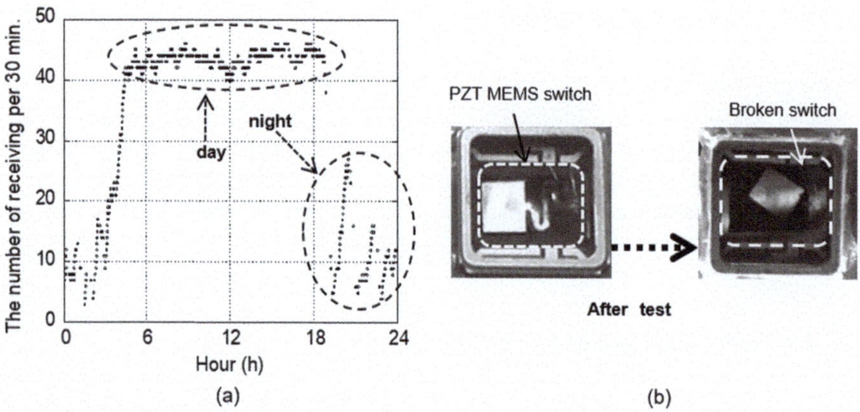

Fig. 11.8 Measurement result (**a**) and broken PZT switches after test (**b**)

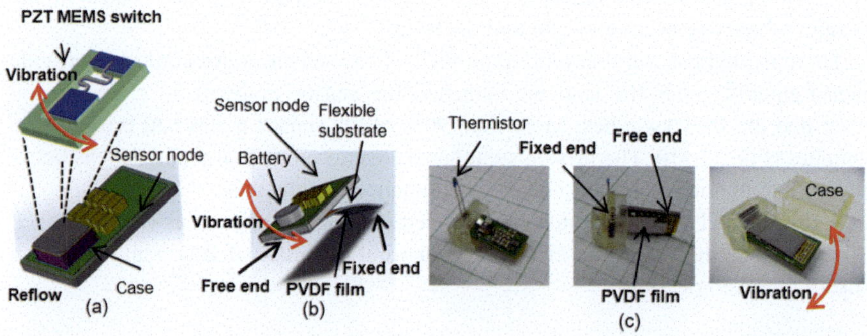

Fig. 11.9 Design and photographs of wireless sensor nodes: (**a**) design of PZT MEMS switch version sensor node, (**b**) design of PVDF film version sensor node, and (**c**) photographs of a fabricated PVDF film version sensor node

11.2.3 PVDF Film Switch Version Sensor Node

11.2.3.1 Design and Fabrication of PVDF Film Version Sensor Node

Sensor nodes attached to chickens should be very small and light, so as not to restrict movement. The PZT MEMS version switches thus reduced lower the volume ratio to the sensor node (Fig. 11.9a), since PZT has a high piezoelectric property. On the other hand, PVDF films are tough and are expected to generate high piezoelectric output voltages under a large stress without causing PVDF films to break down. To take advantage of these features of PVDF films, we suggested that the self-weight of sensor nodes be considered as the mass of the PVDF film cantilevers as shown in Fig. 11.9b (Nogami et al., 2013b). One end of PVDF film is fixed to the case of the sensor node, and the other end is attached to the sensor node. Thus, the PVDF films are cantilevers and the sensor nodes are masses. Since the PVDF films are subjected to force generated by the self-weight of the sensor nodes, high piezoelectric output voltages are expected.

Figure 11.9c shows photographs of the fabricated PVDF film switch version wireless sensor nodes. We bent a PVDF film (Tokyo Sensor 3-1003702-7) twice, and attached it to a flexible substrate. One end of the flexible substrate was soft-soldered to the sensor node, and the other end was fixed to the case. The PVDF films were 10 × 18 mm^2 in size and 110 μm in thickness. Also, d_{31} is 23 × 10^{-12} C/N and g_{31} was 216 × 10^{-3} V m/N. The weight of the sensor node is 5.5 g including the battery. The battery is Sr726W (Sony), whose capacity is 35 mAh, which is sufficient to drive 1 μW sensor node for 2 years.

11.2.3.2 Experimental Procedure

The sensor nodes are sealed in cases and attached to chickens using a band as shown in Fig. 11.10. The thermistor on the sensor case is attached to the inside of the chicken's wing, and s attached to the chicken's wing to wind up with a band. It was easy to attach and took within 10 s.

Figure 11.11 indicates experimental environment in the chicken coop. We logged the sensor data using a Yagi antenna and a notebook PC.

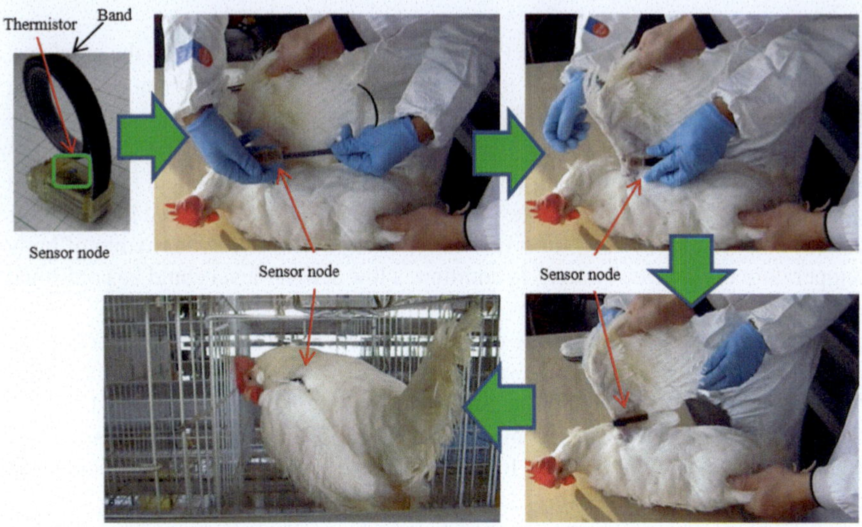

Fig. 11.10 Steps of attaching sensor nodes to chickens. The thermistor on the sensor case is attached to the inside of the chicken's wing to wind up with a band

Fig. 11.11 Experimental environment in chicken coop

11.2.3.3 Measurement Result

Immediately after attaching these PVDF film version sensor nodes to the chickens, many of the chickens pecked at these sensor nodes. These sensor nodes did not break when the chickens pecked them. Five days of measurement data for one chicken are shown in Fig. 11.12. The x-axis is the time of receiving data, the left y-axis is body temperature, and the right y-axis is the number of data receptions in a 1-h period (chicken activity). The red dots are body temperature, and the blue dots are the number of data receptions (chicken activity). The circadian rhythm of body temperature can be clearly seen. In addition, a low activity level could be confirmed in the nighttime.

Figure 11.13 shows the results of the measurement of the body temperature or activity using PVDF film version sensor node for 1 day. We can confirm the difference between the day and night numbers of data receptions. Also, after 81 days, a daily change in activity can be confirmed, as on the first day. Thus, we consider that we have successfully fabricated a tough wireless sensor node for chickens, utilizing the features of PVDF films.

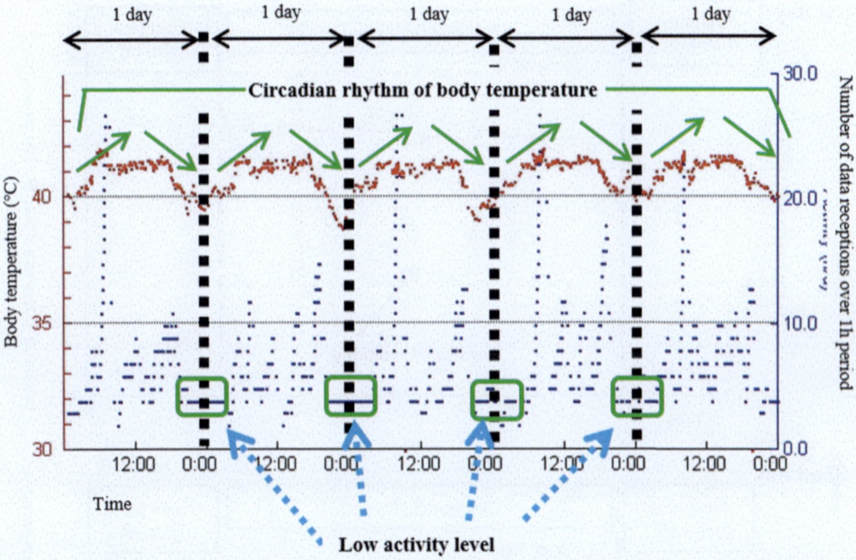

Fig. 11.12 Measurement data for 5 days of one chicken. The circadian rhythm of body temperature could be confirmed

11.3 Cow Health Monitoring

11.3.1 First-Cow's-Tail Sensor Node

11.3.1.1 Design of Compact Wireless Sensor Nodes

The weight of the wearable wireless sensor nodes must be less than 10 g in order for the sensor nodes to be attached to growing cows' tails. Figure 11.14 shows a photograph of the compact sensor node, including a single unit with a sensor, a MCU, and a battery (Nogami et al., 2013a). The three-axis accelerometer H30CD (HITACHI Metals, Ltd.) with 5 μA current consumption during the sleep mode state is used. H30CD can measure temperature with an accuracy of ±3 °C. The transceiver IC M430F2232 (Texas, Ins.) with a 16 bit analog-digital converter (ADC) and a TXC 101 (RF Monolithics, Ins.) are used. The operating frequency of this transceiver is 315 MHz because the permeability of the radio waves in humans and animals. The communication distance is about 50 m without obstructions. The size and weight of the sensor nodes are 18 × 18 × 7 mm^3 and 2.1 g without batteries. The battery was a CR1632 button cell (4.5 g), and the total weight of the sensor nodes is 6.6 g.

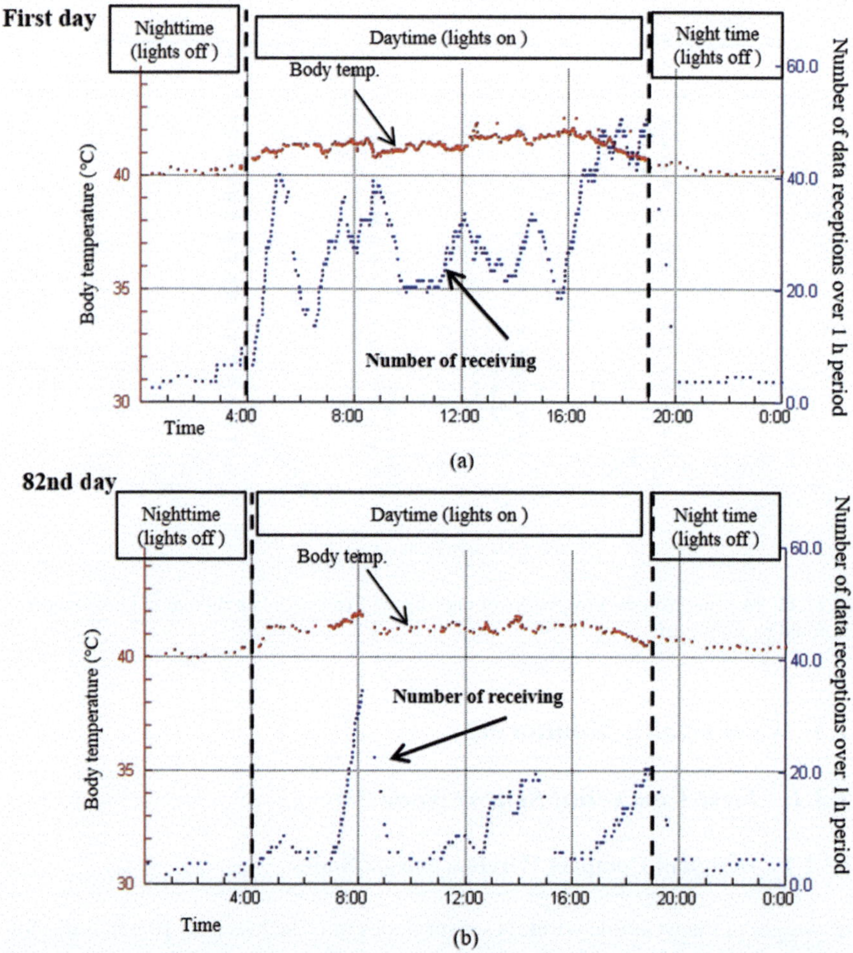

Fig. 11.13 Results of measurement of activity of a chicken using the PVDF film switch version sensor node for 1 day. The x-axis is the time of receiving data, the left y-axis is body temperature, and the right y-axis is the number of data receptions in a 1-h period (chicken activity). (**a**) Data on the first day. (**b**) Data on the 82nd day

11.3.1.2 Experimental Procedure

The experimental object is a calf (Fig. 11.15a). Because calves are much smaller than adult cattle, the size of the wearable sensor nodes is required to be smaller. The smaller sensor nodes can also be attached to larger animals. The measurement location is the base of the tail, as shown in Fig. 11.15b, since this location has a higher skin temperature than other parts. The wireless sensor nodes were attached to the base of the tail with sport tape (Fig. 11.15c).

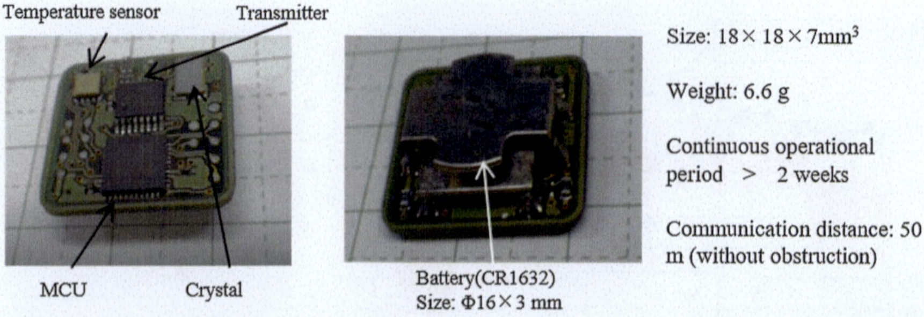

Fig. 11.14 Photographs of compact sensor nodes, including a single unit with a sensor, a MCU, and a battery

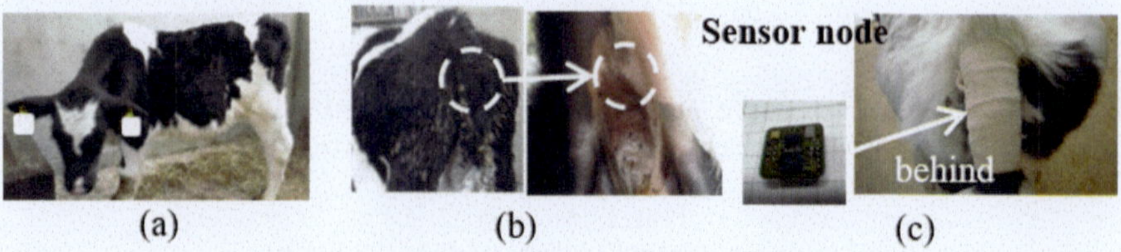

Fig. 11.15 Test calf (**a**) and measurement location (**b**), (**c**)

11.3.1.3 Measurement Result

Figure 11.16 shows measurement results for 16 days. In this experiment, we measured the rectal temperature of the calf twice a day for comparison with the sensor temperature. The red dots are sensor temperature, and the blue squares are rectal temperature in Fig. 11.16a. The sensor temperature is about 3 °C lower than the rectal temperature. We can observe that the sensor temperature is high when the rectal temperature is high. Similarly, we note that the sensor temperature is low when the rectal temperature is low. However, the problem is that the sensor temperature is very low in some cases (dashed circles in Fig. 11.16a). These data increase with the number of days.

Figure 11.16b indicates the relationship between the daily maximum rectal temperature and the daily maximum sensor temperature for 16 days. A strong correlation is indicated. Thus, it is important to measure skin temperature correlated with rectal temperature.

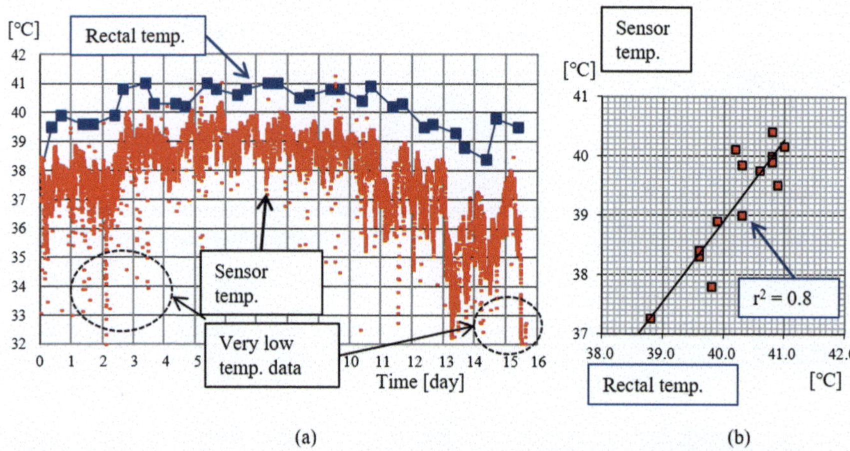

Fig. 11.16 Measurement results of (**a**) continuous monitoring over 16 days and (**b**) relationship between daily maximum sensor temperature and daily maximum rectal temperature over 16 days

11.3.2 Second- and Third-Cow's-Tail Sensor Nodes

11.3.2.1 Design of Wireless Sensor Nodes with Separate Temperature Unit

We designed the temperature unit of the wireless sensor to be pressed against the base of a calf's tail. Because the first-cow's-tail sensor nodes indicated very low temperature in some cases, the frequency of such low values increased over time (dashed circles in Fig. 11.16a). The cause could be a gap between the skin of the calf and the temperature sensor. The gap was caused by the 7 mm thickness of the sensor nodes.

Second-cow's-tail sensor nodes were divided into 2 units, as shown in Fig. 11.17 (Nogami et al., 2014). One unit was the temperature unit, and the other was the main unit. The temperature unit had a temperature sensor mounted on a flexible substrate, and its thickness was only 0.8 mm. This temperature unit enables the temperature sensor to be pressed against the base of the calf's tail.

The main unit includes the data processing device, the data transmission device, and the battery. The data processing device F930 (Silicon Labs, Inc.) and the data transmission device TRC103 (RF Monolithics, Inc.) were used. The battery was a 3V CR2032 button cell. The operating frequency of the transceiver was chosen to be 900 MHz frequency band because the permeability of radio waves at that frequency is high for humans and animals. The measurement and transmission intervals were set as 1 min, resulting in a power consumption of 0.18 mW. The lifetime of the sensor nodes was longer than 5 months.

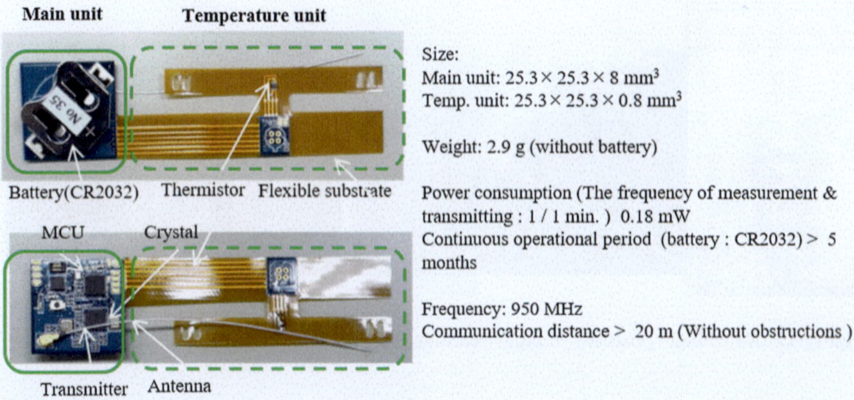

Fig. 11.17 Design of second-cow's-tail sensor node

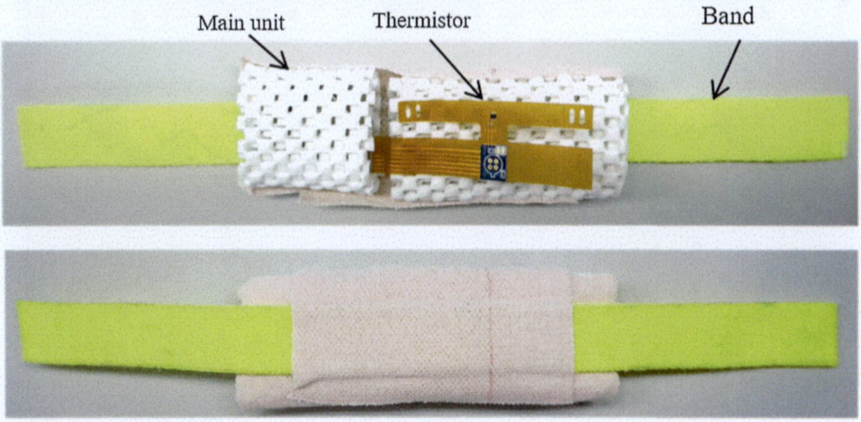

Fig. 11.18 Photographs of the wireless sensor nodes with a band

11.3.2.2 Experimental Procedure

Figure 11.18 shows the photographs of the sensor nodes used in the experiment. With the band, the sensor nodes can be easily attached to the base of a cow's tail. First, we confirmed the difference in skin temperature between the upper and lower positions of the cow's tail (Fig. 11.19). To investigate the measurement stability of the sensor nodes and the number of operating days, we conducted about 14 long-term tests for about 1 month.

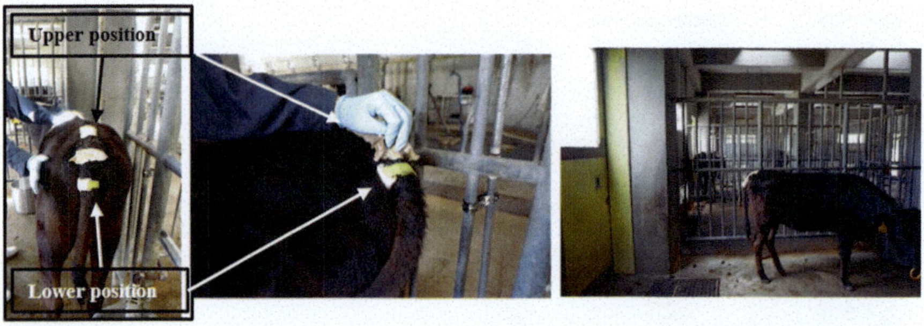

Fig. 11.19 Attachment positions of the wireless sensor nodes and experimental environment

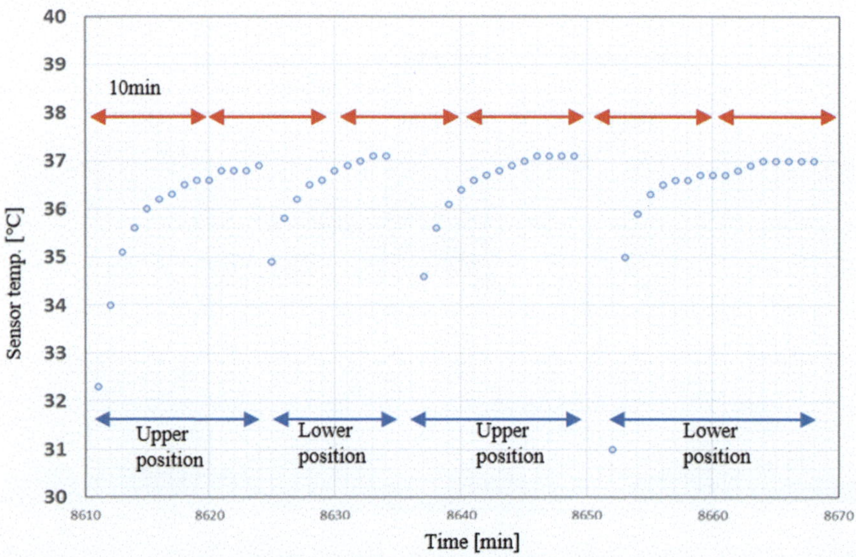

Fig. 11.20 Measured skin temperatures at upper and lower positions

11.3.2.3 Measurement Result and the Number of the Operating Days

There was no skin temperature difference between the upper and lower positions as shown in Fig. 11.20. The temperature reached 37 °C about 10 min after attaching the sensors to the upper and lower positions.

Figure 11.21 shows the measurement results of two calves for the 5-day period. The measurement temperature is shown by the lower line, and the maximum temperature every 30 min is shown by the upper line. The sensor temperature was not below 33 °C in Fig. 11.21. Also, the variability of the anteroposterior data decreased. With first-cow's-tail sensor nodes, the temperature data often indicated values below 33 °C, and the anteroposterior data varied widely. From these results, it can be seen that additional measurement stability is provided by the thinness of the temperature

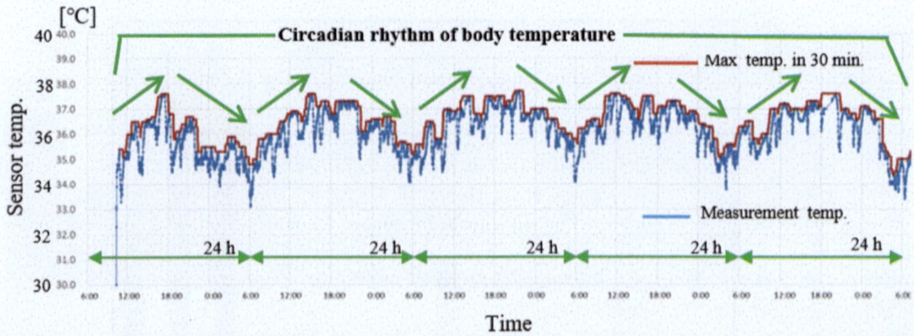

Fig. 11.21 Measured skin temperatures over 5 days. The circadian rhythm of body temperature was clearly confirmed

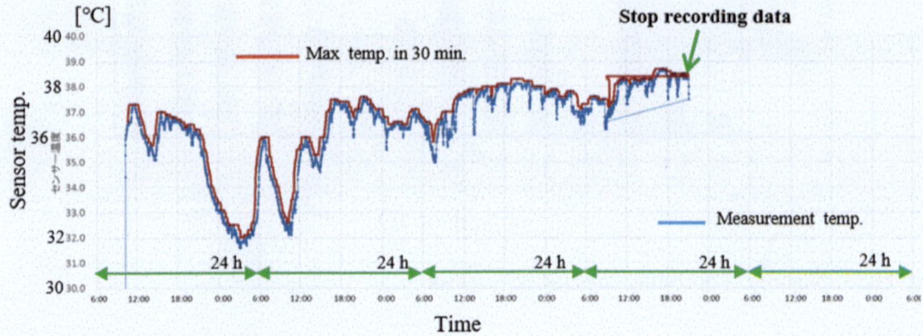

Fig. 11.22 Failure results of one cow

unit (0.8 mm). As a result of the additional measurement stability, we can clearly confirm the circadian rhythm of body temperature, which varied between 34 °C and 38 °C.

Figure 11.22 shows failure results of one cow. On the fourth day, data were not recorded. Figure 11.23 shows the number of operating days for each sensor. Only 4 out of 14 sensor nodes could be used for more than 30 days. The sensor nodes that failed to record temperature data had a damaged temperature sensor part. The flexible substrate improved the temperature measurement stability owing to its thinness, but lacked strength.

11.3.2.4 Design of Third-Cow's-Tail Sensor Nodes

We designed third-cow's tail sensor nodes with a lead wire instead of a flexible substrate. The thickness of the temperature sensor part (a thermistor), which has good adhesion to the skin, is 0.4 mm. In addition, the lead wire is more durable than the wire on the flexible substrate (Miura et al., 2017). If a temperature sensor breaks, it can be easily replaced with a new temperature sensor.

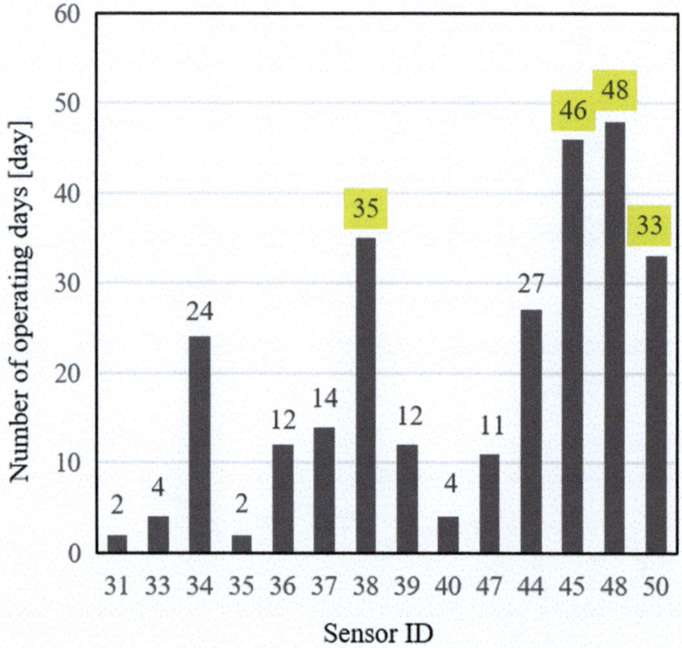

Fig. 11.23 Number of operating days for each sensor

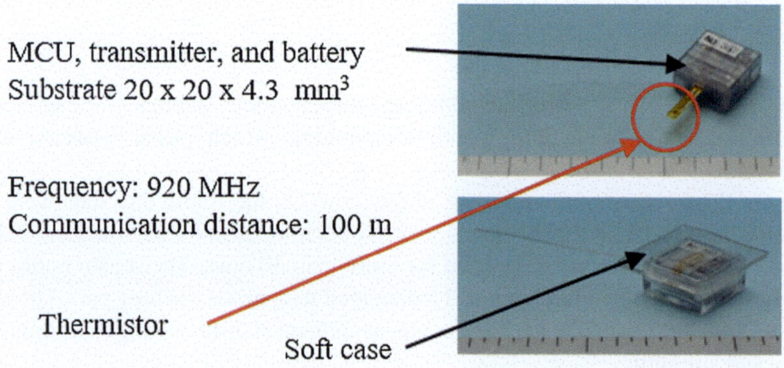

Fig. 11.24 Photographs of the third-cow's-tail sensor nodes

Figure 11.24 shows photographs of the third-cow's-tail sensor nodes. The component consists of a main unit and a thermistor (103JT-025, SEMITEC Corporation, Tokyo, Japan). The main unit contains the micro control unit, the transmitter, and the battery. The main unit measures 20.0 × 20.0 × 4.3 mm^3.

The operating frequency of the transceiver was chosen to be 920 MHz. The distance to the receiver was approximately 100 m without obstacles. The measurement and transmission intervals were set to 2 min, and the lifetime of the sensor nodes was >6 months using the CR2032 button cell.

11.3.3 Cow Coat Sensor Nodes

11.3.3.1 Design of Cow Cote Sensor Nodes

In cow houses, cow coats are used to maintain the body temperature of cows, especially in the winter season. We have designed wireless temperature sensor nodes that can be attached to the cow coat as shown in Fig. 11.25. Wireless sensor nodes consist of a signal processing circuit, a wireless transmitter, and a temperature unit mounted on a flexible substrate. The flexible substrate is soldered onto ribbons with copper wire power lines. The ribbonlike wireless temperature sensor nodes are lightweight and highly bendable and are attached to the cow coats (Takamatsu et al., 2016).

11.3.3.2 Experimental Procedure

Figure 11.26 shows a photograph of the cow coat with ribbonlike wireless temperature sensor nodes. The ribbonlike wireless temperature sensor nodes had 12 temperature sensor units attached to the inside of the cow coat. The cow coat with ribbonlike wireless temperature sensor nodes was worn by one calf.

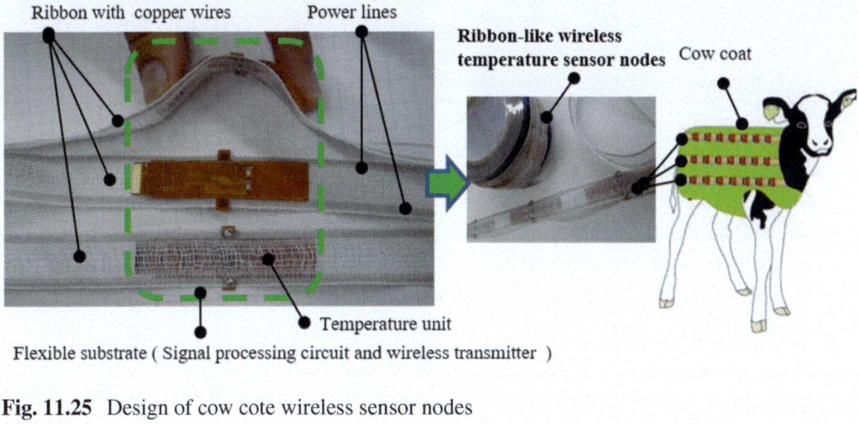

Fig. 11.25 Design of cow cote wireless sensor nodes

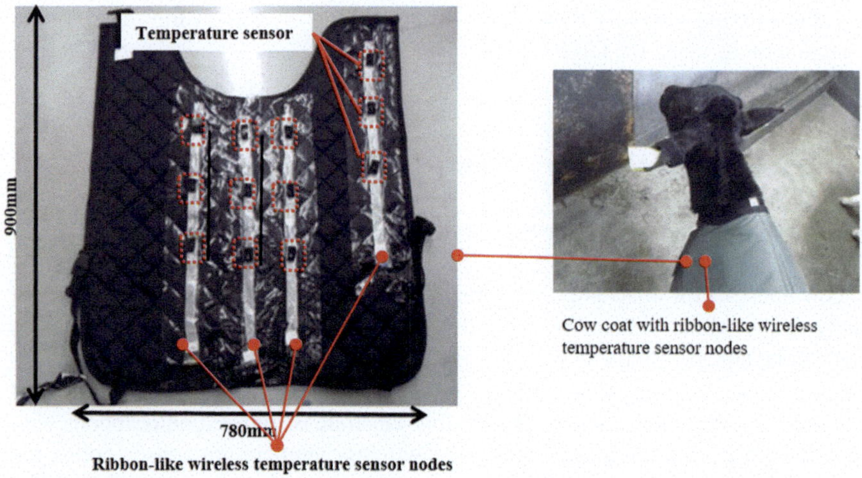

Fig. 11.26 Fabricated cow cote sensor nodes

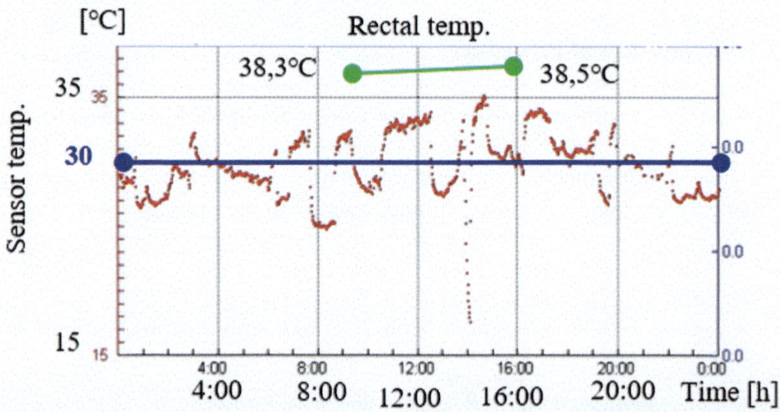

Fig. 11.27 Measurement results of cow cote sensor nodes

11.3.3.3 Measurement Result

Figure 11.27 shows the measurement results of the cow cote sensor nodes. The data from the sensor with the highest measured temperature among the 12 sensors is shown. The sensor temperature was about 30 °C. The circadian rhythm of body temperature could be confirmed.

11.4 Application to Giraffes

A zoo is for recreation, environmental education, the conservation of animal species, and the research of animal ecology. However, the research and study of animal ecology require a great deal of effort. In this section, we describe our wearable sensor nodes and the health monitoring system applied in a zoo.

Figure 11.28 shows a giraffe and the wireless sensor node used in the experiment. The wireless sensor node was the same as the third-cow's-tail sensor node. The sensor nodes were housed into protective gear to measure skin temperature at the ankle.

Figure 11.29 shows the measurement temperature result of the giraffe for 5 days. The upper line is the skin temperature of the giraffe, and the lower line is the outdoor temperature. As in the measurement with cows, the circadian rhythm of skin temperature could be confirmed. The small size, low power consumption, and exchangeable sensor part made application to the giraffe possible.

11.5 Concluding Remarks

Wearable sensor nodes are very useful in an animal health monitoring system to measure biological information of individuals. These wearable sensor nodes must be light and small so that they do not interfere with animal behavior. In addition, low power consumption is required to avoid frequent battery replacement. In these

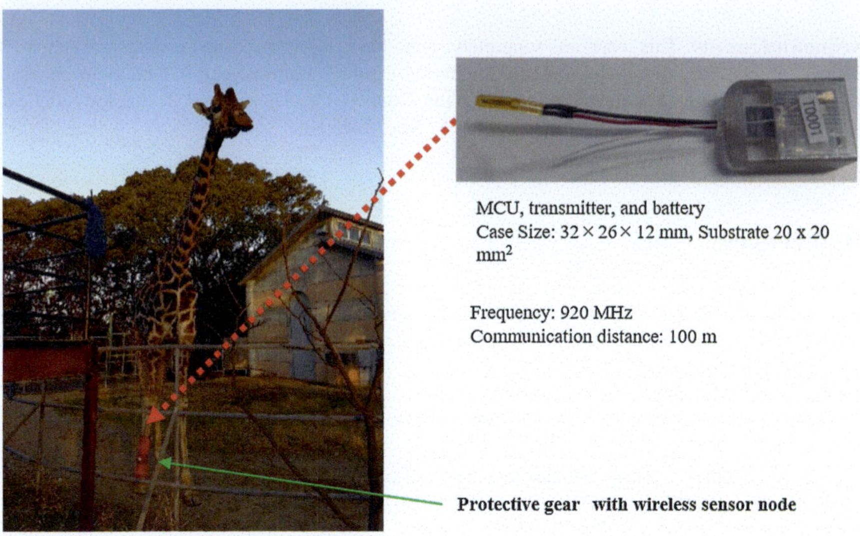

Fig. 11.28 Giraffe and wireless sensor node used in the experiment

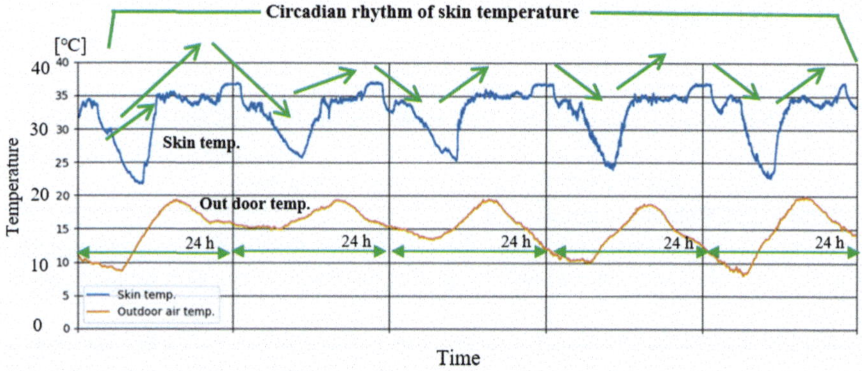

Fig. 11.29 Measurement temperature result of the giraffe for 5 days

aspects, MEMS sensors or event-driven measurement systems with MEMS sensors are highly suitable as described in the second section.

In the second and third sections, we described how wireless sensor nodes must have high durability in order to attach them to animals. PVDF film switch version wireless sensor nodes must be durable enough to withstand impact forces and not break over a long term. On the other hand, skin temperature sensors must have good adhesion, and they must be thin. In this case, it is important to design replaceable sensors. Alternatively, flexible and durable MEMS technologies such as ribbonlike sensor nodes are needed.

In the future, wearable sensor nodes and health monitoring systems may be applied in various situations such as livestock farming, zoos, and pets. In the fourth section, we presented the application of our sensor node to a giraffe in a zoo.

Acknowledgments This work was supported by the Japan Society for the Promotion of Science (JSPS) through the "Funding Program for World-Leading Innovative R&D on Science and Technology (FIRST Program)," initiated by the Council for Science and Technology Policy (CSTP), Grant-in-Aid for Scientific Research (A) No. 24248044.

References

Adil, H., Koser, A. A., & Gupta, A. (2020). Sleep monitoring using laser blood flowmeter. In *2020 IEEE International Students' conference on electrical, electronics and computer science (SCEECS)* (pp. 1–4). IEEE.

Andersson, L. M., Okada, H., Miura, R., Zhang, Y., Yoshioka, K., Aso, H., & Itoh, T. (2016). Wearable wireless estrus detection sensor for cows. *Computers and Electronics in Agriculture, 127*, 101–108.

Hadjidj, A., Souil, M., Bouabdallah, A., Challal, Y., & Owen, H. (2013). Wireless sensor networks for rehabilitation applications: Challenges and opportunities. *Journal of Network and Computer Applications, 36*(1), 1–15.

Iwasaki, W., Sathuluri, R. R., Niwa, O., & Miyazaki, M. (2015). Influence of contact force on electrochemical responses of redox species flowing in nitrocellulose membrane at micropyramid Array electrode. *Analytical Sciences, 31*(7), 729–732.

Khem, B., Hirai, Y., Yamakawa, T., Mori, Y., Inoue, E., Okayasu, T., & Mitsuoka, M. (2018). Effects of different application methods of fertilizer and manure on soil chemical properties and yield in whole crop rice cultivation. *Soil Science and Plant Nutrition, 64*(3), 406–414.

Kobayashi, T., Oyama, S., Takahashi, M., Maeda, R., & Itoh, T. (2008). Microelectromechanical systems-based electrostatic field sensor using Pb (Zr, Ti) O3 thin films. *Japanese Journal of Applied Physics, 47*(9S), 7533.

Kobayashi, T., Okada, H., Masuda, T., Maeda, R., & Itoh, T. (2010). A digital output piezoelectric accelerometer using a Pb (Zr, Ti) O3 thin film array electrically connected in series. *Smart Materials and Structures, 19*(10), 105030.

Miura, R., Yoshioka, K., Miyamoto, T., Nogami, H., Okada, H., & Itoh, T. (2017). Estrous detection by monitoring ventral tail base surface temperature using a wearable wireless sensor in cattle. *Animal Reproduction Science, 180*, 50–57.

Nishihara, K., Iwasaki, W., Nakamura, M., Higurashi, E., Soh, T., Itoh, T., Okada, H., Maeda, R., & Sawada, R. (2013). Development of a wireless sensor for the measurement of chicken blood flow using the laser Doppler blood flow meter technique. *IEEE Transactions on Biomedical Engineering, 60*(6), 1645–1653.

Nogami, H., Okada, H., Miyamoto, T., Maeda, R., & Itoh, T. (2013a). Wearable and compact wireless sensor nodes for measuring the temperature of the base of a Calf's tail. *Sensors and Materials, 25*(9), 577–582.

Nogami, H., Okada, H., Takamatsu, S., Kobayashi, T., Maeda, R., & Itoh, T. (2013b). Unique activity-meter with piezoelectric poly (vinylidene difluoride) films and self weight of the sensor nodes. *Japanese Journal of Applied Physics, 52*(9S1), 09KD15.

Nogami, H., Okada, H., Miyamoto, T., Maeda, R., & Itoh, T. (2014). Wearable wireless temperature sensor nodes appressed to base of a Calf's tail. *Sensors and Materials, 26*, 539–545.

Okada, H., Itoh, T., Suzuki, K., & Tsukamoto, K. (2009a). Wireless sensor system for detection of avian influenza outbreak farms at an early stage. In *Sensors, 2009 IEEE* (pp. 1374–1377). IEEE.

Okada, H., Kobayashi, T., Masuda, T., & Itoh, T. (2009b). Ultra-low power event-driven wireless sensor node using piezoelectric accelerometer for health monitoring. *Japanese Journal of Applied Physics, 48*(7R), 070222.

Takamatsu, S., Yamashita, T., Murakami, T., Masuda, A., & Itoh, T. (2016). Fabrication and evaluation of LED-embedded ribbons for highly flexible lighting applications in rooms. *Microsystem Technologies, 22*(5), 1079–1087.

Takeshita, T., Yoshida, M., Takei, Y., Ouchi, A., Hinoki, A., Uchida, H., & Kobayashi, T. (2019). Relationship between contact pressure and motion artifacts in ECG measurement with electrostatic flocked electrodes fabricated on textile. *Scientific Reports, 9*(1), 1–10.

Yamashita, T., Takamatsu, S., Okada, H., Itoh, T., & Kobayashi, T. (2016). Ultra-thin piezoelectric strain sensor array integrated on a flexible printed circuit involving transfer printing methods. *IEEE Sensors Journal, 16*(24), 8840–8846.

Yun, C. B., & Min, J. (2011). Smart sensing, monitoring, and damage detection for civil infrastructures. *KSCE Journal of Civil Engineering, 15*(1), 1–14.

Zhang, Y., Toda, A., Okada, H., Kobayashi, T., Itoh, T., & Maeda, R. (2012). New wafer-scale MEMS fabrication of 3D silicon/metal cantilever array sensor. In *2012 IEEE 25th international conference on micro electro mechanical systems (MEMS)* (pp. 297–300). IEEE.

Zhang, L., Lu, J., Nogami, H., Okada, H., Itoh, T., & Arai, S. (2018). Solid-state pH sensor prototype for real-time monitoring of the rumen pH value of Japanese cows. *Microsystem Technologies, 24*(1), 457–463.

Index

Z. Yang (ed.), *Advanced MEMS/NEMS Fabrication and Sensors*,
https://doi.org/10.1007/978-3-030-79749-2

The manufacturer's authorised representative in the EU is Springer Nature Customer Service Centre GmbH, Europaplatz 3, 69115 Heidelberg, Germany. If you have any concerns regarding our products, please contact ProductSafety@springernature.com

Printed and bound by CPI Group (UK) Ltd, Croydon, CR0 4YY
30/09/2026
02225248-0004